Biobased Polymers for Environmental Applications

Biobased Polymers for Environmental Applications

Editors

Andrea Lazzeri
Maria Beatrice Coltelli
Patrizia Cinelli

MDPI • Basel • Beijing • Wuhan • Barcelona • Belgrade • Manchester • Tokyo • Cluj • Tianjin

Editors

Andrea Lazzeri
Department of Civil and
Industrial Engineering
University of Pisa
Pisa
Italy

Maria Beatrice Coltelli
Department of Civil and
Industrial Engineering
University of Pisa
Pisa
Italy

Patrizia Cinelli
Department of Civil and
Industrial Engineering
University of Pisa
Pisa
Italy

Editorial Office
MDPI
St. Alban-Anlage 66
4052 Basel, Switzerland

This is a reprint of articles from the Special Issue published online in the open access journal *Polymers* (ISSN 2073-4360) (available at: www.mdpi.com/journal/polymers/special_issues/Biobased_Polym_Environ_Appl).

For citation purposes, cite each article independently as indicated on the article page online and as indicated below:

LastName, A.A.; LastName, B.B.; LastName, C.C. Article Title. *Journal Name* **Year**, *Volume Number*, Page Range.

ISBN 978-3-0365-7669-5 (Hbk)
ISBN 978-3-0365-7668-8 (PDF)

© 2023 by the authors. Articles in this book are Open Access and distributed under the Creative Commons Attribution (CC BY) license, which allows users to download, copy and build upon published articles, as long as the author and publisher are properly credited, which ensures maximum dissemination and a wider impact of our publications.

The book as a whole is distributed by MDPI under the terms and conditions of the Creative Commons license CC BY-NC-ND.

Contents

About the Editors vii

Preface to "Biobased Polymers for Environmental Applications" ix

Laura Aliotta, Alessandro Vannozzi, Ilaria Canesi, Patrizia Cinelli, Maria-Beatrice Coltelli and Andrea Lazzeri
Poly(lactic acid) (PLA)/Poly(butylene succinate-co-adipate) (PBSA) Compatibilized Binary Biobased Blends: Melt Fluidity, Morphological, Thermo-Mechanical and Micromechanical Analysis
Reprinted from: *Polymers* 2021, 13, 218, doi:10.3390/polym13020218 1

Laura Aliotta, Vito Gigante, Maria-Beatrice Coltelli and Andrea Lazzeri
Volume Change during Creep and Micromechanical Deformation Processes in PLA–PBSA Binary Blends
Reprinted from: *Polymers* 2021, 13, 2379, doi:10.3390/polym13142379 23

Laura Aliotta, Alessandro Vannozzi, Patrizia Cinelli, Maria-Beatrice Coltelli and Andrea Lazzeri
Essential Work of Fracture and Evaluation of the Interfacial Adhesion of Plasticized PLA/PBSA Blends with the Addition of Wheat Bran By-Product
Reprinted from: *Polymers* 2022, 14, 615, doi:10.3390/polym14030615 41

Vito Gigante, Luca Panariello, Maria-Beatrice Coltelli, Serena Danti, Kudirat Abidemi Obisesan and Ahdi Hadrich et al.
Liquid and Solid Functional Bio-Based Coatings
Reprinted from: *Polymers* 2021, 13, 3640, doi:10.3390/polym13213640 61

Luca Panariello, Maria-Beatrice Coltelli, Simone Giangrandi, María Carmen Garrigós, Ahdi Hadrich and Andrea Lazzeri et al.
Influence of Functional Bio-Based Coatings Including Chitin Nanofibrils or Polyphenols on Mechanical Properties of Paper Tissues
Reprinted from: *Polymers* 2022, 14, 2274, doi:10.3390/polym14112274 83

Laura Aliotta, Alessandro Vannozzi, Daniele Bonacchi, Maria-Beatrice Coltelli and Andrea Lazzeri
Analysis, Development, and Scaling-Up of Poly(lactic acid) (PLA) Biocomposites with Hazelnuts Shell Powder (HSP)
Reprinted from: *Polymers* 2021, 13, 4080, doi:10.3390/polym13234080 103

Karina Roa, Yesid Tapiero, Musthafa Ottakam Thotiyl and Julio Sánchez
Hydrogels Based on Poly([2-(acryloxy)ethyl] Trimethylammonium Chloride) and Nanocellulose Applied to Remove Methyl Orange Dye from Water
Reprinted from: *Polymers* 2021, 13, 2265, doi:10.3390/polym13142265 121

Yike Meng, Yuan Wang and Chuanyue Wang
Phosphorus Release and Adsorption Properties of Polyurethane–Biochar Crosslinked Material as a Filter Additive in Bioretention Systems
Reprinted from: *Polymers* 2021, 13, 283, doi:10.3390/polym13020283 141

Lili Ren, Zhihui Yang, Lei Huang, Yingjie He, Haiying Wang and Liyuan Zhang
Macroscopic Poly Schiff Base-Coated Bacteria Cellulose with High Adsorption Performance
Reprinted from: *Polymers* 2020, 12, 714, doi:10.3390/polym12030714 159

Lin Yu, Shiman Liu, Weiwei Yang and Mengying Liu
Analysis of Mechanical Properties and Mechanism of Natural Rubber Waterstop after Aging in Low-Temperature Environment
Reprinted from: *Polymers* **2021**, *13*, 2119, doi:10.3390/polym13132119 . . . . **173**

About the Editors

Andrea Lazzeri

Prof. Andrea Lazzeri started his career with a degree in Chemical Engineering from the University of Pisa, Pisa, Italy, in 1982. He obtained his PhD in Advanced Materials at the University of Cranfield in the UK in 1991. Starting in February 1999, he carried out teaching and research activities as a visiting scientist at the Department of Mechanical Engineering and Department of Chemical Engineering at the Massachusetts Institute of Technology (MIT), Cambridge, USA, where he worked with Professors A.S. Argon, R. Cohen and M. Boyce. In 2022, he was appointed as an Associate Professor in the sector of Materials Science and Technology (ING-IND/22) at the University of Pisa; in 2008, he was appointed as a Full Professor at the same university, where he teaches several courses in the field of materials science and engineering, biobased polymers and composites, and plastic processing and polymer characterization.

During the last two decades, Prof. Lazzeri's major research projects have investigated the processing and mechanical properties of polymer-matrix and ceramic-matrix composites, nanostructured materials and biobased/biodegradable polymers for environmental applications. He has worked as a coordinator of four European projects: Forbioplast, Helm, NanoCathedral and Cemwave (ongoing). He has published over 190 scientific papers with around 6000 citations (H-index 41).

Maria Beatrice Coltelli

Maria-Beatrice Coltelli is an Associate Professor at the Department of Civil and Industrial Engineering, University of Pisa, Pisa, Italy, in the sector of Materials Science and Technology (ING-IND/22). She obtained her degree in Chemistry in 2001 and her PhD "Galileo galilei" in Chemical Sciences in 2005, both at the Department of Chemistry and Industrial Chemistry at the University of Pisa. As a post-doctoral researcher, she was involved in the scientific activities of numerous European and regional projects. Moreover, she developed a spin-off company, SPIN-PET, owned by the University of Pisa. She also worked as a technician in a laboratory for material testing (PONTLAB). In September 2013, she began her activity as a researcher at the Department of Civil and Industrial Engineering until 2019, when she was appointed as an Associate Professor.

Her research studies have focused on polymer science, in particular polymer blends and composites. She previously took part in research projects in the field of plastic recycling and plastics from renewable sources, with particular attention on reactive processing and bio-nanocomposites. She also taught a course on Managing and Treatment of Solid Wastes. Currently, she is engaged in didactic activity related to courses in Chemical Engineering (Materials Science and Engineering and Polymeric Materials Science and Technology) and Materials and Nanotechnology (Reactive Processing and Recycling of Polymers) and is involved in the scientific activities of several European projects. She is an author or a co-author of more than 100 international scientific papers with 2700 citations, 12 book chapters, 3 patents, and 1 book. She co-edited one book and six Special Issues in the field of sustainable materials.

Patrizia Cinelli

Prof. Patrizia Cinelli is an Associate Professor in Fundamentals of Chemical Technologies (03/B2, CHIM/07) at the Department of Civil and Industrial Engineering, University of Pisa, Pisa, Italy. She graduated in Chemistry in 1995 at the University of Florence, Florence, Italy, under the supervision of Prof. Dante Gatteschi. In 1999, she obtained her PhD in Chemistry at the University of

Pisa, with research focusing on biodegradable and sustainable polymers for application in agriculture under the supervision of Prof. Emo Chiellini, which was partly performed at the United States Department of Agriculture (USDA), Peoria, IL, USA, where she spent a total of three stages (six months in duration) in 1998, 2000 and 2001. She worked as a researcher for the Interuniversity Consortium of Materials Science and Technology, Florence, Italy, and for the Pisa Division at the Institute for Physical and Chemical Processes of the National Research Council (CNR), Italy. She was a visiting scientist at the University of Almeria, Spain, and at the INTEMA-CONICET, Mar del Plata, Argentina. She is a co-author of over 110 papers (Scopus H-index of 34), 12 book chapters, and 8 patents. She has participated in over 25 EC projects, including the FP5 and the current Horizon 2020. She acted as a coordinator of the BBIJU project, ECOFUNCO "ECO sustainable multiFUNctional biobased COatings with enhanced performance and end of life options", GA 837863, and a technical manager of the BBIJU project RECOVER "Development of innovative biotic symbiosis for plastic biodegradation and synthesis to solve their end of life challenges in the agriculture and food industries", GA 887648.

Prof. Patrizia Cinelli has over 25 years of research experiences in sustainable and biodegradable polymers and in biomass valorization, in particular in production; chemical, thermal and mechanical characterization; and evaluation of sustainability based on biodegradation tests and life cycle assessment.

Preface to "Biobased Polymers for Environmental Applications"

Biobased polymers are attracting a great deal of attention because the extension of their use to replace fossil-based products is promoting a material selection approach oriented toward carbon neutrality. The improved modulation of their thermomechanical properties through the design of blends, biocomposites, and bionanocomposites, in addition to the exploitation of reactive processing techniques, provides such materials with high performances and peculiar functionalities. Moreover, the end-of-life management of biobased polymers is easier than the management of fossil-based polymers because they can be recyclable and, depending on their chemical structures and geometrical features, are also compostable or degradable in a controlled environment. Hence, on the one hand, biobased polymers offer the possibility of differentiating processability and properties to match the requirements of different applications; on the other hand, they offer the possibility of designing the best end-of-life scenarios for specific products. Although these beneficial new alternatives require a great deal of efforts to be integrated into current industrial production and waste management systems, many research activities have been dedicated to this topic. Moreover, biobased polymers are used for interesting environmental applications, such as depollution. The present reprint, which includes five papers from two European projects (BIONTOP and ECOFUNCO), gathers research and review papers dedicated to replacing fossil-based materials with their biobased counterparts with suitable properties, considering not only the structural and functional properties but also the environmental benefits of these new alternative products.

This reprint includes 10 papers. The first three papers describe research activities on poly(lactic acid) (PLA)-based blends and their composites with wheat bran that were carried out under the framework of the BIONTOP project ("Novel Packaging Films and Textiles with Tailored End of Life and Performance Based on Bio-based Copolymers and Coatings", 837761) funded by the European Commission BBI-JU program. The melt fluidity, morphological, thermo-mechanical, creep and fracture properties of these promising biodegradable materials were considered and discussed.

The reprint includes two contributions (a review and a research article) written in the contest of the ECOFUNCO project ("ECO Sustainable Multi FUNctional Biobased COATings with Enhanced Performance and End of Life Options", 837863) funded by the European Commission BBI-JU program. The review explores possible innovations in the field of solid and liquid biobased coatings, whereas the research article examines the modification of paper tissues with biobased additives for sustainable personal care applications.

A research paper on the production of PLA-based composites to valorize hazelnut shell powder, which represents an agro-food waste, is also included in the reprint. Moreover, the reprint presents research papers investigating different biobased materials for environmental depollution, including hydrogel-based, bacterial cellulose-based and biochar-based materials. Finally, a research paper on the behavior of natural rubber during aging is presented.

Andrea Lazzeri, Maria Beatrice Coltelli, and Patrizia Cinelli
Editors

Article

Poly(lactic acid) (PLA)/Poly(butylene succinate-co-adipate) (PBSA) Compatibilized Binary Biobased Blends: Melt Fluidity, Morphological, Thermo-Mechanical and Micromechanical Analysis

Laura Aliotta [1,2,*], Alessandro Vannozzi [1,2], Ilaria Canesi [3], Patrizia Cinelli [1,2,3], Maria-Beatrice Coltelli [1,2,*] and Andrea Lazzeri [1,2,3]

[1] Department of Civil and Industrial Engineering, University of Pisa, 56122 Pisa, Italy; alessandrovannozzi91@hotmail.it (A.V.); patrizia.cinelli@unipi.it (P.C.); andrea.lazzeri@unipi.it (A.L.)
[2] National Interuniversity Consortium of Materials Science and Technology (INSTM), 50121 Florence, Italy
[3] Planet Bioplastics s.r.l., Via San Giovanni Bosco 23, 56127 Pisa, Italy; ilariacanesi@planetbioplastics.com
* Correspondence: laura.aliotta@dici.unipi.it (L.A.); maria.beatrice.coltelli@unipi.it (M.-B.C.); Tel.: +39-050-2217856 (M.-B.C.)

Abstract: In this work poly(lactic) acid (PLA)/poly(butylene succinate-co-adipate) (PBSA) biobased binary blends were investigated. PLA/PBSA mixtures with different compositions of PBSA (from 15 up to 40 wt.%) were produced by twin screw-extrusion. A first screening study was performed on these blends that were characterized from the melt fluidity, morphological and thermo-mechanical point of view. Starting from the obtained results, the effect of an epoxy oligomer (EO) (added at 2 wt.%) was further investigated. In this case a novel approach was introduced studying the micromechanical deformation processes by dilatometric uniaxial tensile tests, carried out with a videoextensometer. The characterization was then completed adopting the elasto-plastic fracture approach, by the measurement of the capability of the selected blends to absorb energy at a slow rate. The obtained results showed that EO acts as a good compatibilizer, improving the compatibility of the rubber phase into the PLA matrix. Dilatometric results showed different micromechanical responses for the 80–20 and 60–40 blends (probably linked to the different morphology). The 80–20 showed a cavitational behavior while the 60–40 a deviatoric one. It has been observed that while the addition of EO does not alter the micromechanical response of the 60–40 blend, it profoundly changes the response of the 80–20, that passed to a deviatoric behavior with the EO addition.

Keywords: poly(lactic) acid (PLA); poly(butylene succinate-co-adipate) (PBSA); binary blends; micromechanical analysis; mechanical tests

1. Introduction

Plastic materials are used constantly in everyday life thanks to their versatility, low cost and huge range of properties. Consequently, the increment of nondegradable plastic waste has remarkably increased so that 150 million tons of plastic per year are consumed worldwide. This fact, combined with the threat of oil depletion, has led in the last decades to the development of biodegradable plastics based on renewable and nonrenewable resources. In fact, the use in specific applications of biodegradable plastics can limit the environmental problems correlated to plastic disposal [1–3].

Nowadays different biodegradable polymers are commercially available on the market. Among them poly(lactic) acid (PLA) is the most interesting, due to its low production cost (compared to other biodegradable polymers), good mechanical properties and easy processability (PLA can be manufactured in conventional extrusion, injection molding, blown film extrusion, cast extrusion, thermoforming and three-dimensional (3D)-printing) [4–7]. However, some PLA drawbacks, such as its brittleness, low toughness and poor heat resistance when employed at a temperature above its glass transition temperature, must be improved to extend its applications [8].

The improvement of PLA toughness (higher ductility and impact resistance) is thus fundamental and for this purpose the simplest approach is physical blending with a more ductile polymer. Among the more investigated biodegradable polymers, those that show a good starting ductility and that can be easily blended with PLA (mainly to increase its toughness) are: poly(butylene succinate) (PBS) [9–11], poly(butylene succinate co-adipate) (PBSA) [8,12,13], poly(caprolactone) (PCL) [14–16], poly(butylene adipate-co-terephthalate) (PBAT) [17–20]. Among these possible combinations of biodegradable blends, in this paper the attention has been focused on blending PLA with PBSA. In fact, depending on end-of-life (EOL) options, PBSA possesses a better eco-efficiency compared with the other biopolymers before mentioned [21]; moreover, its availability is very high thanks to the production capacity of around 100,000 tons per year [22]. PBSA is produced by a polycondensation reaction of 1,4-butanediol with succinic acid and adipic acid, that produces a completely aliphatic polyester having high flexibility, excellent impact strength, as well as thermal and chemical resistance and good biodegradability [11,23]. The use of PBSA alone is impracticable in rigid items due to its low stiffness, strength and melting point, but thanks to its low glass transition temperature its behavior is quite similar to a rubber and therefore lends itself very well to physical blending with PLA in order to increase its toughness by the well-known rubber toughening mechanism [20,24,25]. Thus, several research groups recently evidenced the possibility of modulating PLA brittleness by blending with PBSA [26–33].

However, most polymer blends are immiscible due to an unfavorable enthalpy mixing and consequently they form separated phases [34–36]. The morphology evolution of a biphasic system depends on the blend composition, processing conditions, rheological properties and interfacial tension of the two constituents [37–40]. Different morphologies (droplets, co-continuous, double emulsion) can be achieved by tailoring the ratio of PLA with respect to the rubbery polymer and consequently, it is possible to control the mechanical performance of the final material [41–43].

- Regarding binary blends, the compatibility/miscibility issues must be considered. It is noteworthy that the introduction of chain extenders that are able to reconnect cleaved chains, increases the molecular weight (consequently increasing the melt strength) [17,44,45]. Different types of chain extenders, also available on the market, have been extensively investigated and reported in literature such as: multi-functional epoxides [46], diisocyanate compounds [47], dianhydride [48], bis-oxazolines, tris(nonyl-phenyl) and phosphate (TNPP) [49]. The introduction of chain extenders also provides a better control of the polymer degradation [17,19,50–52] during the process and at the same time enhances the extrusion and injection foamability [52,53]. Moreover, the use of chain extenders can also improve the compatibility between the two phases constituting a binary blend because, especially in the interfacial region, the chain extender can react with both the polymers resulting in the formation of block copolymers acting as effective in situ generated compatibilizers. Chain extenders containing epoxy groups are the most suitable in this case, in fact they are able to react opening the epoxy group ring and creating covalent bonds [45] with both the hydroxyl and carboxylic groups of the polyester chain-ends. The high number of epoxy groups per macromolecule grants efficiency in limiting the decrease of viscosity during processing typical of polyesters that are generally affected by hydrolysis due to atmospheric humidity. For this reason, a multifunctional epoxy oligomer (EO) consisting of styrene, acrylic and glycidyl acrylate units, has been chosen for the binary PLA/PBSA polyester blends. Al-Itry et al. [52] and Wang et al. [54] studied the positive compatibilization effect of a similar EO in a PLA/PBAT system. Lascano et al. [55] explained that EO addition can be advantageous also in PLA/PBSA binary blends (thus very similar to the ones studied in this paper) because it reacts either with the hydroxyl terminal groups of PLA and PBSA, leading to a compatibilization effect and an effective toughening. However, the investigated blends contained up to 30% of PBSA and were not investigated in terms of their failure mechanism and melt fluidity.

In this study a systematic work was carried out by twin screw extruder producing PLA/PBSA blends with different compositions up to 40% of PBSA. These blends were characterized from morphological, melt fluidity, thermal and mechanical points of view. On the basis of the results obtained, the blends showing the best toughness improvement (elongation at break and impact resistance enhancement) were selected and the effect of EO (added at 2 wt.%) as compatibilizer was further investigated. In the second part of the work the micromechanical analysis (which represents a novelty in the study of PLA/PBSA blends) was performed with the help of an optical extensometer, capable of registering both axial and transversal elongation during the tensile tests. In this way it was possible to record the volume variation and to correlate the volume increment to the micromechanical deformation processes (debonding, cavitation, voids growth, etc.). The study was completed with the measurement of the capability of the selected best blends to absorb energy at a slow rate; this measurement was carried out by adopting the elasto-plastic fracture approach based on the ESIS load separation criterion.

2. Materials and Methods

2.1. Materials

The materials used in this work are listed below:

- Poly(lactic) acid (PLA) Luminy LX175 produced by Total Corbion PLA. This biodegradable PLA is derived from natural resources, appears as white pellets and contains about 4% of D-lactic acid units. It is a general-purpose extrusion grade PLA that can be used alone or to produce formulated blends or composites; it can be easily processed on conventional equipment for film extrusion, thermoforming or fiber spinning (density: 1.24 g/cm^3; melt flow index (MFI) (210 °C/2.16 kg): 6 g/10 min).
- Poly(butylene succinate-co-adipate) (PBSA), trade name BioPBS FD92PM, purchased from Mitsubishi Chemical Corporation, is a copolymer of succinic acid, adipic acid and butandiol. It is a soft and flexible semicrystalline polyester suitable for both blown and cast film extrusion (density of 1.24 g/cm^3; MFI (190 °C, 2.16 kg): 4 g/10 min).
- The epoxy oligomer (EO) used in the work is Joncryl ADR 4468 produced by BASF. It is an oligomeric chain extender having about 20 epoxy groups per macromolecule that reacts with the terminal groups of polycondensates, increasing the melt viscosity (Mw: 7250; density: 1.08 g/cm^3; epoxy equivalent weight: 310 g/mol).

EO appears as solid flakes that can be easily fed into the lateral feeder of the extruder. It was added to the PLA/PBSA formulations that showed the best toughness and flexibility.

2.2. Blends and Sample Preparation

Different blends containing increasing amounts of PBSA (from 5 wt.% to 40 wt.%), were produced in pellets according to the compositions reported in Table 1. To produce the granules, a semi-industrial Comac EBC 25HT (L/D = 44) (Comac, Cerro Maggiore, Italy) twin screw extruder was used. Before extrusion all solid materials were dried in a Piovan DP 604–615 dryer (Piovan S.p.A., Verona, Italy). PLA granules were introduced into the main extruder feeder, while PBSA was fed with a specific lateral feeder. In fact, after setting the weight percentage to be added, this feeder allows to a constant concentration in the melt during the extrusion to be obtained. EO was added only in the formulations that showed enhanced mechanical properties and it was fed separately with another lateral feeder. A fixed quantity of EO equal to 2 wt.% was chosen on the basis of literature works [17,54,56–59] but also considering the minimum quantity settable in the extruder that guaranteed a constant feeding without dosage problems. The blends containing EO are indicated in brackets and followed by the letter J in Table 1.

Table 1. Blend names and composition.

Blend Name	PLA–PBSA wt.%	EO wt.%
PLA	100–0	0
95–5	95–5	0
90–10	90–10	0
85–15	85–15	0
80–20	80–20	0
(80–20)J	80–20	2
60–40	60–40	0
(60–40)J	60–40	2

The temperature profile of the extruder (11 zones) used for blends preparation was: 150/175/180/180/180/185/185/185/185/185/190 °C, with the die zone at 190 °C for the blends containing up to 20 wt% of PBSA. Due to the major quantity of PBSA (having a lower melting temperature than PLA) decreasing the viscosity, for the blends containing 40 wt.% of PBSA a slightly decrease of 5 °C to the entire temperature profile was carried out; for the 60–40 and (60–40)J blends the temperatures in the zones from 1 to 11 were: 150/170/175/175/175/180/180/180/180/180/185 °C, with the die zone at 185 °C. The screw rate was 300 rpm with a total mass flow rate of 20 kg/h.

The extruded strands were cooled in a water bath at room temperature and reduced in pellets by an automatic knife cutter. All pellets were finally dried again at 60 °C.

After the extrusion, the pellets of the different blends were injection molded using a Megatech H10/18 injection molding machine (TECNICA DUEBI s.r.l., Fabriano, Italy) to obtain two types of specimens: dogbone specimens for tensile tests according to ISO 527-1A (width: 10 mm, thickness: 4 mm, length: 80 mm) and parallelepiped Charpy specimens for Charpy impact test according to ISO 179 (width: 10 mm, thickness: 4 mm, length 80 mm). After the injection molding the Charpy specimens were V-notched in the middle by a V-notch manual cutter (V-notch: 2 mm at 45°).

The samples injection molding parameters are reported in Table 2.

Table 2. Injection molding conditions.

Main Injection Molding Parameters	PLA	95–5	90–10	85–15	80–20	(80–20)J	60–40	(60–40)J
Temperature profile (°C)				180/185/190				
Mold temperature (°C)	70	70	60	60	55	65	55	65
Injection holding time (s)				5				
Cooling time (s)			15				25	
Injection pressure (bar)	90	90	80	80	80	120	80	100

The injection molding was performed using the same temperature profile for all blends. An increase in the cooling time was necessary with the increase of the PBSA content. The mold temperature was also lowered progressively with the increasing amount of PBS from 70 °C for pure PLA to 55 °C for 80–20 and 60–40 blends.

The addition of EO produces a significant increment of the melt viscosity, which in turn raises the pressure requested for filling completely the mold. Consequently, for the blends containing EO the increased viscosity was balanced by increasing both the mold temperature and the injection pressure.

2.3. Torque and Melt Flow Rate Analysis

The torque analysis is useful to obtain an indirect estimation of viscosity variations during the extrusion. Torque values were obtained by introducing about 6 g of the extruded granules in a MiniLab II twin-screw mini-compounder (HAAKE, Vreden, Germany). This equipment is able to compound the molten material and at the same time it records the torque values during the extrusion. The extrusion (performed at 190 °C and 100 rpm) was monitored for 1 min and every 10 s an assessment of the torque value was recorded. The measurements were carried out three times and the average value was reported.

The melt flow behavior of the blends was also investigated using a Melt Flow Tester M20 (CEAST, Torino, Italy) equipped with an encoder. The encoder, following the movement of the piston, is able to acquire the melt volume rate (MVR) of the polymer blends. Before the test, granules of the blends obtained from the extrusion process were dried in a ventilated oven (set at 60 °C) for one day. The melt flow rate (MFR), defined as the weight of the molten polymer passing through a capillary (having a specific length and diameter) in 10 min under a specific weight (according the ISO 1133:2005), was also recorded. In particular, the standard ISO 1133D custom TTT was used with a customized procedure: the sample was preheated without the weight for 40 s at 190 °C, then the weight of 2.160 kg was released on the piston and after 5 s a blade cut the spindle starting the real test. At this point the MVR was recorded, every 3 s, by the encoder. All the MVR data were reported with their standard deviation thanks to the CEAST Visuamelt software of the equipment. The MFR values' standard deviations were calculated by considering the results obtained by the measurements.

2.4. Mechanical Characterization

Both tensile and dilatometry tests were carried out on ISO 527-1A dog-bone specimens using an MTS Criterion model 43 universal tensile (MTS Systems Corporation, Eden Prairie, MN, USA) testing machine, at a crosshead speed of 10 mm/min, equipped with a 10 kN load cell and interfaced with a computer running MTS Elite Software. Tests were conducted 3 days after the injection molding process and during this time the specimens were stored in a dry keeper (SANPLATEC Corp., Osaka, Japan) at controlled atmosphere (room temperature and 50% humidity).

For tensile tests at least ten specimens were tested and for each blend composition the average values of the main mechanical properties were reported.

Regarding dilatometry, because of the large number of formulated blends, the tests were carried out only for two selected compositions. At least five specimens for each selected material were tested at room temperature and also in this case the tests were carried out after 3 days from the injection molding process. Transversal and axial specimen elongations were recorded, during the tensile test, using a video extensometer (GenieHM1024 Teledyne DALSA camera) interfaced with a computer running ProVis software (Fundamental Video Extensometer). Furthermore, the data in real-time were transferred to MTS Elite software in order to measure not only the axial and transversal strains but also the load value.

The volume strain was calculated, assuming equal the two lateral strain components, according to the following Equation [60–62]:

$$\frac{\Delta V}{V_0} = (1+\varepsilon_1)(1+\varepsilon_2)^2 - 1 \tag{1}$$

where the volume variation is ΔV, the starting volume is V_0, ε_1 is the axial (or longitudinal) strain and ε_2 is the lateral strain.

The impact tests were performed using V-notched ISO 179 parallelepiped specimens on a Instron CEAST 9050 machine (INSTRON, Canton, MA, USA) equipped with a 15 J Charpy pendulum. At least ten specimens for each blend were tested at room temperature. The impact tests were also carried out 3 days after the injection molding process.

To evaluate the energy stored by the sample before the fracture, three-point bending tests were also carried out. In this case, the tests were performed only on the most significant formulations. The already cited MTS universal testing machine was used. The methodology adopted to calculate the fracture energy at the starting point of crack propagation (J_{Ilim})

follows the ESIS TC4 load separation protocol [63,64]. According to this protocol, the tests must be carried out at 1 mm/min crosshead speed on 80 mm × 10 mm × 4 mm SENB specimens (that is the same parallelepiped specimen typology adopted for Charpy impact test) cut in two different ways: "sharp" (half notched samples) and "blunt" (drilled in the center with a 2 mm diameter hole and then cut for half width). To obtain the sharp notch (5 mm), during the cutting process, compressed air was used in order to avoid the "notch closing" phenomenon caused by excessive overheating generated by the cutter. A "sacrificial specimen" placed under the "good one" was used to guarantee a correct notch of the sample without closure (qualitatively evaluated with a "passing" paper) and to avoid plastic deformation around it. At least five specimens were tested for each selected blend. Thus the J_{lim} was calculated following the Load Separation Criterion procedure [65–69] for which it is necessary to construct a load separation parameter curve, obtained from the load (named P) vs. displacement (named u) during the three-point bending tests. The load vs. displacement curves must be obtained for both types of specimens used (sharp and blunt). In fact, in the sharp specimens the fracture is able to propagate, while in the blunt specimens crack growth cannot occur. At this point it must be defined the S_{sb} curve that represents the variation of the load separation parameter and is defined as follows:

$$S_{sb} = \frac{P_s}{P_b}\bigg|u_{pl} \tag{2}$$

where the subscripts s and b indicate the sharp and the blunt notched specimens, respectively. The plastic displacement is denoted as u_{pl} and it is expressed as:

$$u_{pl} = u - P \cdot C_0 \tag{3}$$

in which u is the total displacement and C_0 is the initial elastic specimen compliance. It must be pointed out that for ductile polymers (like those investigated in this paper), fracture initiation is a complex and progressive process that is characterized by the slow development of the crack front across the thickness of fracture transition [20,67,69]. This limit point is the pseudo-initiation of fracture. Thus, once that limit point was defined, the corresponding J_{lim} can be calculated as:

$$J_{lim} = \frac{2 \cdot U_{lim}}{b \cdot (w - a_0)} \tag{4}$$

where U_{lim} is the elastic behavior limit point, b is the sample thickness, w is the sample width and a_0 is the initial crack length.

2.5. Optical Analysis

Morphological analysis was carried out on cryo-fractured Charpy samples by FEI Quanta 450 FEG scanning electron microscope (SEM) (Thermo Fisher Scientific, Waltham, MA, USA). To avoid charge build up, the sample surfaces were sputtered (on a LEICA EM ACE 600 High Vacuum Sputter Coater, Wetzlar, Germany) with a thin surface layer of platinum. Image-J software was used to analyze the SEM images and to calculate the number average radius (R_n), the volume average radius (R_v) and the size distribution (SD) of the dispersed phase droplets. At least 150 droplets were counted to calculate R_v, R_n and SD parameters according the following Equations [70]:

$$R_n = \frac{\sum_i n_i R_i}{n_i} \tag{5}$$

$$R_v = \frac{\sum_i n_i R_i^4}{n_i R_i^3} \tag{6}$$

$$SD = \frac{R_v}{R_n} \tag{7}$$

The fracture surface of the tensile specimen broken during the dilatometric tests offered reliable information about the micromechanical deformations that occurred during the tensile tests. Consequently, some specimens were cold fractured along the tensile direction. In this case the specimens were coated with a thin layer of platinum prior to microscopy to avoid charge build up.

2.6. Thermal Characterization

Thermal properties of PLA and PLA/PBSA blends were investigated by calorimetric analysis (Q200 TA- DSC). Nitrogen, set at 50 mL/min, was used as purge gas for all measurements. Indium was used as a standard for temperature and enthalpy calibration of DSC. The materials used for DSC analysis have been cut from the ISO 5271-A dog-bone injection mold specimens. In order to evaluate if an eventual crystallization occurred during the specimen injection molding (affecting the mechanical behavior of the materials), the thermal properties were evaluated considering only the first DSC heating run. The sampling was carried out taking the material in the same region of the specimens to avoid differences ascribable to different cooling rates in the specimen thickness. The samples, with mass between 11.5 and 15 mg, were sealed inside aluminum pans before measurement. PBS granules were also analyzed in order to better understand how its thermal properties could affect the thermograms of the binary blends.

The samples were quickly cooled from room temperature to $-50\ °C$ and kept at this temperature for 1 min. Then the samples were heated at $10\ °C/min$ to $200\ °C$ to delete the thermal history then a second cooling scan from $-70\ °C$ to $190\ °C$, at $10\ °C/min$, was carried out. Melting temperature (T_m) and cold crystallization temperature (T_{cc}) of the blends were recorded at the maximum of the melting peak and at the minimum of the cold crystallization peak, respectively. The enthalpies of melting and cold crystallization were determined from the corresponding peak areas in the thermograms. Where possible the PLA and PBSA crystallinity were calculated according the following Equation:

$$X_{cc,\ PLA(or\ PBSA)} = \frac{\Delta H_{m,\ PLA\ (or\ PBSA)} - \Delta H_{cc,\ PLA\ (or\ PBSA)}}{\Delta H°_{m,\ PLA\ (or\ PBSA)} \cdot wt.\%\ PLA\ (or\ PBSA)} \quad (8)$$

where X_{cc}, is the crystallinity fraction of PLA or PBSA, ΔH_m and ΔH_{cc} are the melting and cold crystallization enthalpies respectively, while $\Delta H°_m$ is the theoretical melting heat of 100% crystalline polymer. For PLA a $\Delta H°_m$ value of 93 J/g [71] and for PBSA a $\Delta H°_m$ value of 142 J/g were considered [8].

The heat deflection temperature or heat distortion temperature (HDT) corresponds to the temperature at which the polymeric material deforms under a specified load. This property is fundamental during the design and production of thermoplastic components. The HDT is also strictly correlated to the polymer crystallinity, in fact it is noteworthy that a highly crystalline polymer has an HDT value higher than its amorphous counterpart [72]. For this purpose, the determination of the deflection temperature under load (HDT) was carried out on a CEAST HV 3 (INSTRON, Canton, MA, USA) in accordance with ISO 75 (method A). A flexural stress of 1.81 MPa and a bath heating rate of $120\ °C/h$ were used. The sample size was 80 mm $\times$ 10 mm $\times$ 4 mm. When the sample bar deflects by 0.34 mm, the corresponding bath temperature represents the HDT (Type A) value. At least five measurements were carried out and the average value was reported.

3. Results and Discussions

3.1. Melt Fluidity, Morphology and Thermal Properties of PLA/PBSA Blends

From the MFR and torque results, reported in Figure 1, it can be observed that the increase of PBSA content resulted in a decrease in torque values and an increment in MFR.

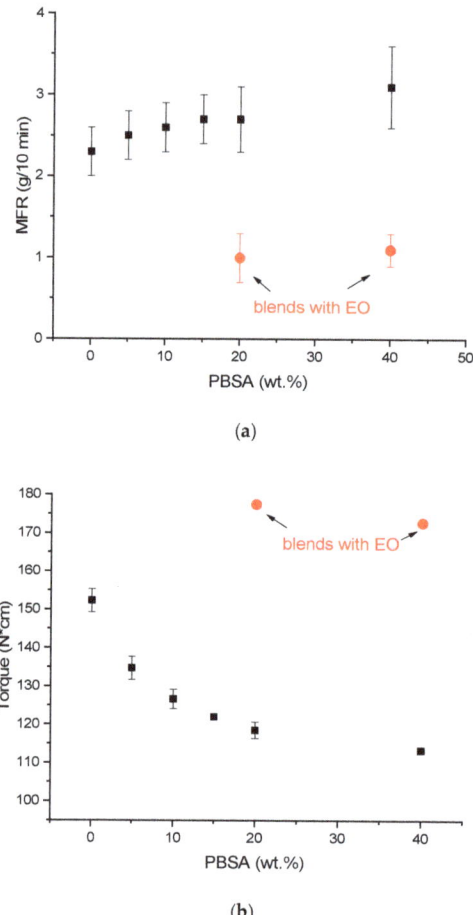

Figure 1. (**a**) Trend of MFR as a function of PBSA content; (**b**) trend of torque as a function of PBSA content.

In fact, the pure PLA showed a torque of 152.3 ± 3.0 N*cm and an MFR of 2.3 ± 0.3 g/10 min, but the pure PBSA was less viscous, showing a torque of 104.0 ± 7.1 N*cm and a MFR of 7.1 ± 0.7 g/10 min. Hence the fluidity in the melt and thus the processability of the produced granules can be modulated as a function of composition. This trend is in accordance with the necessity to appropriately modify the extrusion and injection molding for the different compositions, as explained in Section 2.2. Interestingly, the blends containing EO showed an increased value of torque and a significantly lower value of MFR, because of the increase in molecular weight due to the branching reactions of the EO. A simplified scheme of the reactions occurring in the melt between PLA, PBSA and EO is reported in Figure 2.

The lowest MFR value was recorded for the (80/20)J blend, thus the one that required an increase in the mold temperature and injection pressure during the injection molding process (as previously observed and reported in Table 2).

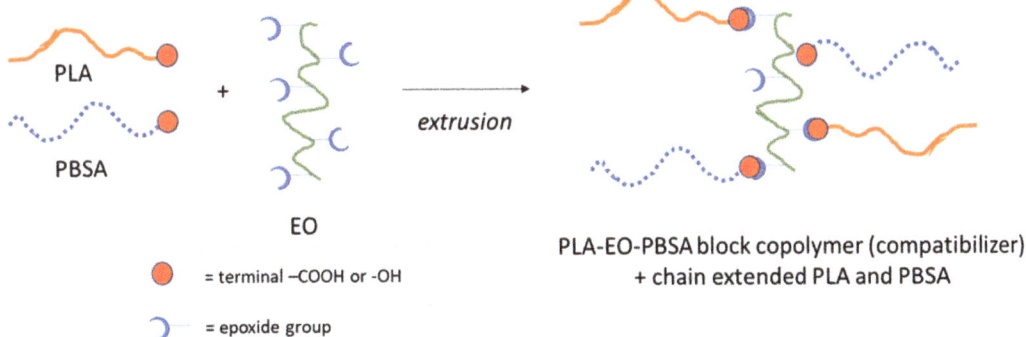

Figure 2. Simplified scheme of the reactions occurring in the melt between PLA, PBSA and EO.

The main tensile properties (Young's modulus, yield stress, stress at break, elongation at break and Charpy impact resistance) determined from the tensile stress–strain curves and Charpy impact test, are reported in Table 3.

Table 3. Tensile and impact properties for each formulation containing an increasing PBSA content.

Blend Name	Young's Modulus (GPa)	Yield Stress (MPa)	Stress at Break (%)	Elongation at Break (%)	Charpy Impact Resistance (kJ/m^2)
PLA	3.63 ± 0.12	/	62.7 ± 1.7	3.7 ± 0.3	2.7 ± 0.3
95–5	3.07 ± 0.1	/	59.4 ± 0.1	4.0 ± 0.2	2.8 ± 0.3
90–10	3.02 ± 0.13	57.4 ± 0.7	23.3 ± 2.3	45.8 ± 10.9	3.5 ± 0.2
85–15	2.91 ± 0.10	55.3 ± 0.7	21.3 ± 1.1	57.9 ± 18.9	4.4 ± 0.2
80–20	2.72 ± 0.10	51.2 ± 0.6	19.7 ± 0.6	86.7 ± 31.5	4.6 ± 0.2
60–40	1.99 ± 0.12	39.9 ± 0.8	21.5 ± 0.8	192.8 ± 31.1	9.2 ± 0.5

The mechanical behavior of the blends changed, passing from neat PLA to binary PLA/PBSA blends. PLA alone had the typical mechanical response of a fragile material with a high stiffness and tensile strength but a low Charpy impact resistance and elongation at break. Neat PLA failed just after the elastic region (no yielding point is observed), as typical in a brittle fracture. The PBSA addition maintained the material stiffness at an acceptable level but at the same time improved its flexibility for contents higher than 5 wt.% of PBSA. In fact, 95–5 blend still showed a fragile behavior without yielding, and a low elongation at break, low Charpy impact resistance and still high stress at break. From 5 up to 40 wt.% of PBSA, the increment in elongation at break and Charpy impact resistance was almost proportional to PBSA content (Figure 3a). A marked decrement of stress at break was registered due to the elastic characteristic conferred by the growing PBSA addition. The variation of stress at break, also considering the values of the deviations reported, can be considered negligible. However, the slight stress at break increase for the 60–40 blend could be attributable to the change in the morphology of the blend which became co-continuous. Hence, also the more ductile PBSA phase results were continuous. Then, in this co-continuous blend both phases fully contributed to the blend mechanical response in all directions and resulted in a more effective stress transfer.

The Young's Modulus decreased by increasing the PBSA content in monotonic way (Figure 3b). Hence, in the investigated range a wide modulation of properties is possible by varying the blend composition.

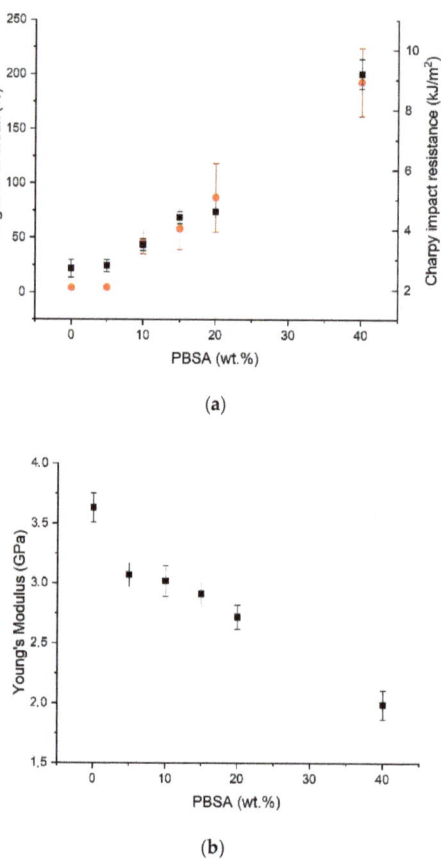

Figure 3. (**a**) Trend of elongation at break (red circles) and Charpy Impact strength (black squares) as a function of PBSA content; (**b**) trend of Young's Modulus as a function of PBSA content.

The results of the mechanical tests are closely related to the materials morphology (Figure 4).

The PBSA, until 20 wt.% in the blends, appeared as a spherical/ellipsoidal dispersed phase in the PLA matrix, indicating the poor miscibility between PLA and PBSA. In some cases, it was possible to observe voids at the interface (debonding). The cause of this debonding could be related to the cryofractured samples, where dilatational stresses were generated to the mismatch of the thermal coefficients between the PBSA particles and the PLA matrix during cooling before fracture, as was also observed for PLA–PBAT system [20]. The 60–40 mixture confirmed what is generally observed in literature [38,41] where a change in morphology occurs (from PBSA dispersed particles to a co-continuous structure); so the remarkable improvement in elongation at break, especially of the impact strength observed, is explained with the achievement of the inversion point region.

A typical two-phase structure, where discrete droplets of the minor phase are dispersed in the matrix, was observed in the samples with PBSA content up to 20%. For these blends it was possible to calculate the R_n, R_v and SD values, given by the ratio between R_v and R_n, (reported in Table 4) according to Equations (5)–(7), respectively.

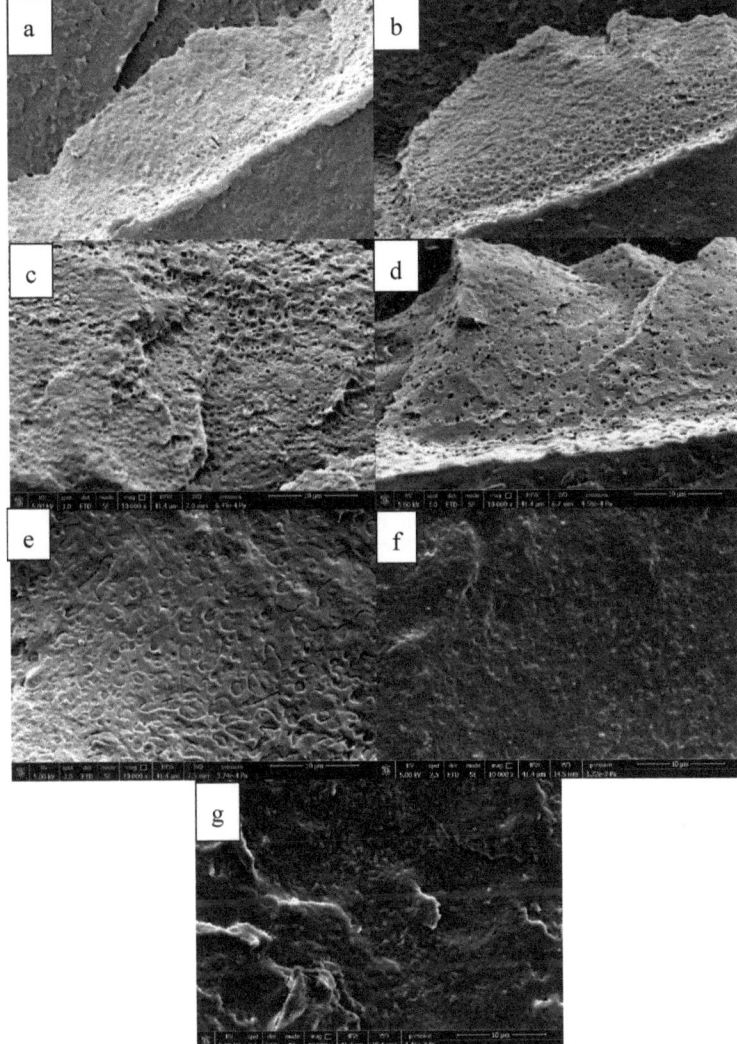

Figure 4. SEM micrographs of morphology phase development of PLA/PBSA blends: (**a**) 95–5; (**b**) 90–10; (**c**) 85–15; (**d**) 80–20; (**e**) 60–40; (**f**) (80/20)J; (**g**) (60/40)J.

Table 4. R_n, R_v and SD values of the dispersed PBSA phase for PLA/PBSA blends at different compositions up to 20 wt.% of PBSA content.

Blend Name	Rn (μm)	Rv (μm)	SD
95–5	0.41	0.54	1.32
90–10	0.42	0.58	1.38
85–15	0.54	0.78	1.44
80–20	0.55	0.80	1.45

The average size of the domain was between 1 and 2 μm indicating that PBSA and PLA were thermodynamically incompatible [41]. A slight increment of R_v values was observed with the increase in the PBSA content. This behavior is in agreement with the coalescence theory [73] for which, during the mixing process, the dispersed phase can collide with each other and coalesce, forming bigger droplets. The probability of droplet collision will be more accentuated by increasing the PBSA content, with the number of PBSA droplet collisions that will be proportional to the square of the PBSA concentration [74]. However, this size distribution increment is not as pronounced as could be expected. This behavior was encountered by other authors [70] and can be ascribed to the high shear rate and extensional flow during the extrusion process that limited the particle coalescence favoring the break-up of dispersed droplets. Moreover, the two polymers, being both polyesters, are characterized by a good chemical affinity in full agreement with the evolution of mechanical properties as a function of PBSA content. However, the dispersed phase dimensions are in accordance with the general rule regarding rubber toughening, which states that the dispersed phase must be distributed as small domains in the polymer matrix [75,76]. This aspect, combined with a sufficient interfacial adhesion, would increase the elongation at break and the impact properties of the final material in accordance with the results obtained from the mechanical tests.

The 60–40 blend showed a co-continuous morphology where the dispersed phase coalesced until it formed bigger structures throughout the whole blend [77–80]. Consequently, due to the irregularity of the shape assumed by this structure, the calculation of R_v, R_n and SD parameters was not feasible. The analysis of the micrographs has anyway shown that the dimension of the two interpenetrating phases is quite low, probably thanks to effective processing and this explain the possibility of a good modulation of properties by acting on composition. Moreover the presence of EO improved the compatibility of the phases: a decrease in phase dimensions can be noticed both in the (80/20)J and in (60/40)J blend (Figure 4f,g) as well as an increased interfacial adhesion [44,81–83] making difficult the observation of these interfaces during the analysis.

The first heating thermograms obtained from DSC analysis and reported in Figure 5a,b show the thermal history of the samples produced by injection molding.

The analysis of these data was preferred to get correlation with mechanical results, measured on injection molded specimens. Moreover, this analysis can provide useful information regarding the peculiar injection molding process adopted for the different blends. In fact, the crystallization occurring in the mold had a significant role in allowing a rapid and efficient ejection of the specimen by the machine without any distortion. Moreover, thermal properties related to the use of EO were yet investigated by Nunes et al. [84] in PLA. EO determined a decrease in the crystallinity in pure PLA and in blends with PBAT because of the introduced disorder due to the branching of chains. Lascano et al. [55] did not determine the crystallinity of their PLA/PBSA but the thermograms they reported were in good agreement with Nunes et al. [84] observations. These considerations can explain the observed necessity to increase the molding temperature in the injection molding of blends containing EO, to counterbalance the reduced tendency to crystallization with a decreased undercooling favoring the extension of the crystalline fraction.

According to literature [85,86], PBSA has a triple melting peak centered at around 87.14 °C; the melting behavior can occur with a melting peak numbers that depend on the processing conditions. Consequently, it was not possible to measure with accuracy the PBSA melting enthalpy and the PLA cold crystallization enthalpy due to the overlap between the PBSA melting endotherms, the PLA cold crystallization exotherms (that became more marked with increasing the PBSA content) and the enthalpic relaxation peak occurring in correspondence of the PLA glass transition temperature and ascribed to the specimen's aging.

With the PBSA addition also the PLA double melting behavior started to become observable. These double peaks were correlated to the remelting of newly formed crystallite during heating. Crystallites of disordered α' form (with low melting temperature) recrystallize in a more ordered α form having a higher melting temperature. In any case, the right melting peak of PLA was deemed to be the one which was in a higher temperature range [10]. According to the melting recrystallization model [87] the small and imperfect crystals would be transformed into more stable and perfect crystals. However, at high cooling rates (like that reached during the injection molding process), the granules passed to the recrystallization region so fast that there was insufficient time for the molten material to reorganize into new ordered crystals generating consequently low melting crystallites [13]. In fact, it is known in literature that the heating rate influences markedly the conversion from α'-to α-form for PLA as well as the presence of PBSA double or triple- melting peaks [83,88,89]. The absence of PBSA cold crystallization in the blends and also in the pure material can be ascribed to different factors. The most important is probably correlated to the PBSA macromolecules having a very fast crystallization rate during cooling and this leading to an absence (or very low quantitative) of amorphous domains that could recrystallize again during heating [8].

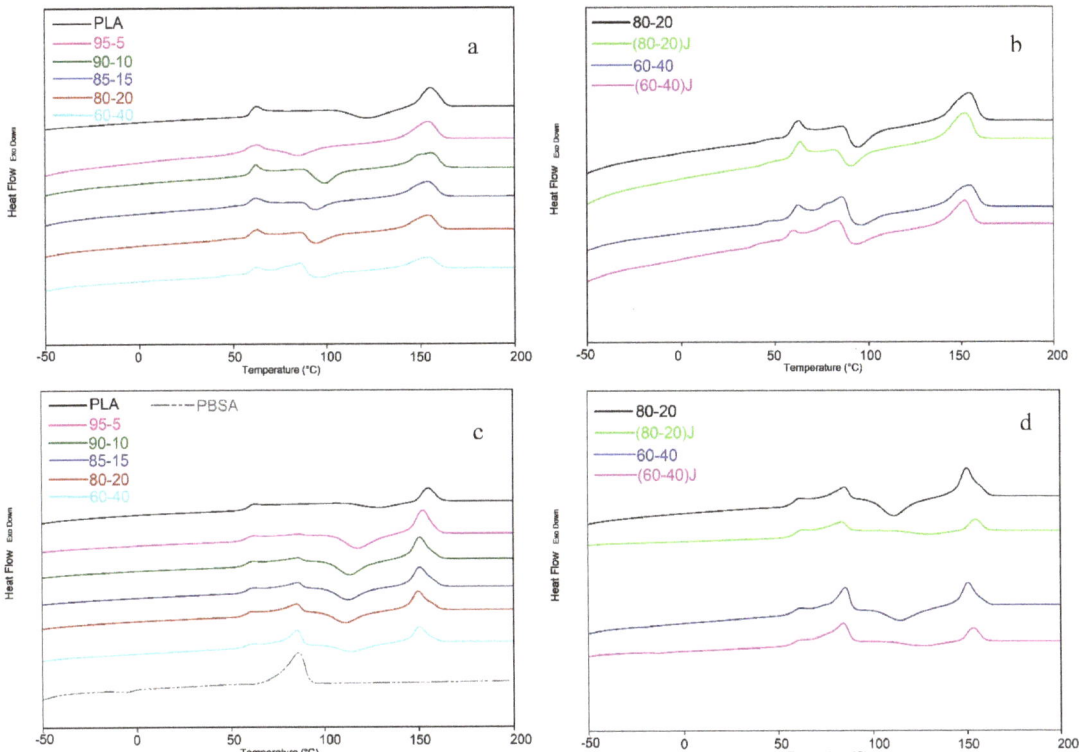

Figure 5. DSC thermograms (**a**) first heating for PLA/PBSA blends; (**b**) first heating for the selected blends with EO; (**c**) second heating for PLA/PBSA blends; (**d**) second heating for the selected blends with EO. Heat flow, reported in J/g, is expressed in arbitrary units to allow the curves shifting.

For all the blends, the cold crystallization temperature of PLA was shifted to a lower temperature if compared to pure PLA (Figure 5a–d); in agreement with Lascano et al. [55], the PBSA melting contributes to the increase in the PLA chain motion allowing PLA chains to arrange into packed structures at lower temperatures. Consequently, the cold crystallization temperature and the melting temperature of PLA decreased with the PBSA addition. However, the most marked decrement of the PLA cold crystallization temperature was observed for the 95–5 blend (especially in the first heating scan, related to the injection molding cycle that could enhance the PBSA nucleating ability because of the high temperature of the mold and high holding time) where the PBSA dispersed phase, present in low content but with a low particle size, seemed to act as nucleating agent [10] thanks to the effect of the extended interface. In fact, in good agreement, by increasing the PBSA content, the PLA cold crystallization temperature rose slightly but still remained lower than that of pure PLA.

An enthalpic relaxation peak above the glass transition temperature due to the aging [90] could be observed in the first heating scan (Figure 5a,b) for all blends, but it was reduced in blends containing EO because of their more disordered structure hindering chain relaxation. However, for both heating scans, it could be observed that the glass transition temperature of PLA remained almost unchanged with PBSA content, confirming the restricted miscibility of the two biopolymers as reported by Lee et al. [13].

In order to better understand the HDT obtained data (Figure 6), showing an almost constant value in all the examined blends, although it was not possible to determine the crystallinity values of the injection molded specimens considering the first heating scan, an estimation of the crystallinity was made on the second run. In this case, having deleted the thermal history, the single PBSA granules were also investigated. In fact, in the second run, thanks to the disappearance of the aging peak and to the lower cooling rate with respect to injection molding, the peaks of PBSA and PLA show an almost null overlap (Figure 5c,d).

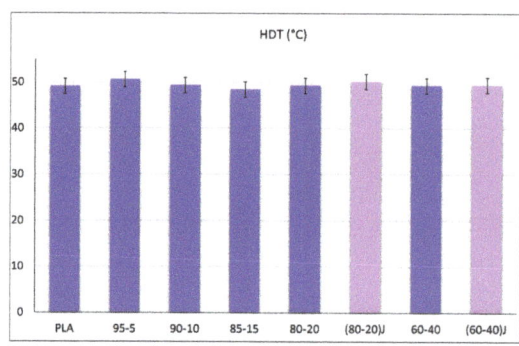

Figure 6. HDT values for all blend compositions.

In Figure 7 the crystallinity percentage values for PLA, PBSA and the sum of them (that is, the total crystallinity of the sample) are reported. The enhancement of PLA crystallinity with PBSA content, suggests that the PBSA droplets, dispersed into the PLA matrix effectively act as crystallization nuclei for PLA accelerating the crystallization during the heating process [10]. For all the blends the maximum total crystallization, seen as the sum of PLA and PBSA crystallinity percentage, was between 27 and 32%. Hence only small variations of the crystalline fraction were present in the different blends and this was reflected in the HDT values (Figure 6) which were approximately the same for all the examined blends. In fact, it is well known that the HDT of neat PLA is at about its T_g. The HDT is significantly affected by the degree of crystallization [91]. The low crystallization rate of PLA made it essentially amorphous under the injection molding conditions adopted in this work. Although a quantitative evaluation of the samples crystallinity was not possible for all blend compositions, it could be supposed that the maximum degree of sample crystallinity was lower than that measured during the second heating run. Consequently, the low crystallinity (due to the low crystallization rate) reached by the PLA under practical molding conditions was reflected in a null variation of the HDT that remained almost unchanged for all the blends analyzed (Figure 6).

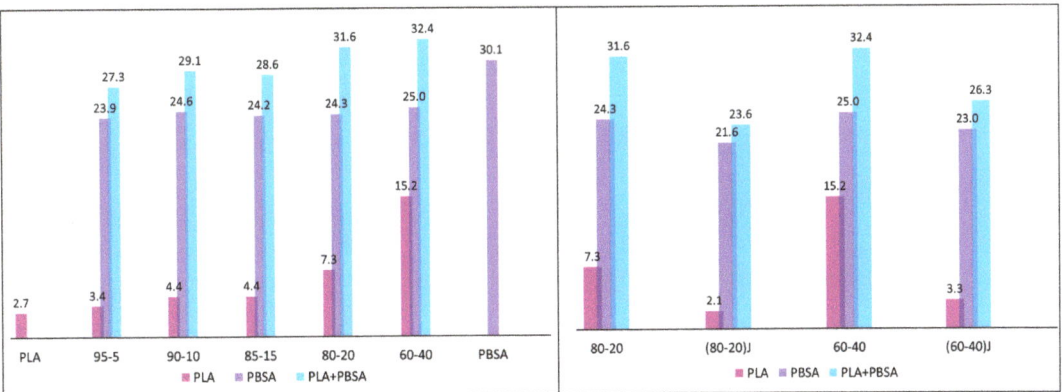

Figure 7. Crystallinity values for PLA/PBSA blends in the second heating scan.

3.2. Effect of the EO on Mechanichal and Failure Behaviour of PLA/PBSA Blends

EO was added (at 2 wt.%) to 80–20 and 60–40 blends. These two formulations were selected because they showed the highest elongation at break and Charpy impact resistance values.

From a morphological point of view, it can be observed in Figure 3, that EO acted as a compatibilizer, working as a bridge between the PLA and PBSA phases, leading to a reduction of the interfacial tension and thus leading to better adhesion and dispersion between the two phases [55,92,93].

The interaction created by EO in the PLA–PBSA blends can also be observed in the torque values that for (80–20)J and (60–40)J were higher than the torque value of pure PLA (Figure 1). The MFR, on the other hand, reached the lowest values when EO was added. It is known that the torque increment is related to the molten polymer viscosity increase, which is caused by a molecular weight increment in polyesters [17]. Furthermore, the chain extender produces an increment of the melt viscosity creating interactions between the matrix and the dispersed phase. This marked viscosity variation with respect to the viscosity values recorded for corresponding blends without EO (80–20 and 60–40), led to a significant variation of the injection molding conditions (as can be observed in Table 2) where the mold temperature and the injection pressure were increased.

From a thermal point of view, it can be observed (Figure 5) that the EO addition caused a slight decrement of T_g that could be ascribable to the increased compatibility between the two polymers [17]. A slight shift of the PLA cold crystallization temperature was also registered. On the other hand, the introduction of EO limited the chain mobility and the shift of PLA cold crystallization temperature did not occur in the presence of the chain extender. As could be expected, the chain extender depressed severely the PLA crystallization while the PBSA crystallization did not seem to be much influenced by the EO addition (Figure 7).

However, from the results of HDT (Figure 6), it could be detected that the EO addition did not cause a significant variation of the HDT values and it did not significantly affect the final crystallinity of the material. In any case the increase in molecular weight and branching activity of EO on PLA counterbalanced the decrease in crystallinity in this phase, thanks to the increased resistance of the amorphous phase achieved thanks to the formation of new inter-macromolecular bonds.

The enhanced compatibility induced by EO was overwhelmingly reflected in the mechanical properties (Figure 8). EO caused a relevant increment of the mechanical properties. Not only an improvement in elongation at break and Charpy impact resistance occurred, but also a marked increment in stress at break was recorded.

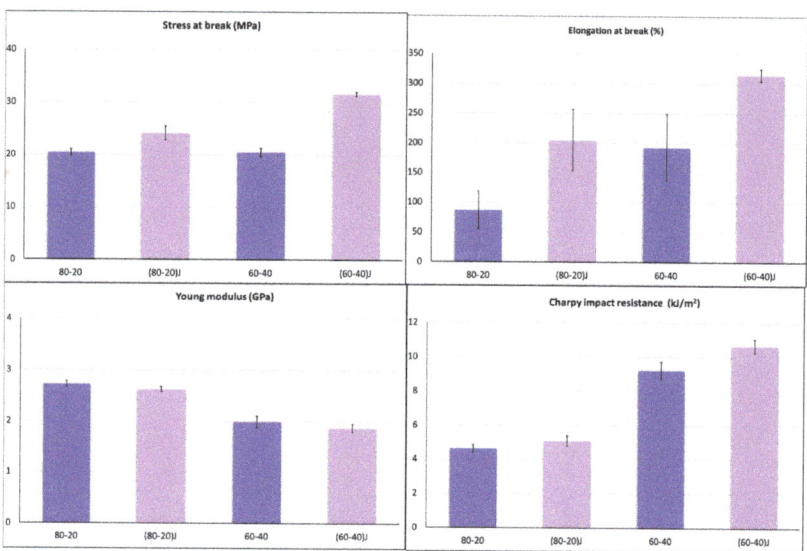

Figure 8. Comparison of the main mechanical properties for blends with and without Joncryl.

Essentially, the increase in stress at break and elongation at break may be related to the molecular weight increase, due to the polyester reaction with EO (Figure 2). On the other hand, the toughness increment is more related to the morphology of the system and, as can be expected, better toughness is reached for the (60–40)J blend where a co-continuous structure combined to a better phase compatibility was observed.

To better understand the effective toughness enhancement obtained with the EO addition, the J_{Ilim} was calculated adopting the elasto-plastic fracture mechanism approach. The J_{Ilim} values correspond to the energy absorbed by the specimen at the moment of the crack propagation in slow-rate test conditions.

Comparing the J_{Ilim} values with the "brittle" G value of pure PLA (equal to 2.97 kJ/m^2 [94,95]) all blends showed very interesting values (Figure 9). The fracture energy released at the beginning of crack propagation was markedly increased with the EO addition for the (80–20)J blend, while for the 60–40 composition the J_{Ilim} value remained almost unchanged with the EO addition. Probably, the 60–40 blend being co-continuous, the effect of EO addition was less marked; the co-continuous nature of the blend led it to have a starting high value of J_{Ilim} and Charpy impact resistance that were not significantly altered by the presence of EO.

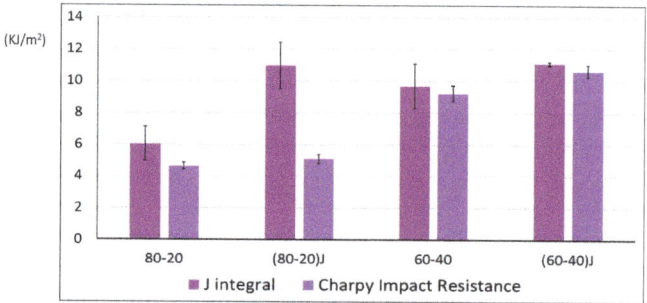

Figure 9. Comparison between J_{Ilim} and Charpy impact resistance values (both in KJ/m^2) for blends with and without Joncryl.

The results of dilatometric tensile tests are reported in Figure 10 where the volume variation (calculated by Equation (1)) is reported as function of axial elongation (the tests were carried out until the deformation of the specimens remained homogeneous).

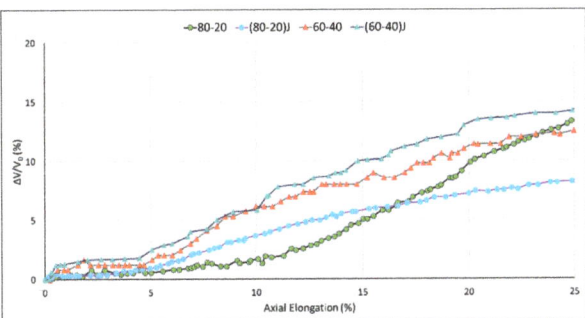

Figure 10. Volume strain–strain curves for blends with and without EO.

The characterization of the dilatational response of a material when it is subjected to an applied stress can lead to an appreciated deformation mechanism in the bulk of the material. According to the type of volume strain–strain curve obtained, it is possible to distinguish between: a cavitational response, a dilational response and deviatoric response [96]. It can be observed that except for the 80–20 blend, where a cavitational response is observed, for the other blends a deviatoric response was registered.

As a consequence, for the 80–20 blend, in the large deformation limit, the hydrostatic tensile stress will cause cavitation type mechanisms that will lead to a rapid volume increase. In particular, when the stress approaches the yield stress value, the cavitation-type mechanism will produce voiding that will cause the rapid increase in the volume strain, and the coalescence of these voids will bring to the final rupture of the specimen [60]. The SEM micrographs at the cryo-fractured surface of tensile specimens along the draw direction (Figure 11) confirmed the presence of many small and big elongated voids grown along the tensile direction due to cavitation for the 80–20 blend.

The volume dilatation response for the other blends ((80–20)J, 60–40 and (60–40)J)), displayed a deviatoric behavior; for these materials, when the stress value approaches stress at yield, the material will continue to deform by changing shape and not volume, leading to a shear yielding as the main failure mechanism. The SEM images confirmed that a different deformation mechanism occurred where the material underwent deformation along the draw direction without voiding formation.

Similar to what was observed for the J_{lim} value, the EO addition in the 80–20 mixture, significantly changed the micromechanical behavior of the material. This marked change can probably be ascribed to the capacity of the chain extender to modify the macromolecule structure and create interactions between the matrix and the dispersed phase, modifying the initial interface adhesion and leading to a different micromechanical response.

On the other hand, the effect of EO to the micromechanical behavior of the 60–40 blend could be considered negligible. The dilatometric curves of the 60–40 and (60–40)J blends were very similar with a very slight increment of the volume variation for the (60–40)J blend. In this case the micromechanical behavior was probably ascribable to the different morphology of the 60–40 blend that, being co-continuous, did not show dispersed PBSA particles into the PLA matrix.

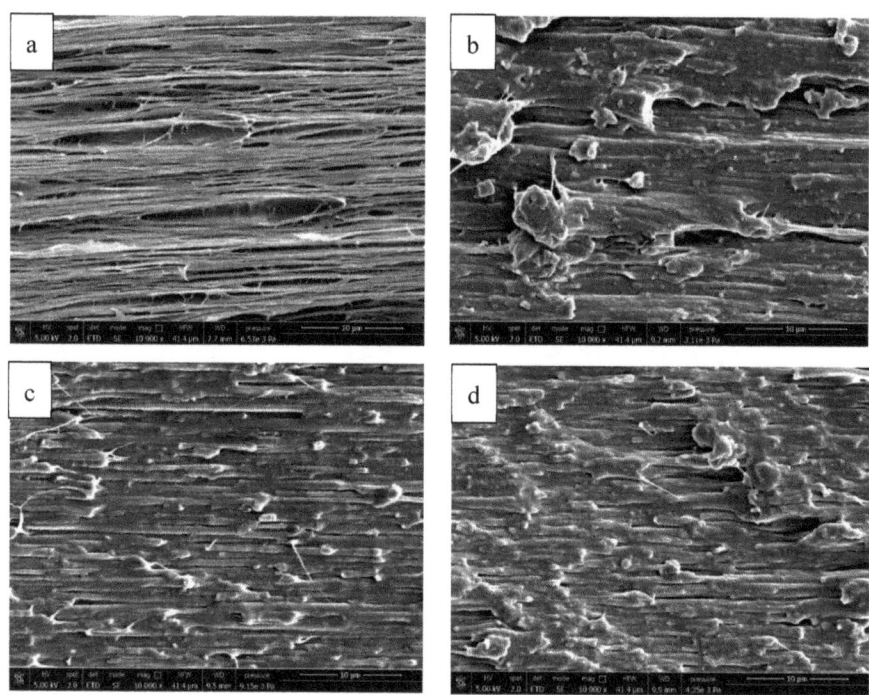

Figure 11. SEM micrographs made at the surface of tensile specimen cryo-fractured along the draw direction: (**a**) 80–20; (**b**) (80/20)J; (**c**) 60–40; (**d**) (60/40)J.

4. Conclusions

In this study, PLA/PBSA binary blends with different PBSA contents (from 5 up to 40 wt.%) were investigated from a rheological, thermal and mechanical point of view. To the blends (80–20 and 60–40), that presented a higher value of impact resistance and elongation at break, 2 wt.% of an epoxy oligomer (EO) as compatibilizer was added. The two blends also presented different morphologies: the 80–20 blend possesses a morphology where the PBSA particles are dispersed within the PLA matrix, while the 60–40 blend shows a co-continuous structure.

From the thermal properties point of view, no significant variations were observed caused by the EO addition, whereas, as could be expected, the interaction created by EO produced an increment of the melt viscosity.

However, the most interesting results were observed from the micro and macro mechanical analysis. The mechanical results showed that the Charpy impact resistance value (high speed test) and the J_{lim} (energy absorbed from the specimen at the moment of the crack propagation in slow-rate test conditions) are higher for the 60–40 blends having a co-continuous morphology. However, for this 60/40 blend the EO addition does not significantly alter the mechanical response. In fact, dilatometric tests show that for the 60/40 blend the micromechanical response is of the deviatoric type and the morphology, investigated at the surface of cryo-fractured tensile specimen along the draw direction, was comparable.

On the other hand, the 80/20 blend was significantly affected by the presence of EO. The micromechanical response passed from a cavitational response to a deviatoric one, probably induced by the marked change in the dispersed phase morphology with the EO addition and the higher complexity of macromolecule structure generated by the reactive processing.

The systematic study described in this paper was thus useful for a better knowledge of thermo-mechanical behavior of PLA blends containing up to 40% of PBSA.

Moreover, the EO used in the present paper resulted in being quite useful in effectively modulating melt viscosity and this is necessary not only for adapting the material to several processing methodologies, but also as compatibility enhancer, impacting positively the mechanical performance. However, this additive is not biobased and biodegradable. Hence in the future it could be important to replace it with biobased and biodegradable ones, to grant a full circularity of the material by keeping into account both its origin and final disposal.

Author Contributions: Conceptualization, M.-B.C. and L.A.; experimental work, A.V., I.C. and L.A.; theoretical background, L.A. and A.L.; data curation and elaboration, A.V., I.C., L.A. and M.-B.C.; writing—original draft preparation, L.A. and A.V.; writing—review and editing, M.-B.C.; supervision, P.C. and A.L.; project administration and funding acquisition, M.-B.C. and P.C. All authors have read and agreed to the published version of the manuscript.

Funding: This research was funded by the Bio-Based Industries Joint Undertaking under the European Union Horizon 2020 research program (BBI-H2020), BIONTOP project, grant number G.A 837761.

Institutional Review Board Statement: Not applicable.

Informed Consent Statement: Not applicable.

Data Availability Statement: The data presented in this study are available on request from the corresponding author.

Acknowledgments: Belen Monje Martinez of AIMPLAS, coordinator of BIONTOP project, is thanked for helpful suggestions regarding the present manuscript.

Conflicts of Interest: The authors declare no conflict of interest.

References

1. Vroman, I.; Tighzert, L. Biodegradable polymers. *Materials* **2009**, *2*, 307–344. [CrossRef]
2. Luckachan, G.E.; Pillai, C.K.S. Biodegradable Polymers—A Review on Recent Trends and Emerging Perspectives. *J. Polym. Environ.* **2011**, *19*, 637–676. [CrossRef]
3. La Mantia, F.P.; Morreale, M.; Botta, L.; Mistretta, M.C.; Ceraulo, M.; Scaffaro, R. Degradation of polymer blends: A brief review. *Polym. Degrad. Stab.* **2017**, *145*, 79–92. [CrossRef]
4. Gross, R.A.; Kalra, B. Biodegradable Polymers for the Environment. *Science* **2002**, *297*, 803–807. [CrossRef] [PubMed]
5. Aliotta, L.; Gigante, V.; Coltelli, M.; Cinelli, P.; Lazzeri, A.; Seggiani, M. Thermo-Mechanical Properties of PLA / Short Flax Fiber Biocomposites. *Appl. Sci.* **2019**, *9*, 3797.
6. Harris, A.M.; Lee, E.C. Improving mechanical performance of injection molded PLA by controlling crystallinity. *J. Appl. Polym. Sci.* **2008**, *107*, 2246–2255. [CrossRef]
7. Nofar, M.; Maani, A.; Sojoudi, H.; Heuzey, M.C.; Carreau, P.J. Interfacial and rheological properties of PLA/PBAT and PLA/PBSA blends and their morphological stability under shear flow. *J. Rheol.* **2015**, *59*, 317–333. [CrossRef]
8. Bureepukdee, C.; Suttireungwong, S.; Seadan, M. A study on reactive blending of (polylactic acid) and poly(butylene succinate co adipate). *IOP Conf. Ser. Mater. Sci. Eng.* **2015**, *87*, 012070. [CrossRef]
9. Nofar, M.; Tabatabaei, A.; Sojoudiasli, H.; Park, C.B.; Carreau, P.J.; Heuzey, M.-C.; Kamal, M.R. Mechanical and bead foaming behavior of PLA-PBAT and PLA-PBSA blends with different morphologies. *Eur. Polym. J.* **2017**, *34*, 231–244. [CrossRef]
10. Yokohara, T.; Yamaguchi, M. Structure and properties for biomass-based polyester blends of PLA and PBS. *Eur. Polym. J.* **2008**, *44*, 677–685. [CrossRef]
11. Hassan, E.; Wei, Y.; Jiao, H.; Yu, M. Dynamic Mechanical properties and Thermal stability of Poly(lactic acid) and poly(butylene succinate) blends composites. *J. Fiber Bioeng. Inform.* **2013**, *6*, 85–94. [CrossRef]
12. Pradeep, S.A.; Kharbas, H.; Turng, L.S.; Avalos, A.; Lawrence, J.G.; Pilla, S. Investigation of thermal and thermomechanical properties of biodegradable PLA/PBSA composites processed via supercritical fluid-assisted foam injection molding. *Polymers* **2017**, *9*, 22. [CrossRef] [PubMed]
13. Lee, S.; Lee, J.W. Characterization and processing of Biodegradable polymer blends of poly(lactic acid) with poly(butylene succinate adipate). *Korea Aust. Rheol. J.* **2005**, *17*, 71–77.
14. Takayama, T.; Todo, M.; Tsuji, H.; Arakawa, K. Improvement of fracture properties of PLA/PCL polymer blends by control of phase structures. *Kobunshi Ronbunshu* **2006**, *63*, 626–632. [CrossRef]

15. Semba, T.; Kitagawa, K.; Ishiaku, U.S.; Kotaki, M.; Hamada, H. Effect of compounding procedure on mechanical properties and dispersed phase morphology of poly(lactic acid)/polycaprolactone blends containing peroxide. *J. Appl. Polym. Sci.* **2007**, *103*, 1066–1074. [CrossRef]
16. Harada, M.; Lida, K.; Okamoto, K.; Hayashi, H.; Hirano, K. Reactive Compatibilization of Biodegrdable Poly(lactic acid)/Poly(caprolactone) Blends with reacive Processing Agents. *Polym. Eng. Sci.* **2008**, 1359–1368. [CrossRef]
17. Arruda, L.C.; Magaton, M.; Bretas, R.E.S.; Ueki, M.M. Influence of chain extender on mechanical, thermal and morphological properties of blown films of PLA/PBAT blends. *Polym. Test.* **2015**. [CrossRef]
18. Kumar, M.; Mohanty, S.; Nayak, S.K.; Rahail Parvaiz, M. Effect of glycidyl methacrylate (GMA) on the thermal, mechanical and morphological property of biodegradable PLA/PBAT blend and its nanocomposites. *Bioresour. Technol.* **2010**, *101*, 8406–8415. [CrossRef]
19. Muthuraj, R.; Misra, M.; Mohanty, A.K. Biodegradable Poly(butylene succinate) and Poly(butylene adipate-co-terephthalate) Blends: Reactive Extrusion and Performance Evaluation. *J. Polym. Environ.* **2014**, *22*, 336–349. [CrossRef]
20. Gigante, V.; Canesi, I.; Cinelli, P.; Coltelli, M.B.; Lazzeri, A. Rubber Toughening of Polylactic Acid (PLA) with Poly(butylene adipate-co-terephthalate) (PBAT): Mechanical Properties, Fracture Mechanics and Analysis of Ductile-to-Brittle Behavior while Varying Temperature and Test Speed. *Eur. Polym. J.* **2019**, *115*, 125–137. [CrossRef]
21. Changwichan, K.; Silalertruksa, T.; Gheewala, S.H. Eco-efficiency assessment of bioplastics production systems and end-of-life options. *Sustainability* **2018**, *10*, 952. [CrossRef]
22. European Bioplastics. Bioplastics Market Data. Available online: https://www.european-bioplastics.org/market/ (accessed on 25 September 2020).
23. Gigante, V.; Coltelli, M.; Vannozzi, A.; Panariello, L.; Fusco, A.; Trombi, L.; Donnarumma, G.; Danti, S.; Lazzeri, A. Flat Die Extruded Biocompatible Poly(Lactic Acid). *Polymers* **2019**, *11*, 1857. [CrossRef] [PubMed]
24. Bucknall, C.B.; Heather, P.S.; Lazzeri, A. Rubber toughening of plastics. *J. Mater. Sci.* **2000**, *24*, 2255–2261. [CrossRef]
25. Lazzeri, A.; Bucknall, C.B. Recent developments in the modeling of dilatational yielding in toughened plastics. *Toughening Plast. Adv. Model. Exp.* **2000**, *264*, 14–35.
26. Nofar, M.; Salehiyan, R.; Ciftci, U.; Jalali, A.; Durmuş, A. Ductility improvements of PLA-based binary and ternary blends with controlled morphology using PBAT, PBSA, and nanoclay. *Compos. Part B Eng.* **2020**, *182*, 107661. [CrossRef]
27. Meng, L.; Yu, L.; Khalid, S.; Liu, H.; Zhang, S.; Duan, Q.; Chen, L. Preparation, microstructure and performance of poly(lactic acid)-Poly(butylene succinate-co-butyleneadipate)-starch hybrid composites. *Compos. Part B Eng.* **2019**, *177*, 107384. [CrossRef]
28. Garcia-Campo, M.; Quiles-Carrillo, L.; Sanchez-Nacher, L.; Balart, R.; Montanes, N. High toughness poly(lactic acid) (PLA) formulations obtained by ternary blends with poly(3-hydroxybutyrate) (PHB) and flexible polyesters from succinic acid. *Polym. Bull.* **2018**, *76*, 1839–1859. [CrossRef]
29. Supthanyakula, R.; Kaabbuathongb, N.; Chirachanchai, S. Random poly(butylene succinate-co-lactic acid) as a multi-functional additive for miscibility, toughness, and clarity of PLA/PBS blends. *Polymer* **2016**, *105*, 1–9. [CrossRef]
30. Su, S.; Kopitzky, R.; Tolga, S.; Kabasci, S. Polylactide (PLA) and Its Blends with Poly(butylene succinate) (PBS): A Brief Review. *Polymers* **2019**, *11*, 1193. [CrossRef]
31. Hamad, K.; Kaseem, M.; Ayyoob, M.; Joo, J.; Deri, F. Polylactic acid blends: The future of green, light and tough. *Prog. Polym. Sci.* **2018**, *85*, 83–127. [CrossRef]
32. Wang, R.; Wang, S.; Zhang, Y.; Wan, C.; Ma, P. Toughening modification of PLLA/PBS blends via in situ compatibilization. *Polym. Eng. Sci.* **2009**, *49*, 26–33. [CrossRef]
33. Kfoury, G.; Raquez, J.-M.; Hassouna, F.; Odent, J.; Toniazzo, V.; Ruch, D.; Dubois, P. Recent advances in high performance poly(lactide): From "green" plasticization to super-tough materials via (reactive) compounding. *Front. Chem.* **2013**, *1*, 32. [CrossRef] [PubMed]
34. Cardinaels, R.; Moldenaers, P. Morphology development in immiscible polymer blends. In *Polymer Morphology: Priciples, Characterization and Properties*; Guo, Q., Ed.; John Wiley and Sons: Hoboken, NJ, USA, 2016; Chapter 19, pp. 348–373.
35. Thomas, S.; Grohens, Y.; Jyotishkumar, P. (Eds.) *Characterization of Polymer Blends Miscibility, Morphology and Interfaces*; Wiley-VCH Verlag GmbH & Co. KG: Weinheim, Germany, 2014.
36. Harrats, C.; Thomas, S.; Groeninckx, G. (Eds.) *Micro- and Nanostructured Multiphase Polymer Blend Systems: Phase Morphology and Interfaces*; CRC Press, Taylor and Francis: Boca Raton, FL, USA, 2006.
37. Hu, L.; Vuillaume, P.Y. Chapter 7—Reactive Compatibilization of Polymer Blends by Coupling Agents and Interchange Catalysts. In *Compatibilization of Polymer Blends*; Ajitha, A.R., Thomas, S., Eds.; Elsevier: Amsterdam, The Netherlands, 2020; pp. 205–248.
38. Homklin, R.; Hongsriphan, N. Mechanical and thermal properties of PLA/PBS cocontinuous blends adding nucleating agent. *Energy Procedia* **2013**, *34*, 871–879. [CrossRef]
39. Dhibar, A.K.; Kim, J.K.; Khatua, B.B. Cocontinuous phase morphology of asymmetric compositions of polypropylene/high-density polyethylene blend by the addition of clay. *J. Appl. Polym. Sci.* **2011**, *119*, 3080–3092. [CrossRef]
40. Ito, E.N.; Pessan, L.A.; Covas, J.A.; Hage, E., Jr. Analysis of the morphological development of PBT/ABS blends during the twin screw extrusion and injection molding processes. *Int. Polym. Process.* **2003**, *18*, 376–381. [CrossRef]
41. Wu, D.; Yuan, L.; Laredo, E.; Zhang, M.; Zhou, W. Interfacial properties, viscoelasticity, and thermal behaviors of poly(butylene succinate)/polylactide blend. *Ind. Eng. Chem. Res.* **2012**, *51*, 2290–2298. [CrossRef]

42. Macosko, C.W. Morphology Development and Control in Immiscible Polymer Blends. *Macromol. Symp.* **2000**, *149*, 171–184. [CrossRef]
43. Ajitha, A.R.; Thomas, S. Chapter 1—Introduction: Polymer blends, thermodynamics, miscibility, phase separation, and compatibilization. In *Compatibilization of Polymer Blends*; Ajitha, A.R., Thomas, S., Eds.; Elsevier: Amsterdam, The Netherlands, 2020; pp. 1–29.
44. Coltelli, M.B.; Bronco, S.; Chinea, C. The effect of free radical reactions on structure and properties of poly(lactic acid) (PLA) based blends. *Polym. Degrad. Stab.* **2010**, *95*, 332–341. [CrossRef]
45. Corre, Y.-M.; Duchet, J.; Reignier, J.; Maazouz, A. Melt strengthening of poly(lactic acid) through reactive extrusion with epoxy-functionalized chains. *Rheol. Acta* **2011**, *50*, 613–629. [CrossRef]
46. Han, C.D. *Rheology and Processing of Polymeric Materials: Volume 1: Polymer Rheology*; Oxford University Press on Demand: Oxford, UK, 2007; Volume 1, ISBN 0195187822.
47. Meng, Q.; Heuzey, M.-C.; Carreau, P.J. Control of thermal degradation of polylactide/clay nanocomposites during melt processing by chain extension reaction. *Polym. Degrad. Stab.* **2012**, *97*, 2010–2020. [CrossRef]
48. Wu, C.-S. Utilization of peanut husks as a filler in aliphatic–aromatic polyesters: Preparation, characterization, and biodegradability. *Polym. Degrad. Stab.* **2012**, *97*, 2388–2395. [CrossRef]
49. Bhatia, A.; Gupta, R.K.; Bhattacharya, S.N.; Choi, H.J. Compatibility of biodegradable poly(lactic acid) (PLA) and poly(butylene succinate) (PBS) blends for packaging application. *Korea Aust. Rheol. J.* **2007**, *19*, 125–131. [CrossRef]
50. Gu, S.Y.; Zhang, K.; Ren, J.; Zhan, H. Melt rheology of polylactide/poly(butylene adipate-co-terephthalate) blends. *Carbohydr. Polym.* **2008**, *74*, 79–85. [CrossRef]
51. Lin, S.; Guo, W.; Chen, C.; Ma, J.; Wang, B. Mechanical properties and morphology of biodegradable poly(lactic acid)/poly(butylene adipate-co-terephthalate) blends compatibilized by transesterification. *Mater. Des.* **2012**, *36*, 604–608. [CrossRef]
52. Al-Itry, R.; Lamnawar, K.; Maazouz, A. Improvement of thermal stability, rheological and mechanical properties of PLA, PBAT and their blends by reactive extrusion with functionalized epoxy. *Polym. Degrad. Stab.* **2012**, *97*, 1898–1914. [CrossRef]
53. Pilla, S.; Kim, S.G.; Auer, G.K.; Gong, S.; Park, C.B. Microcellular extrusion-foaming of polylactide with chain-extender. *Polym. Eng. Sci.* **2009**, *49*, 1653–1660. [CrossRef]
54. Wang, X.; Peng, S.; Chen, H.; Yu, X.; Zhao, X. Mechanical properties, rheological behaviors, and phase morphologies of high-toughness PLA/PBAT blends by in-situ reactive compatibilization. *Compos. Part B Eng.* **2019**, *173*, 107028. [CrossRef]
55. Lascano, D.; Quiles-Carrillo, L.; Balart, R.; Boronat, T.; Montanes, N. Toughened poly (lactic acid)-PLA formulations by binary blends with poly(butylene succinate-co-adipate)-PBSA and their shape memory behaviour. *Materials* **2019**, *12*, 622. [CrossRef]
56. Chaiwutthinan, P.; Leejarkpai, T.; Kashima, D.P.; Chuayjuljit, S. Poly(Lactic Acid)/Poly(Butylene Succinate) Blends Filled with Epoxy Functionalised Polymeric Chain Extender. *Adv. Mater. Res.* **2013**, *664*, 644–648. [CrossRef]
57. Elhassan, A.S.M.; Saeed, H.A.M.; Eltahir, Y.A.; Xia, Y.M.; Wang, Y.P. Modification of PLA with Chain Extender. *Appl. Mech. Mater.* **2014**, *716–717*, 44–47. [CrossRef]
58. Nakayama, D.; Wu, F.; Mohanty, A.K.; Hirai, S.; Misra, M. Biodegradable Composites Developed from PBAT/PLA Binary Blends and Silk Powder: Compatibilization and Performance Evaluation. *ACS Omega* **2018**, *3*, 12412–12421. [CrossRef] [PubMed]
59. Coltelli, M.-B.; Gigante, V.; Vannozzi, A.; Aliotta, L.; Danti, S.; Neri, S.; Gagliardini, A.; Morganti, P.; Panariello, L.; Lazzeri, A. Poly(lactic) acid (PLA) based nano-structured functional films for personal care applications. In Proceedings of the AUTEX 2019—19th World Textile Conference on Textiles at the Crossroads, Ghent, Belgium, 11–15 June 2019.
60. Aliotta, L.; Cinelli, P.; Beatrice Coltelli, M.; Lazzeri, A. Rigid filler toughening in PLA-Calcium Carbonate composites: Effect of particle surface treatment and matrix plasticization. *Eur. Polym. J.* **2018**, *113*, 78–88. [CrossRef]
61. Aliotta, L.; Gigante, V.; Acucella, O.; Signori, F.; Lazzeri, A. Thermal, Mechanical and Micromechanical Analysis of PLA/PBAT/POE-g-GMA Extruded Ternary Blends. *Front. Mater.* **2020**. [CrossRef]
62. Lazzeri, A.; Thio, Y.S.; Cohen, R.E. Volume strain measurements on $CACO_3$/polypropylene particulate composites: The effect of particle size. *J. Appl. Polym. Sci.* **2004**, *91*, 925–935. [CrossRef]
63. Bernal, C.R.; Montemartini, P.E.; Frontini, P.M. The use of load separation criterion and normalization method in ductile fracture characterization of thermoplastic polymers. *J. Polym. Sci. Part B Polym. Phys.* **1996**, *34*, 1869–1880. [CrossRef]
64. Baldi, F.; Agnelli, S.; Riccò, T. On the determination of the point of fracture initiation by the load separation criterion in J-testing of ductile polymers. *Polym. Test.* **2013**, *32*, 1326–1333. [CrossRef]
65. Sharobeam, M.H.; Landes, J.D. The load separation criterion and methodology in ductile fracture mechanics. *Int. J. Fract.* **1991**, *47*, 81–104. [CrossRef]
66. Baldi, F.; Agnelli, S.; Riccò, T. On the applicability of the load separation criterion in determining the fracture resistance (JIC) of ductile polymers at low and high loading rates. *Int. J. Fract.* **2010**, *165*, 105–119. [CrossRef]
67. Agnelli, S.; Baldi, F.; Riccò, T. A tentative application of the energy separation principle to the determination of the fracture resistance (Jic) of rubbers. *Eng. Fract. Mech.* **2019**, *90*, 76–88. [CrossRef]
68. Blackman, B.; Baldi, F.; Castellani, L.; Frontini, P.; Laiarinandrasana, L.; Pegoretti, A.; Rink, M.; Salazar, A.; Visser, H.; Agnelli, S. Application of the load separation criterion in J-testing of ductile polymers: A round-robin testing exercise. *Polym. Test.* **2015**, *44*, 72–81.

69. Agnelli, S.; Baldi, F. *A Testing Protocol for the Construction of the "Load Separation Parameter Curve" for Plastics*; ESIS TC4 communication; ESIS: Delft, The Netherlands, 2013.
70. Gui, Z.Y.; Wang, H.R.; Gao, Y.; Lu, C.; Cheng, S.J. Morphology and melt rheology of biodegradable poly(lactic acid)/poly(butylene succinate adipate) blends: Effect of blend compositions. *Iran. Polym. J.* **2012**, *21*, 81–89. [CrossRef]
71. Fischer, E.W.; Sterzel, H.J.; Wegner, G. Investigation of the structure of solution grown crystals of lactide copolymers by means of chemical reactions. *Kolloid Z. Z. Polym.* **1973**, *251*, 980–990. [CrossRef]
72. Tang, Z.; Zhang, C.; Liu, X.; Zhu, J. The crystallization behavior and mechanical properties of polylactic acid in the presence of a crystal nucleating agent. *J. Appl. Polym. Sci.* **2012**, *125*, 1108–1115. [CrossRef]
73. Bhadane, P.A.; Champagne, M.F.; Huneault, M.A.; Tofan, F.; Favis, B.D. Erosion-dependant continuity development in high viscosity ratio blends of very low interfacial tension. *J. Polym. Sci. Part B Polym. Phys.* **2006**, *44*, 1919–1929. [CrossRef]
74. Xu, X.; Yan, X.; Zhu, T.; Zhang, C.; Sheng, J. Phase morphology development of polypropylene/ethylene-octene copolymer blends: Effects of blend composition and processing conditions. *Polym. Bull.* **2007**, *58*, 465–478. [CrossRef]
75. Lazzeri, A.; Bucknall, C.B. Dilatational bands in rubber-toughened polymers. *J. Mater. Sci.* **1993**, *28*, 6799–6808. [CrossRef]
76. Wu, S. Chain structure, phase morphology, and toughness relationships in polymers and blends. *Polym. Eng. Sci.* **1990**, *30*, 753–761. [CrossRef]
77. Pötschke, P.; Paul, D.R. Formation of Co-continuous Structures in Melt-Mixed Immiscible Polymer Blends. *J. Macromol. Sci. Part C* **2003**, *43*, 87–141. [CrossRef]
78. Harrats, C.; Coltelli, M.B.; Groeninckx, G. Features on the development and stability of phase morphology in complex multicomponent polymeric systems: Main focus on processing aspects. In *Polymer Morphology: Principles, Characterization, and Processing*, 1st ed.; Guo, Q., Ed.; Wiley: Hoboken, NJ, USA, 2016; pp. 418–438.
79. Coltelli, M.-B.; Mallegni, N.; Rizzo, S.; Cinelli, P.; Lazzeri, A. Improved Impact Properties in Poly(lactic acid) (PLA) Blends Containing Cellulose Acetate (CA) Prepared by Reactive Extrusion. *Materials* **2019**, *12*, 270. [CrossRef]
80. Coltelli, M.B.; Harrats, C.; Aglietto, M.; Groeninckx, G. Influence of compatibilizer precursor structure on the phase distribution of low density poly(ethylene) in a poly(ethylene terephthalate) matrix. *Polym. Eng. Sci.* **2008**, *48*, 1424–1433. [CrossRef]
81. Coltelli, M.B.; Toncelli, C.; Ciardelli, F.; Bronco, S. Compatible blends of biorelated polyesters through catalytic transesterification in the melt. *Polym. Degrad. Stab.* **2011**, *96*, 982–990. [CrossRef]
82. Quiles-Carrillo, L.; Montanes, N.; Lagaron, J.M.; Balart, R.; Torres-Giner, S. In Situ Compatibilization of Biopolymer Ternary Blends by Reactive Extrusion with Low-Functionality Epoxy-Based Styrene–Acrylic Oligomer. *J. Polym. Environ.* **2019**, *27*, 84–96. [CrossRef]
83. Rasselet, D.; Caro-Bretelle, A.-S.; Taguet, A.; Lopez-Cuesta, J.-M. Reactive Compatibilization of PLA/PA11 Blends and Their Application in Additive Manufacturing. *Materials* **2019**, *12*, 485. [CrossRef] [PubMed]
84. de CD Nunes, E.; de Souza, A.G.; dos S. Rosa, D. Effect of the Joncryl® ADR Compatibilizing Agent in Blends of Poly(butylene adipate-co-terephthalate)/Poly(lactic acid). *Macromol. Symp.* **2019**, *383*, 1–6. [CrossRef]
85. Wang, Y.; Bhattacharya, M.; Mano, J.F. Thermal analysis of the multiple melting behavior of poly(butylene succinate-co-adipate). *J. Polym. Sci. Part B Polym. Phys.* **2005**, *43*, 3077–3082. [CrossRef]
86. Fenni, S.E.; Wang, J.; Haddaoui, N.; Favis, B.D.; Müller, A.J.; Cavallo, D. Crystallization and self-nucleation of PLA, PBS and PCL in their immiscible binary and ternary blends. *Thermochim. Acta* **2019**, *677*, 117–130. [CrossRef]
87. Wunderlich, B. *Macromolecular Physics*; Elsevier: Amsterdam, The Netherlands, 2012; Volume 2, ISBN 0323148948.
88. Di Lorenzo, M.L. Calorimetric analysis of the multiple melting behavior of poly(L-lactic acid). *J. Appl. Polym. Sci.* **2006**, *100*, 3145–3151. [CrossRef]
89. Aliotta, L.; Cinelli, P.; Coltelli, M.B.; Righetti, M.C.; Gazzano, M.; Lazzeri, A. Effect of nucleating agents on crystallinity and properties of poly(lactic acid) (PLA). *Eur. Polym. J.* **2017**, *93*, 822–832. [CrossRef]
90. Mallegni, N.; Phuong, T.V.; Coltelli, M.B.; Cinelli, P.; Lazzeri, A. Poly(lactic acid) (PLA) based tear resistant and biodegradable flexible films by blown film extrusion. *Materials* **2018**, *11*, 148. [CrossRef]
91. Ostrowska, J.; Sadurski, W.; Paluch, M.; Tyński, P.; Bogusz, J. The effect of poly(butylene succinate) content on the structure and thermal and mechanical properties of its blends with polylactide. *Polym. Int.* **2019**, *68*, 1271–1279. [CrossRef]
92. Eslami, H.; Kamal, M.R. Effect of a chain extender on the rheological and mechanical properties of biodegradable poly(lactic acid)/poly[(butylene succinate)-co-adipate] blends. *J. Appl. Polym. Sci.* **2013**, *129*, 2418–2428. [CrossRef]
93. Supthanyakul, R.; Kaabbuathong, N.; Chirachanchai, S. Poly(l-lactide-b-butylene succinate-b-l-lactide) triblock copolymer: A multi-functional additive for PLA/PBS blend with a key performance on film clarity. *Polym. Degrad. Stab.* **2017**, *142*, 160–168. [CrossRef]
94. Nascimento, L.; Gamez-Perez, J.; Santana, O.O.; Velasco, J.I.; Maspoch, M.L.; Franco-Urquiza, E. Effect of the Recycling and Annealing on the Mechanical and Fracture Properties of Poly(Lactic Acid). *J. Polym. Environ.* **2010**, *18*, 654–660. [CrossRef]
95. Todo, M.; Takayama, T. Fracture mechanisms of biodegradable PLA and PLA/PCL blends. *Biomater. Chem.* **2011**.
96. Naqui, S.I.; Robinson, I.M. Tensile dilatometric studies of deformation in polymeric materials and their composites. *J. Mater. Sci.* **1993**, *28*, 1421–1429. [CrossRef]

Article

Volume Change during Creep and Micromechanical Deformation Processes in PLA–PBSA Binary Blends

Laura Aliotta [1,2,*], Vito Gigante [1,2], Maria-Beatrice Coltelli [1,2] and Andrea Lazzeri [1,2,*]

[1] Department of Civil and Industrial Engineering, University of Pisa, 56122 Pisa, Italy; vito.gigante@dici.unipi.it (V.G.); maria.beatrice.coltelli@unipi.it (M.-B.C.)
[2] National Interuniversity Consortium of Materials Science and Technology (INSTM), 50121 Florence, Italy
* Correspondence: laura.aliotta@dici.unipi.it (L.A.); andrea.lazzeri@unipi.it (A.L.)

Abstract: In this paper, creep measurements were carried out on poly(lactic acid) (PLA) and its blends with poly(butylene succinate-adipate) (PBSA) to investigate the specific micromechanical behavior of these materials, which are promising for replacing fossil-based plastics in several applications. Two different PBSA contents at 15 and 20 wt.% were investigated, and the binary blends were named 85-15 and 80-20, respectively. Measurements of the volume strain, using an optical extensometer, were carried out with a universal testing machine in creep configuration to determine, accompanied by SEM images, the deformation processes occurring in a biopolymeric blend. With the aim of correlating the creep and the dilatation variation, analytical models were applied for the first time in biopolymeric binary blends. By using an Eyring plot, a significant change in the curves was found, and it coincided with the onset of the cavitation/debonding mechanism. Furthermore, starting from the data of the pure PLA matrix, using the Eyring relationship, an apparent stress concentration factor was calculated for PLA-PBSA systems. From this study, it emerged that the introduction of PBSA particles causes an increment in the apparent stress intensity factor, and this can be ascribed to the lower adhesion between the two biopolymers. Furthermore, as also confirmed by SEM analysis, it was found that debonding was the main micromechanical mechanism responsible for the volume variation under creep configuration; it was found that debonding starts earlier (at a lower stress level) for the 85-15 blend.

Keywords: binary blends; creep; micromechanics

1. Introduction

Physical polymeric blends, constituted by a rubbery phase embedded in a more rigid matrix, are a long-lasting route for the preparation of new materials with a modulated balance of properties [1]. The advent of biobased and biodegradable polymers could positively affect the environmental profile of products, thanks to an improved carbon neutrality with respect to fossil-based counterparts and a more environmentally friendly end of life [2–4]. The increased interest towards biopolymers has resulted in a revival of blending technology and, in order to exploit their potential and enter new markets, the study of these materials is at the center of scientific research [5]. Improved knowledge of the mechanical behavior of biobased materials is fundamental for better exploiting their peculiar properties and comparing them with fossil-based ones, favoring the replacement of the latter in several application sectors.

In this context, it is beneficial to investigate micromechanical deformation processes that occur in physical polymeric blends, knowing that external stresses can be the cause of the starting point of numerous micromechanical deformation processes that play a critical role regarding the failure of pure heterogeneous systems [6]. The two basic micromechanical deformation processes in pure polymers are shear yielding and crazing [7], while in multiphase systems, it can also be active debonding and cavitation [8]. The debonding mechanism involves the formation of cavities/voids at the interface between the rubber

phase and the matrix, while void formation occurring internally to the rubber particles is known as cavitation [9]. The expansion of the cavities occurs when the volumetric strain energy is greater than that required for the creation of the void surface area [10]. The parameter governing the cavitation or debonding mechanism is the value of the polymer/rubber interfacial adhesion: high adhesion values contribute to the internal cavitation of the rubber particles, while low values contribute to the debonding mechanism [11]. It is important to state that micromechanical deformations are competitive processes, the prevailing one is determined by the inherent properties of the matrix polymer and by local stress distribution [12]. The role and importance of void formation within or around the rubber particles in polymer blends are still not clear, but they have been at the center of academic interest in material science [13–16]. On the basis of energy balances, Lazzeri and Bucknall [17,18] developed the idea that cavitation/debonding of rubber particles arises at the crack tip and can be the cause of dilatational shear yielding and/or crazing in the matrix. The following matrix distortion is not homogeneous and becomes highly localized due to the formation of bands of voids and sheared material called "dilatational shear bands" or more simply "dilatational bands". The effect of cavitation is a local decreasing of the bulk modulus and hydrostatic stress components near the void and a corresponding growth of the stress deviatoric component. Higher elastic energy may then cause a faster advance of shear bands and, thus, a larger plastic zone form is attained [19].

In order to understand the micromechanical deformation processes that occur in particular polymeric blends, measurements and analysis of the volume change during uniaxial tensile or creep tests can lead to a better understanding of the deformation phenomena [20,21]. In particular, the theories existing in literature state that during shear yielding, the volume of the sample remains constant, while crazing, debonding, and cavitation are characterized by the increase in volume strain [22]. To make a "quantitative evaluation" of the deformation mechanisms effective in rubber-toughened systems, dilatometric studies of tensile creep have been carried out as function of the tensile stress or strain prior to fracture [23].

During uniaxial tensile testing, after that the rubber particles generate voids (by cavitation and or debonding), the voids elongate as the specimen extends. Generally, debonding or cavitation occurs before yielding [24]. The void growth mechanism is a second and successive deformation process that causes great differences in volume variation. Since this process occurs after yielding and it is in common to both cavitation and debonding phenomena, very interesting are creep studies in which different stress levels (below the yield stress) are investigated in order to identify the stress level for which the debonding or cavitation process begins.

In tensile creep experiments, the onset of dilatational yielding goes with a rapid increase in deformation; for this reason, the void volume will increase with strain in addition to the volume change of the matrix itself [20]. To avoid the drawbacks of mechanical extensometers, which include range limitations, recently video-controlled tensile testing equipment has been developed by G'Sell et al. [25,26] to optically evaluate the volumetric strain.

The abovementioned important concepts have been tested and developed on conventional polymeric systems; however, to the best of our knowledge, they have not been extensively explored on biopolymeric blends. In particular, coupling between the dilatometric volume measurement with an optical system performed in real time during creep tests of a biopolymeric blend is a novelty.

The present paper, starting from one of the author's own experience [27,28], aimed to explore the micromechanical deformation mechanisms of blends based on polylactic acid (PLA) and polybutylene succinate adipate (PBSA). These blends were recently considered for their potential use in packaging and personal care/sanitary applications [29], hence, in sectors where products have a short life thus highly contributing to the production of enormous amounts of waste [30]. In this paper, two formulations were studied in which PBSA was introduced at 15 and 20 wt.%, respectively.

Krishnan et al. [31], LeBarbe [32], and Nagarjan et al. [33] published exhaustive reviews regarding the problem of PLA toughening and the necessity of increasing the ductility without losing too many of the characteristics that make PLA interesting, i.e., high elastic modulus, good processability, high tensile strength (without forgetting biodegradability and biobased content).

In any case, while different researchers have investigated PLA toughened with biodegradable rubber, in which they stated that certainly the size of the dispersed particles, together with the quality of interfacial adhesion, determines the final toughening effect in PLA [34–40], only a few papers studied the short-time creep behavior of PLA-based blends [41–43], and none of them addressed the issue of relating micromechanical deformation phenomena with volumetric dilatometric variations from creep tests. For this reason, in this work, measurements of the volume strain, using an optical extensometer, were conducted with a universal testing machine in creep configuration to determine, accompanied by SEM images, the micromechanical deformation processes involved in a biopolymeric blend system.

Analytical models were also applied in order to correlate the creep to the dilatation variation; in particular, the Andrade equation [8] was applied and the b parameter for the polymeric systems was calculated. Plotting the $\log b$ against the applied stress, using an Eyring plot, a significant change in the curves was found, and it coincided with the onset of the cavitation/debonding mechanism. Furthermore, knowing the data of the pure matrix, from the Eyring relationship, the apparent stress concentration factor was calculated for the PLA binary blends with 15 and 20 wt.% of PBSA.

2. Theoretical Analysis and Theoretical Background

Creep rupture of a polymer is the result of combined events (such as viscoelastic deformation, primary and secondary bond rupture, shear yielding, crazing, void formation and growth) and fibril breakdown with intrinsic and extrinsic flaws, leading to fracture. In the case of polymer blends or composites, the interfacial strength and morphology must also be taken into account [44]. According to the creep curves (*creep strain (ε) vs. creep time (t)*), in polymers, four stages can be considered [45]: (I) the first stage of instantaneous deformation (ε_0), (II) the second stage named primary creep (ε_1), (III) the third stage named secondary or transient creep (ε_2) in which the creep rate reaches a steady-state value, and the fourth stage (IV) (ε_4) in which the creep rate increases abruptly and the final creep rupture occurs. Transient creep of many materials (including polymeric materials) obeys Andrade's law in which creep strain is proportional to the cube root of time according to the following equation [46–48]:

$$\varepsilon(t) = \varepsilon(0) + bt^{\frac{1}{3}} \qquad (1)$$

where $\varepsilon(0)$ is the time independent instantaneous elongation due to the elastic or plastic deformation of polymer once the external load is applied; b is a function of stress and temperature. The validity of the Andrade's approach can be easily verified by plotting in the region of the transient creep, $\varepsilon(t)$ against $t^{1/3}$, and checking if the data align on a straight line. If the data fits the model, the b parameter can be thus obtained, because it is equal to the angular coefficient of the straight line.

If the b parameter is known, it can be related to the activation volume according to the following Eyring relationship [48]:

$$b = Q \exp\left(\frac{\gamma V \sigma}{k_B T}\right) \qquad (2)$$

where Q is a constant, γ is a stress concentration factor, V is the activation volume for the deformation process, k_B Boltzmann's constant, and T the temperature. Making the assumption that in the absence of rubber particles (or other stress-concentrating additives like rigid fillers), $\gamma = 1$ and the activation volume can be easily calculated for the pure matrix. Once the activation volume is obtained, the variation in the apparent stress concentration factor induced by the addition of the rubber particles to the matrix can be obtained following the procedure explained in [48].

In [8], it was shown that Eyring plots of log b against applied stress were linear for the pure polyamide (PA66), but for the rubber toughened polyamide (RTPA66), it showed a sharp increase in $d \log b/d\sigma$, where significant dilatation begins. Matching the results of creep tests and scanning electron micrographs, it was concluded that this cavitation accelerates shear yielding in the nylon matrix. The main explanation for this behavior was correlated with the energy-balance model for cavitation combined with the modified version of Gurson's equation for dilatation at yielding. According to the energy-balance model, the critical volume deformation Δ_v^c above which a particle can cavitate can be determined by Equation (3) [18]:

$$\Delta_v^c = 4\left(\frac{4\Gamma}{3K_r D}\right)^{3/4} \tag{3}$$

where Γ is the surface energy of the rubber, D the particle diameter, and K_r the rubber bulk modulus. When a rubber-toughened material is subjected to an external load, during the earlier stages of deformation, the hydrostatic component of the stress in the material starts to build up and, at a certain point, the biggest particles will start to cavitate and/or to debond. In this initial stage, voids will appear randomly, but their presence significantly affects the yielding and fracture behavior of polymers. If the particles cavitate, Lazzeri and Bucknall [17,18] proposed a modified version of the Gurson yield function [49] to account for the effects of cavitation on the yielding behavior of rubber-toughened polymers:

$$\sigma_e^2 = \sigma_\phi^2\left[\left(1 - \frac{\mu\sigma_m}{\sigma_\phi}\right) - 2fq_1\cosh\left(\frac{3q_2\sigma_m}{2\sigma_\phi}\right) + (q_1 f)^2\right] \tag{4}$$

where $\sigma_\phi = \sigma_0(1 - q_{1\phi})$ is the effective stress at yield for a rubber-toughened polymer containing a volume fraction ϕ, when the mean normal stress σ_m and the void content f are both zero. In this equation, σ_m is the yield stress of the pure matrix at $\sigma_m = f = 0$. The factors $q_1 = 1.375$ and $q_2 = 0.927$ were introduced to improve the fit between Gurson's predictions and data from numerical analysis [19]. Following dilatational yielding, the measured activation volume V_m increases with the volume fraction of voids, f, according to the following relationship:

$$V_m = V(1 + 2f) \tag{5}$$

This equation shows that the presence of voids significantly affects the rate of yielding as indicated by the increase in apparent activation volume.

If the dominating micromechanical deformation process is the rubber particle debonding, the stress necessary to initiate debonding, the number of debonded particles, and the size of the voids formed can be described by the following equation proposed by Pukanszky and Voros [12]:

$$\sigma^D = -C_1\sigma^T + C_2\left(\frac{W_{AB}E}{R}\right)^{\frac{1}{2}} \tag{6}$$

where σ^D and σ^T are debonding and thermal stresses, respectively, W_{AB} is the reversible work of adhesion, and R denotes the radius of the particle. C_1 and C_2 are constants which depend on the geometry of the debonding process.

3. Materials and Methods

3.1. Materials

The materials used in this work for the binary blends production were:

- Poly(lactic) acid (PLA), trade name Luminy LX175, purchased from Total Corbion PLA. It is a biodegradable PLA, derived from natural resources, that appears as white spherical pellets. According to the datasheet producer, this PLA contains approximately 4% of D-lactic acid units and can be used alone or blended with other polymers or additives for the production of suitable blends and composites. This PLA grade can be processed easily on conventional equipment for film extrusion thermoforming or fiber spinning (density: 1.24 g/cm^3, melt flow index (MFI) (210 °C/2.16 kg): 6 g/10 min);

- Poly(butylene succinate-co-adipate) (PBSA), trade name BioPBS FD92PM, purchased from Mitsubishi Chemical Corporation, is a copolymer of succinic acid, adipic acid, and butandiol. It is a soft and flexible semicrystalline polyester that can be blended in extruder with other polymers but can be also processed by blown and cast film extrusion (density of 1.24 g/cm^3, MFI (190 °C, 2.16 kg): 4 g/10 min).

3.2. Methods

3.2.1. Blends and Samples Preparation

Two different binary blends compositions containing, respectively, 15 and 20 wt.% PBSA (see Table 1 for the blends' names and compositions) were produced in pellets using a Comac EBC 25HT (L/D = 44) (Comac, Cerro Maggiore, Italy) twin screw extruder. The samples were named PLA (pure PLA), 85-15 (blend of PLA and PBSA 85/15 by weight) and 80-20 (blend of PLA and PBSA 80/20 by weight). Before the extrusion the materials were dried for 12 h in a DP 604–615 dryer (Piovan S.p.A., Verona, Italy). PLA granules were introduced into the main extruder feeder, while PBSA granules were fed into a specific side feeder. The temperature profiles of the extruder (11 zones) used for blends preparation were 150/175/180/180/180/185/185/185/185/185/190 °C, with the die zone at 190 °C. A screw rate of 300 rpm and a total mass flow rate of 20 kg/h were set. The strands coming out from the extruder were rapidly cooled in a water bath and then cut into pellets by an automatic knife cutter. All pellets were finally dried again at 60 °C.

Table 1. Injection molding conditions.

Parameters	PLA	85-15	80-20
Temperature profile (°C)	180/185/190	180/185/190	180/185/190
Mold temperature (°C)	70	60	55
Injection holding time (s)	5	5	5
Cooling time (s)	15	15	15
Injection pressure (bar)	90	80	80

After the extrusion, the extruded pellets were injection molded in a Megatech H10/18 injection molding machine (TECNICA DUEBI s.r.l., Fabriano, Italy) to obtain ISO 527-1A dog-bone specimens (width: 10 mm, thickness: 4 mm, length: 80 mm) for tensile tests. From the injection molding parameters (Table 1), it can be observed that the same temperature profile was adopted for all blends as well as the same cooling time was set. The mold temperature was also lowered progressively with the increasing amount of PBSA from 70 °C for pure PLA to 55 °C for 80-20. The PBSA addition causes a decrement in viscosity that results in a lowering of the injection pressure [28].

3.2.2. FT-IR Characterization

Infrared spectra of pure PLA, PBSA, and PLA/PBSA blends were recorded in the 550–4000 cm^{-1} range using a Nicolet 380 Thermo Corporation Fourier Transform Infrared (FTIR) Spectrometer (Thermo Fisher Scientific, Waltham, MA, USA) equipped with smart Itx ATR (attenuated total reflection) accessory with a diamond plate. Two hundred and fifty-six scans at a 4 cm^{-1} resolution were collected. The analysis was performed on the material sampled on the gate region of injection molded specimens.

3.2.3. Mechanical Characterization

Tensile and creep tests were carried out on ISO 527-1A dog-bone specimens using an MTS Criterion model 43 universal testing machine (MTS Systems Corporation, Eden

Prairie, MN, USA) equipped with a 10 kN load cell and interfaced with a computer running MTS Elite Software. Tests were conducted, at room temperature, 3 days after the injection molding process and during this time the specimens were stored in a dry keeper (SANPLATEC Corp., Osaka, Japan) at a controlled atmosphere (room temperature and 50% humidity).

For standard uniaxial tensile tests, at least ten specimens for each blend composition were tested at a constant crosshead speed of 10 mm/min. The average values of the main mechanical properties were reported.

In order to investigate the nature of the deformation process, constant-load creep tests were carried out at room temperature at different stress levels, below the yield stress value, from 10 up to 40 MPa. The initial load for obtaining the set stress was reached by subjecting the specimen to uniaxial test at a speed of 10 MPa/min and, subsequently, maintaining the applied load for 8 h. Once the desired load was reached, the variation in deformation over time was recorded. Furthermore, to estimate the volume change that occurred during the creep test, transversal and axial specimen elongation were recorded with a video extensometer (Genie HM1024 Teledyne DALSA camera) interfaced with a computer running ProVis Software (Fundamental Video Extensometer) which, in turn, is interfaced with the MTS Elite Software. The volume strain ($\Delta V/V_0$) was calculated, assuming equal the two lateral strain components, according to the following equation [20,50,51]:

$$\frac{\Delta V}{V_0} = (1 + \varepsilon_1)(1 + \varepsilon_2)^2 - 1 \qquad (7)$$

where the volume variation is ΔV, the starting volume is V_0, ε_1 is the axial (or longitudinal) strain, and ε_2 is the lateral strain.

In order to verify the reproducibility of the results obtained, at least 3 tests for stress level were carried out for each formulation.

3.2.4. Optical Analysis

The fracture surface of the specimens subjected to creep offers reliable information about the micromechanical deformations that occurred during the creep tests. Consequently, some specimens, appropriately selected at certain stress levels, were cold fractured along the tensile direction. The fracture surfaces, coated prior with a thin layer of platinum to avoid charge build up, were investigated by an FEI Quanta 450 FEG (Thermo Fisher Scientific, Waltham, MA, USA) scanning electron microscope (SEM).

4. Results

After the preparation of the specimens for investigation regarding creep behavior, the blends and the pure polymers were characterized by infrared ATR spectroscopy (Figure 1).

The pure PLA showed an infrared spectrum with main bands at 2996 and 2946 cm^{-1} (stretching C–H), 1746 cm^{-1} (stretching C=O), 1180 cm^{-1} (stretching C–O–C), and 1082 cm^{-1} (stretching O–C–C–) [52,53]. PBSA, being, like PLA, an aliphatic polyester, showed similar bands, but shifted at lower wavenumbers because of the higher macromolecular flexibility of PBSA [54], being 2943, 2867, 1725, 1161, and 1043 cm^{-1}, respectively. The blends PLA/PBSA 85/15 and 80/20 blends showed a spectrum much more similar to pure PLA, but some differences could be noticed. In the blend, the main stretching C=O band at 1746 cm^{-1} showed a shoulder at lower wavenumbers attributable to the presence of the C=O stretching band of PBSA. Moreover, a similar trend could be observed for the band at 1180 cm^{-1}, showing a shoulder at lower wavenumbers due to the overlapping of the 1161 cm^{-1} band of PBSA. The 955 cm^{-1} was more intense in blends because PBSA showed this band, attributable to the –C–OH bending in the carboxylic acid groups of PBS [55] or vinyl esters [56] more intense than PLA. Moreover, the band at 806 cm^{-1} (present in the spectrum of pure PBSA at 840 cm^{-1}) was attributable to the presence of vinyl ester moieties [57]. In general, the ATR evidence resulted in good agreement with the selected compositions of PLA/PBSA blends.

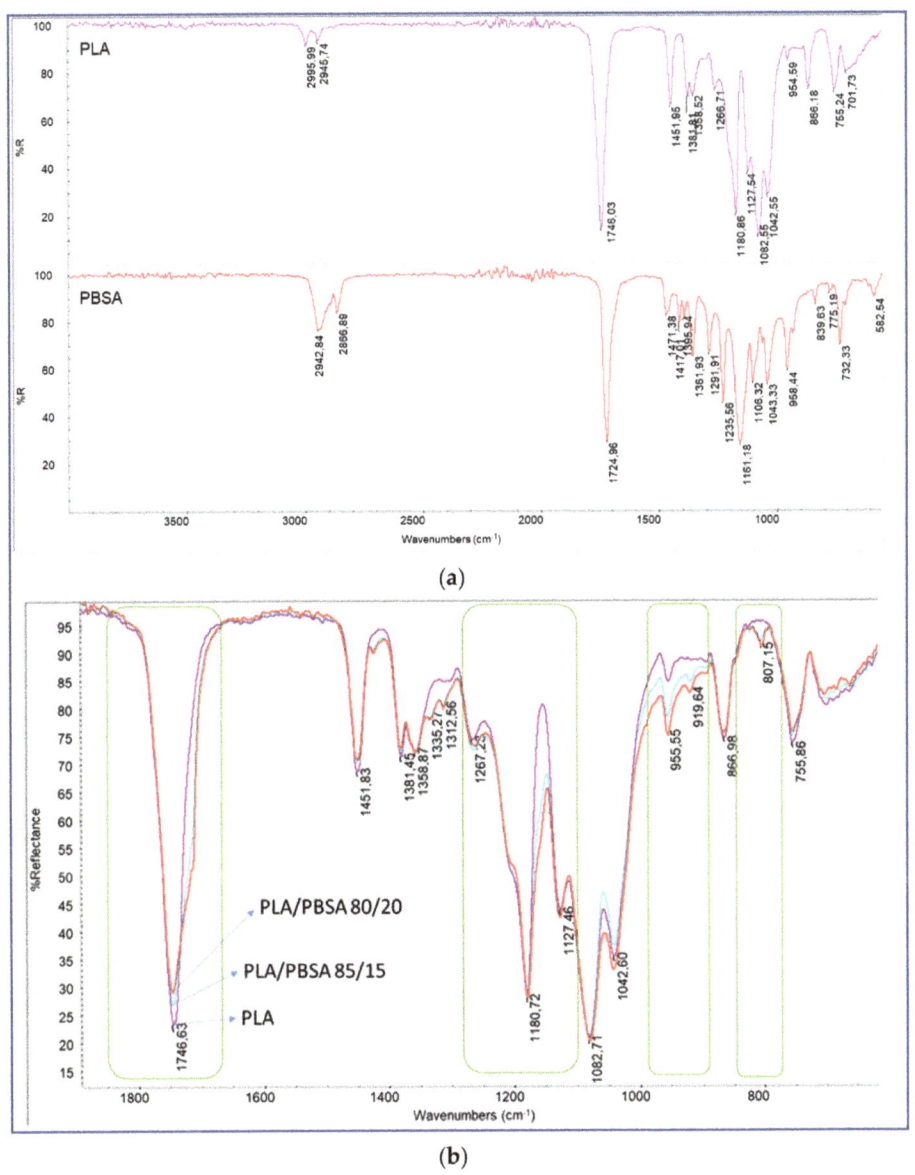

Figure 1. (a) ATR spectra of pure PLA and PBSA; (b) ATR spectra of PLA-PBSA blends.

In particular, the infrared characterization did not show any evident chemical change in the two polymers; this is can be ascribed to the very short processing time. In fact, significant transesterification occurs in the presence of proper catalysts and for processing times longer than 10 min [58]. Moreover, in the literature, it has been observed by Ding et al. [59] that the formation of copolymers between PLA and PBSA during extrusion and injection molding can be considered negligible. In addition, any eventual degradation during processing can be excluded due to the processing temperature adopted that did not overcome 190 °C. For the pure polymers and their blends, in fact, it was observed that the onset temperature for pure PLA was 274 °C, and the temperature at which the maximum degradation rate was reached (inflection point) was 354.5 °C; while for PBSA, these two temperatures were shifted at 301 and 401 °C, respectively, indicating a higher thermal stability for this polymer [60–62].

From the tensile tests (reported in Table 2), it can be observed that the addition of PBSA makes the material more ductile. Pure PLA is a fragile material characterized by an elevated modulus of 3.58 GPa, a high tensile strength of 62 MPa, and a low elongation at break (3.6%). The mechanical results are in accordance to what can be observed in the literature [28,62,63]; the introduction of the rubbery PBSA phase at 15 and 20 wt.% led to a decrement in the Young's modulus and tensile strength counterbalanced by an increment of the elongation at break. These trends are more marked where the PBSA content was higher (20 wt.%). With the addition of PBSA, the material yielded with the appearance of the neck that propagated along the gauge length.

Table 2. Tensile tests results.

Blend Name	Young's Modulus (GPa)	Stress at Break (%)	Elongation at Break (%)	Yield Stress (MPa)	Elongation at Yielding (%)
PLA	3.58 ± 0.04	61.58 ± 0.87	3.57 ± 0.23	-	-
85-15	2.89 ± 0.02	20.56 ± 1.35	62.71 ± 16.63	55.05 ± 0.66	4.18 ± 0.04
80-20	2.75 ± 0.07	19.51 ± 0.49	71.67 ± 11.92	51.07 ± 0.86	4.12 ± 0.11
PBSA	0.25 ± 0.04	23.4 ± 0.55	898.92 ± 21.03	16.6 ± 0.22	29.02 ± 0.23

Standard creep curves at different stress levels are reported in Figure 2.

At stresses of 10 and 20 MPa, all the specimens did not break during the test time interval (set at 8 h). However, when increasing the PBSA content, the creep resistance decreased. In fact, pure PLA did not break even at 30 MPa, while the 85-15 and 80-20 blends began to break (8 h before) already at 30 MPa. Moreover, at the same stress level applied, the time at which breakage occurred decreased with the increase in the rubber content.

The results of the volume variations recorded after that the specimens reached the set stress levels (reported in Figure 3) showed an almost linear variation in the volume change with time. The volume variation at time zero ($\Delta V/V_{0,i}$) corresponded to the intercept of the y-axis in the $\Delta V/V_0$ vs. the axial strain curve. $\Delta V/V_{0,i}$ can be linked to the volume variation caused by the instantaneous elastic deformation of the specimen at the selected stress level applied. Increasing the stress level, the instantaneous volume change increased in accordance with the increment in the instantaneous elongation.

The volume change did not increase proportionally with the rubber content but, on the contrary, a greater volume variation was encountered for the 85-15 blend. This trend can be observed both from Figure 3 but also from the volume variation at time zero reported in Table 3 in which the slopes of the straight lines passing through the experimental data are also reported. For pure PLA, the slope remained almost constant with a slight increment with the stress level. For the 85-15 and 80-20 blends, at higher stresses, the volume strain increased more rapidly with higher slopes values. The introduction of PBSA made a substantial contribution to the dilatational processes that were more marked for the 85-15 composition.

In order to better understand the volume variation curves, it is necessary to investigate deeper the micro-mechanics of the deformation process and how the PBSA addition influences the micromechanical behavior of the binary blend.

For all compositions, it can be observed in Figure 4 that the linear region of the creep curves can be estimated, with a good fitting, by the Andrade equation (Equation (1)); consequently, for all the stress levels applied, the Andrade b parameter for pure PLA and its binary blends can be obtained, making the linear regression of the $\varepsilon(t)$ against $t^{1/3}$ plots.

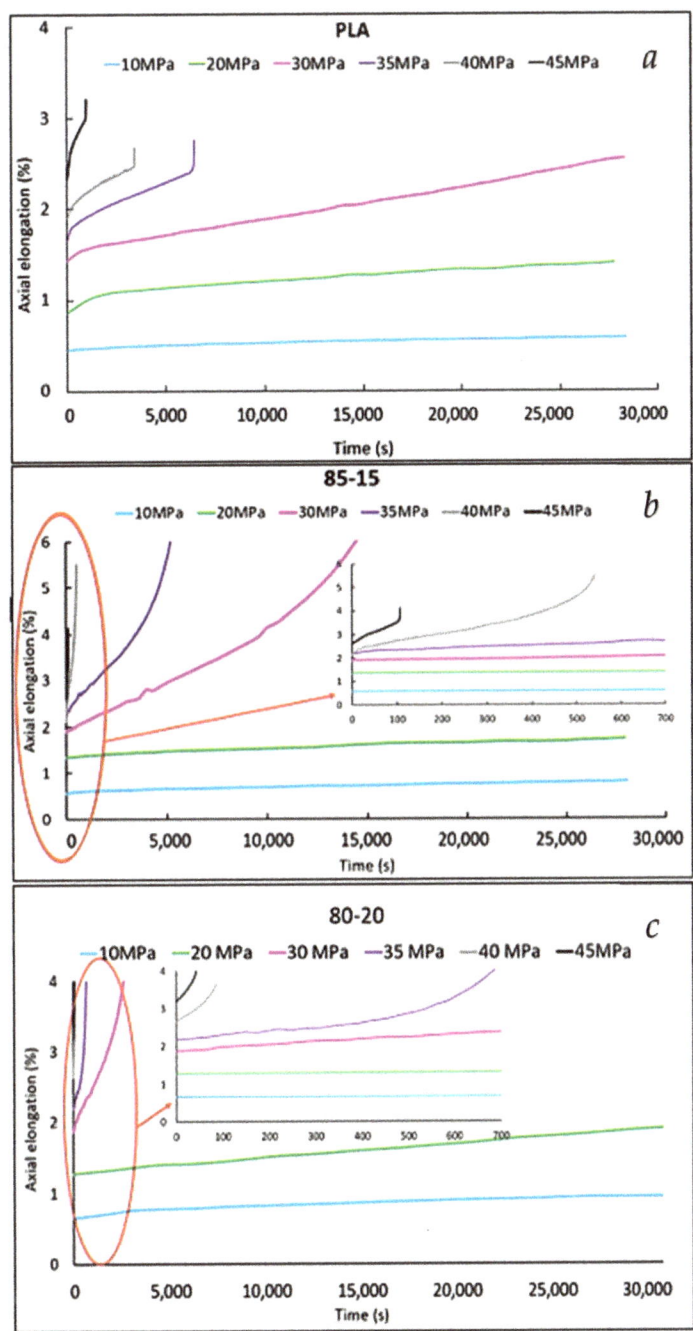

Figure 2. Axial elongation vs. time (creep curves) curves of pure PLA (**a**), 85-15 (**b**), and 80-20 (**c**) for different applied stresses from 10 to 45 MPa.

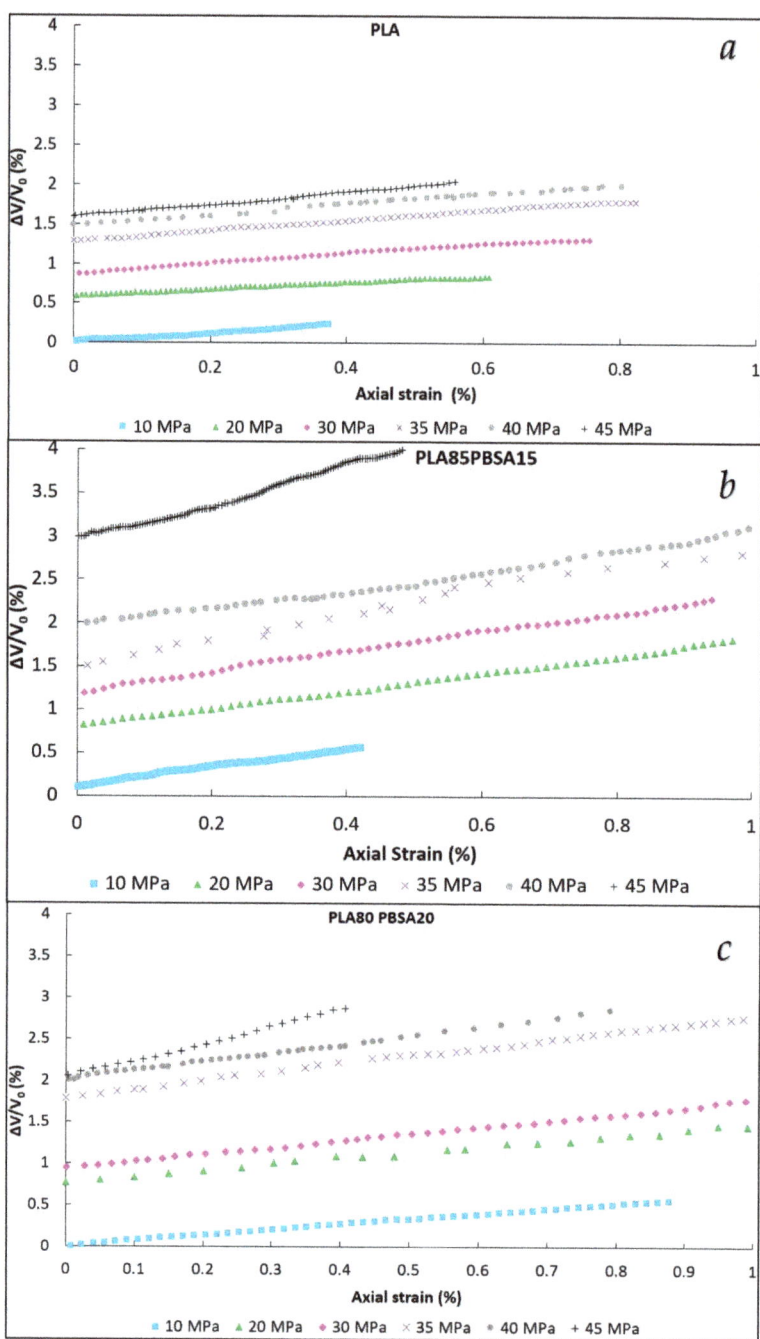

Figure 3. Volume variation vs. axial strain curves (dilatometric curves) of pure PLA (**a**), 85-15 (**b**), and 80-20 (**c**).

Table 3. Volume variation at time zero and the slopes of the ΔV/V0 vs. the axial strain curve at different stress levels.

Blend Name	$\Delta V/V_{0,i}$ (%)	Slope of the $\Delta V/V_0$ vs. Axial Strain Curve	Stress Level (MPa)
PLA	0	0.5	10
	0.59	0.55	20
	0.88	0.61	30
	1.29	0.65	35
	1.43	0.66	40
	1.59	0.78	45
85-15	0.12	1.03	10
	0.8	1.05	20
	1.21	1.14	30
	1.6	1.22	35
	1.94	1.24	40
	2.93	2.22	45
80-20	0.02	0.63	10
	0.76	0.72	20
	0.94	0.83	30
	1.79	1	35
	2.01	1.04	40
	2.03	2.04	45

Applying the Eyring relationship (Equation (2)) between the applied stress, σ, and the b parameter, interesting results can be obtained plotting (Eyring plot) $\log b$ against the applied stress, σ. It can be observed from Figure 3, that the Eyring plot for pure PLA gives a straight line in accordance with the Eyring model for stress activated flow. This result is also in accordance to what was observed for pure PA66 in a rubber-toughened polyamide 6,6 system [48]. For pure matrix, due to the absence of rubber PBSA particles, the stress concentration factor, γ, can be assumed equal to 1, and the activation volume, V, for pure PLA can be easily obtained from the Eyring plot slope. An apparent activation volume of 0.25 nm^3 was obtained.

Under 20 MPa and 30 MPa for the 85-15 and 80-20 blends, respectively, the Eyring plots behaved in a similar manner to pure PLA (following Equation (2)) but with a slightly increased slope. An explanation for this behavior was found to be in accordance with the literature [48] and was ascribed to the same deformation mechanisms that operate in both pure PLA and PLA–PBSA blends having the same value of activation volume, V. The higher slope was ascribed to an increased value of the stress concentration factor, γ. On this basis, using the value of the activation volume found for pure PLA, the stress concentration factor for the PLA–PBSA binary blends can be easily obtained from the Eyring slope until 20 MPa and 30 MPa. For the two blends, 85-15 and 80-20, the stress concentration factor values obtained were very close to each other ($\gamma = 1.15$ for 85-15 and $\gamma = 1.16$ for 80-20); consequently, the γ seemed not to be affected directly by the difference in volume variation observed for the 85-15 and 80-20 blends.

However, a very interesting difference emerged from the Eyring plot (Figure 5) in which the slope changed for the binary blends occurring at two different stress levels. A small increase with no slope change could be detected over the lower range of applied stress (up to approximately 18 MPa), where, according to the literature [48], shear yielding is the predominant deformation mechanism. The stress level for which the change in slope took place can be easily identified using a simple geometric construction (the intersection point is highlighted by the green and orange arrows in Figure 5). The slope change occurred earlier (close to 20 MPa) for the 85-15 blend, while it occurred later (at approximately 25 MPa) for the 80-20 blend. The different intersection points, registered for the two binary blends investigated, were strictly connected to the volume variation differences encountered for the two types of blends. The intersection point allowed for the identification of the stress level for which the micromechanical process of debonding and/or cavitation starts and contributes to deformation. Based on the results observable in Figure 5, the debonding and/or cavitation mechanism was apparently activated earlier (at a lower stress level) for the 85-15 blend.

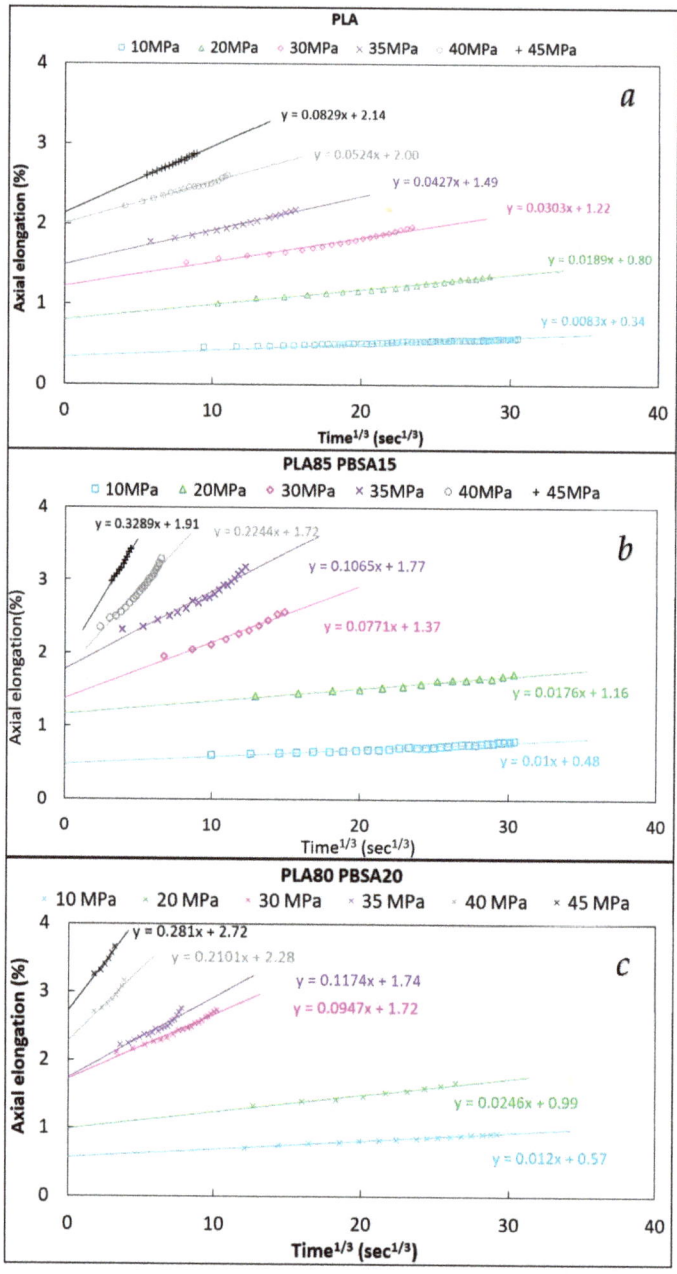

Figure 4. Andrade equation fittings (axial elongation vs. cubic root time) of pure PLA (**a**), 85-15 (**b**), and 80-20 (**c**).

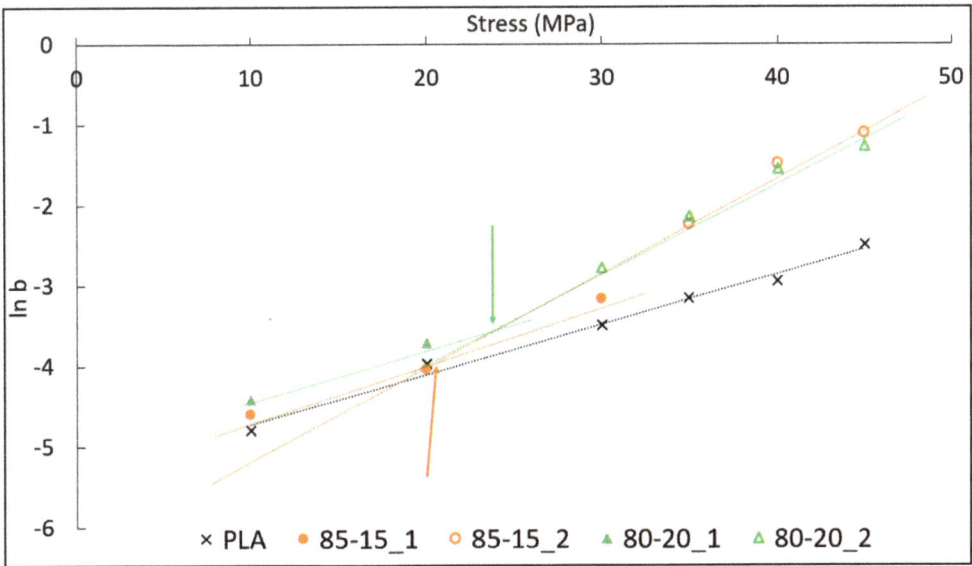

Figure 5. Andrade B parameter in which is highlighted the different slope changes between pure PLA and its blends (85-15 and 80-20).

In fact, it must be taken into account that the creep test was carried out at a constant rate of 10 MPa/min to achieve the set stress levels. Consequently, a slightly different strain rate (of approximately 1mm/min) caused by the higher deformability of the blend that increased with PBSA content was registered. Nevertheless, the application of this method was extremely useful, because it allowed to highlight the connection between the micromechanical processes and their volume variation caused by the introduction of a rubber dispersed phase into a polymeric matrix. The voids generated by the cavitation/debonding mechanism caused the increase in the apparent activation volume, since voids allow plastic flow of the matrix around it more easily than an intact rubber particle. In fact, the bulk modulus of a rubber particle is very high, while for a void it is zero. Thus, the increased local strain rate of the polymer matrix was due to the fact that the material was no longer a "real" continuum on a microscopic scale. The deformation involves a macroscopic volume increase due to the growth of the voids generated around or inside the rubber particles, which is favored by the high level of triaxiality, leading to a nonlinear yield curve. In contrast to the case of a "continuum" polymer, where high levels of triaxiality favor crazing and cleavage mechanisms over shear yielding, for a "porous" polymer, a high triaxiality considerably accelerates plastic flow.

In order to confirm the results obtained and to understand if PBSA particles undergo debonding or cavitation, SEM micrographs (at 8000 X, Figure 6) at the surface of the tensile specimen, cryo-fractured along the draw direction and at different stress levels, were carried out.

On the basis of the results obtained in Figure 5, the most significant stress levels were analyzed. In particular, three stress levels were chosen: (I) 10 MPa, which was the lowest value for which both the binary blends underwent the same deformation mechanism and no deviation in the Eyring's slopes occurred; (II) 30 MPa in which the slope variation occurred for both 85-15 and 80-20; (III) 45 MPa, which was the highest stress level tested and for which the micromechanical deformation process should be more marked.

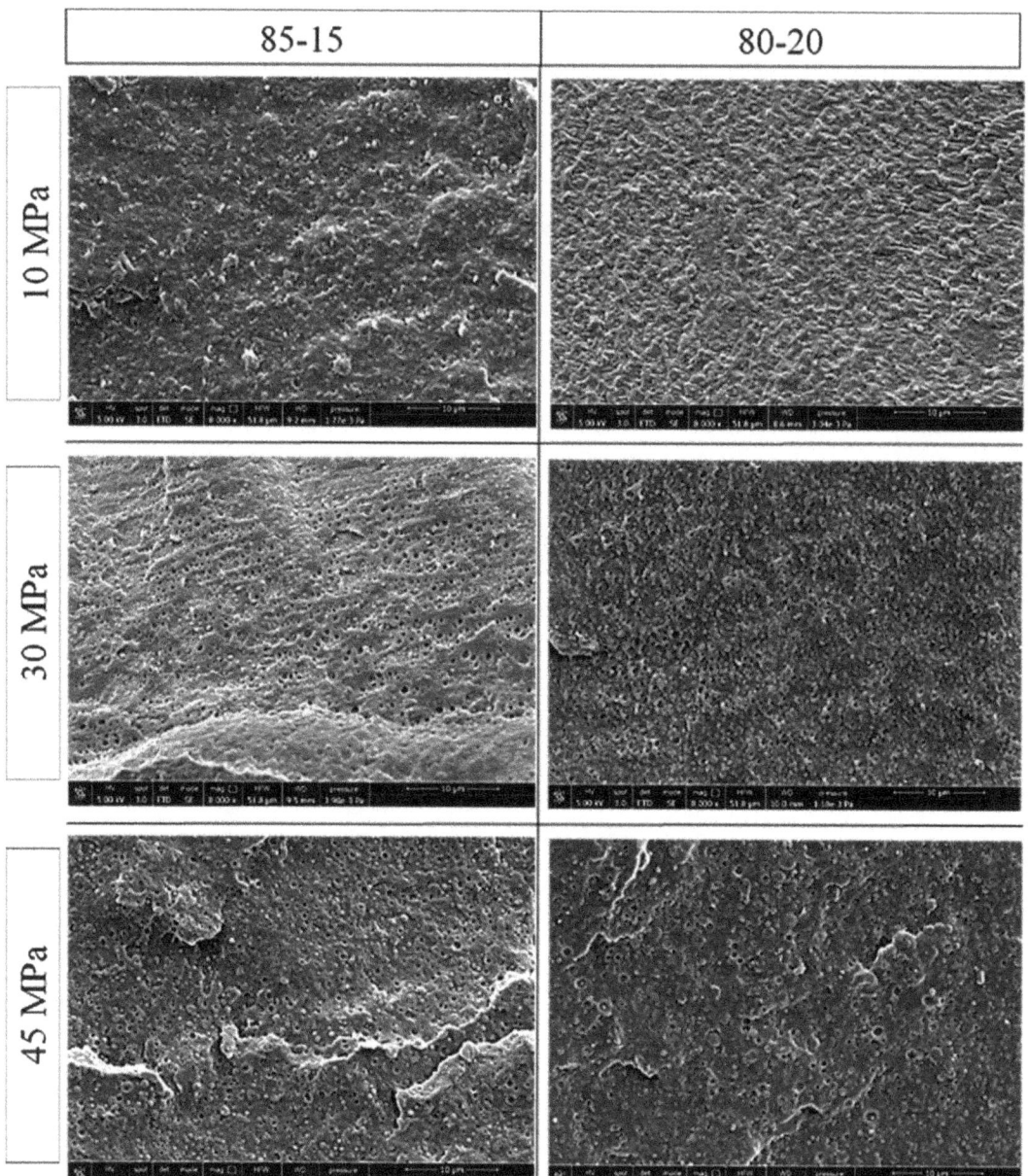

Figure 6. SEM micrographs made at the surface of tensile specimen cryo-fractured along the draw direction at different creep stress levels.

From the SEM images shown in Figure 6, it emerges that the PBSA particle size increases with the PBSA content, in agreement to what is also observed in the literature [28]. At 10 MPa, for both blends, neither cavitation nor debonding of the PBSA particles appeared. At this stress level, only shear yielding took place. The fracture surfaces were characterized by well dispersed PBSA spherical particles that were well attached into the PLA matrix. The matrix showed signs of deformation caused by the creep test with greater matrix deformation occurring for the 80-20 composition in which the rubber amount was higher.

At 30 MPa, it was evident that the micromechanical deformation that occurred was caused by the debonding of the PBSA particles. Clear visible voids around the PBSA particles could be distinguished. At 30 MPa, therefore, despite 80-20 having a higher PBSA content, the 85-15 blend had the greatest number of particles that underwent debonding. This result is in agreement to what emerged from the Eyring's plot; in fact, for the 85-15 composition debonding started at a lower stress level, generating at 30 MPa a greater number of debonded particles. By increasing the stress level of the creep test (up to 45 MPa), debonding continued to be the main micromechanical deformation mechanism for both the binary blends. Increasing the stress level, the quantity of particles that underwent debonding also increased.

5. Conclusions

Over the last years, there has been a growing interest toward biobased and biodegradable polymers that show a more environmentally friendly end of life and, at the same time, are able to improve carbon neutrality when compared to their fossil-based counterparts. In order to improve the physical and mechanical properties of these biopolymers, a blending technique is fundamental. For this purpose, a better knowledge of the micromechanical deformation processes of these new biopolymeric blends is fundamental in order to better exploit their peculiar properties, favoring the replacement of their fossil-based counterparts in several application sectors.

Despite different studies that can be found in the literature on biopolymeric blends, an in-depth investigation of micromechanical mechanisms has not yet been considered. In this study, poly(lactic acid) (PLA) and poly(butylene succinate-adipate) (PBSA) blends were investigated with a PBSA content of 15 and 20 wt.%, respectively (named 85-15 and 80-20). Measurements of the volume strain, using an optical extensometer, were carried out with a universal testing machine in creep configuration trying to determine, accompanied by SEM images, the micromechanical deformation processes involved in the biopolymeric blend systems. Analytical models were also applied to correlate the creep to the dilatation variation; in particular, the Andrade equation was applied and the b parameter for the polymeric systems was calculated. Plotting the $\log b$ against the applied stress, using an Eyring plot, a significant change in the curves was found. The binary blends, in fact, showed a sharp increase in $dlogb/d\sigma$ where significant dilatation began. The point at which the slope change occurred reasonably coincided with the onset of the cavitation/debonding mechanism. For both binary blends, the SEM analysis evidenced that the starting micromechanical deformation mechanism responsible for the volume increment was due to the PBSA debonding.

Another interesting result obtained by applying the Eyring relationship was the calculation (knowing the data of the pure PLA matrix) of the stress concentration factor; it emerged that the second PBSA phase acts as a stress concentrator, probably due to the presence of weak interfaces between PLA and PBSA that are also responsible for the debonding mechanism.

Author Contributions: Conceptualization, A.L. and L.A.; methodology, L.A. and V.G.; validation, L.A. and V.G.; investigation, L.A. and V.G.; resources, A.L. and M.-B.C.; data curation, L.A., M.-B.C. and V.G.; writing—original draft preparation, L.A., V.G., and M.-B.C.; writing—review and editing, M.-B.C., and A.L. All authors have read and agreed to the published version of the manuscript.

Funding: Centre for Instrumentation Sharing—University of Pisa (CISUP) is thanked for its support in the use of FEI Quanta 450 FEG scanning electron microscope. BIONTOP project (grant number: G.A 837761) is thanked for the helpful suggestions regarding the present manuscript.

Institutional Review Board Statement: Not applicable.

Informed Consent Statement: Not applicable.

Data Availability Statement: The data presented in this study are available on request from the corresponding author.

Acknowledgments: Centre for Instrumentation Sharing—University of Pisa (CISUP) is thanked for its support in the use of FEI Quanta 450 FEG scanning electron microscope. BIONTOP project (grant number: G.A 837761) is thanked for the helpful suggestions regarding the present manuscript.

Conflicts of Interest: The authors declare no conflict of interest.

References

1. Gigante, V.; Canesi, I.; Cinelli, P.; Beatrice Coltelli, M.; Lazzeri, A. Rubber toughening of Polylactic acid (PLA) with Poly(butylene adipate-co-terephthalate) (PBAT): Mechanical properties, fracture mechanics and analysis of brittle—Ductile behavior while varying temperature and test speed. *Eur. Polym. J.* **2019**, *115*, 125–137. [CrossRef]
2. Narayan, R. Carbon footprint of bioplastics using biocarbon content analysis and life-cycle assessment. *MRS Bull.* **2011**, *36*, 716–721. [CrossRef]
3. Sohn, Y.J.; Kim, H.T.; Baritugo, K.; Jo, S.Y.; Song, H.M.; Park, S.Y.; Park, S.K.; Pyo, J.; Cha, H.G.; Kim, H. Recent advances in sustainable plastic upcycling and biopolymers. *Biotechnol. J.* **2020**, *15*, 1900489. [CrossRef]
4. Changwichan, K.; Silalertruksa, T.; Gheewala, S.H. Eco-efficiency assessment of bioplastics production systems and end-of-life options. *Sustainability* **2018**, *10*, 952. [CrossRef]
5. Narancic, T.; Verstichel, S.; Reddy Chaganti, S.; Morales-Gamez, L.; Kenny, S.T.; De Wilde, B.; Babu Padamati, R.; O'Connor, K.E. Biodegradable plastic blends create new possibilities for end-of-life management of plastics but they are not a panacea for plastic pollution. *Environ. Sci. Technol.* **2018**, *52*, 10441–10452. [CrossRef] [PubMed]
6. Pukánszky, B.; Van Es, M.; Maurer, F.H.J.; Vörös, G. Micromechanical deformations in particulate filled thermoplastics: Volume strain measurements. *J. Mater. Sci.* **1994**, *29*, 2350–2358. [CrossRef]
7. Bucknall, C.B.; Cote, F.F.P.; Partridge, I.K. Rubber toughening of plastics—Part 9 Effects of rubber particle volume faction on deformation and fracture in HIPS. *J. Mater. Sci.* **1986**, *21*, 301–306. [CrossRef]
8. Wang, F.; Drzal, L.T.; Qin, Y.; Huang, Z. Enhancement of fracture toughness, mechanical and thermal properties of rubber/epoxy composites by incorporation of graphene nanoplatelets. *Compos. Part A Appl. Sci. Manuf.* **2016**, *87*, 10–22. [CrossRef]
9. Argon, A.S.; Cohen, R.E. Toughenability of polymers. *Polymer* **2003**, *44*, 6013–6032. [CrossRef]
10. Bucknall, C.B.; Ayre, D.S.; Dijkstra, D.J. Detection of rubber particle cavitation in toughened plastics using thermal contraction tests. *Polymer* **2000**, *41*, 5937–5947. [CrossRef]
11. Pearson, R.A.; Yee, A.F. Toughening mechanisms in elastomer-modified epoxies. *J. Mater. Sci.* **1989**, *24*, 2571–2580. [CrossRef]
12. Pukánszky, B.; Voros, G. Mechanism of interfacial interactions in particulate filled composites. *Compos. Interfaces* **1993**, *1*, 411–427. [CrossRef]
13. Belayachi, N.; Benseddiq, N.; Naït-Abdelaziz, M.; Hamdi, A. On cavitation and macroscopic behaviour of amorphous polymer-rubber blends. *Sci. Technol. Adv. Mater.* **2008**, *9*, 025008. [CrossRef] [PubMed]
14. Nagarajan, V.; Mohanty, A.K.; Misra, M. Blends of polylactic acid with thermoplastic copolyester elastomer: Effect of functionalized terpolymer type on reactive toughening. *Polym. Eng. Sci.* **2018**, *58*, 280–290. [CrossRef]
15. Shang, M.; Wu, Y.; Shentu, B.; Weng, Z. Toughening of pbt by poe/poe- g-gma elastomer through regulating interfacial adhesion and toughening mechanism. *Ind. Eng. Chem. Res.* **2019**, *58*, 12650–12663. [CrossRef]
16. Agnelli, S.; Baldi, F.; Castellani, L.; Pisoni, K.; Vighi, M.; Laiarinandrasana, L. Study of the plastic deformation behaviour of ductile polymers: Use of the material key curves. *Mech. Mater.* **2018**, *117*, 105–115. [CrossRef]
17. Lazzeri, A.; Bucknall, C.B. Dilatational bands in rubber-toughened polymers. *J. Mater. Sci.* **1993**, *28*, 6799–6808. [CrossRef]
18. Lazzeri, A.; Bucknall, C.B. Recent developments in the modeling of dilatational yielding in toughened plastics. In *Toughening of Plastics: Advances in Modeling and Experiments*; ACS Publications: Washington, DC, USA, 2000; p. 264. ISBN 1947-5918.
19. Bagheri, R.; Pearson, R.A. Role of particle cavitation in rubber-toughened epoxies: 1. Microvoid toughening. *Polymer* **1996**, *37*, 4529–4538. [CrossRef]
20. Aliotta, L.; Cinelli, P.; Coltelli, M.B.; Lazzeri, A. Rigid filler toughening in PLA-Calcium Carbonate composites: Effect of particle surface treatment and matrix plasticization. *Eur. Polym. J.* **2019**, *113*, 78–88. [CrossRef]
21. Aliotta, L.; Gigante, V.; Acucella, O.; Signori, F.; Lazzeri, A. Thermal, Mechanical and Micromechanical Analysis of PLA/PBAT/POE-g-GMA Extruded Ternary Blends. *Front. Mater.* **2020**, *7*, 130. [CrossRef]
22. Coumans, W.J.; Heikens, D. Dilatometer for use in tensile tests. *Polymer* **1980**, *21*, 957–961. [CrossRef]
23. François, P.; Gloaguen, J.M.; Hue, B.; Lefebvre, J.M. Volume strain measurements by optical extensometry: Application to the tensile behaviour of RT-PMMA. *J. Phys. III* **1994**, *4*, 321–329. [CrossRef]
24. Thio, Y.S.; Argon, A.S.; Cohen, R.E.; Weinberg, M. Toughening of isotactic polypropylene with $CaCO_3$ particles. *Polymer* **2002**, *43*, 3661–3674. [CrossRef]
25. G'Sell, C.; Hiver, J.M.; Dahoun, A.; Souahi, A. Video-controlled tensile testing of polymers and metals beyond the necking point. *J. Mater. Sci.* **1992**, *27*, 5031–5039. [CrossRef]
26. Rezgui, F.; Swistek, M.; Hiver, J.M.; G'Sell, C.; Sadoun, T. Deformation and damage upon stretching of degradable polymers (PLA and PCL). *Polymer* **2005**, *46*, 7370–7385. [CrossRef]

27. Aliotta, L.; Vannozzi, A.; Panariello, L.; Gigante, V.; Coltelli, M.-B.; Lazzeri, A. Sustainable Micro and Nano Additives for Controlling the Migration of a Biobased Plasticizer from PLA-Based Flexible Films. *Polymers* **2020**, *12*, 1366. [CrossRef] [PubMed]
28. Aliotta, L.; Vannozzi, A.; Canesi, I.; Cinelli, P.; Coltelli, M.; Lazzeri, A. Poly(lactic acid) (PLA)/Poly(butylene succinate-co-adipate) (PBSA) Compatibilized Binary Biobased Blends: Melt Fluidity, Morphological, Thermo-Mechanical and Micromechanical Analysis. *Polymers* **2021**, *13*, 218. [CrossRef]
29. Coltelli, M.; Aliotta, L.; Vannozzi, A.; Morganti, P.; Fusco, A.; Donnarumma, G.; Lazzeri, A. Properties and Skin Compatibility of Films Based on Poly (Lactic Acid) (PLA) Bionanocomposites Incorporating Chitin Nanofibrils (CN). *J. Funct. Biomater.* **2020**, *11*, 21. [CrossRef]
30. Borrelle, S.B.; Ringma, J.; Law, K.L.; Monnahan, C.C.; Lebreton, L.; McGivern, A.; Murphy, E.; Jambeck, J.; Leonard, G.H.; Hilleary, M.A. Predicted growth in plastic waste exceeds efforts to mitigate plastic pollution. *Science* **2020**, *369*, 1515–1518. [CrossRef]
31. Krishnan, S.; Pandey, P.; Mohanty, S.; Nayak, S.K. Toughening of Polylactic Acid: An Overview of Research Progress. *Polym. Plast. Technol. Eng.* **2016**, *55*, 1623–1652. [CrossRef]
32. Lebarbe, T. *Synthesis of Novel "Green" Polyesters from Plant Oils: Application to the Rubber-Toughening of poly(L-lactide)*; Université de Bordeaux: Bordeaux, France, 2013.
33. Nagarajan, V.; Mohanty, A.K.; Misra, M. Perspective on Polylactic Acid (PLA) based Sustainable Materials for Durable Applications: Focus on Toughness and Heat Resistance. *ACS Sustain. Chem. Eng.* **2016**, *4*, 2899–2916. [CrossRef]
34. Park, C.K.; Jang, D.J.; Lee, J.H.; Kim, S.H. Toughening of polylactide by in-situ reactive compatibilization with an isosorbide-containing copolyester. *Polym. Test.* **2021**, *95*, 107136. [CrossRef]
35. D'Anna, A.; Arrigo, R.; Frache, A. Rheology, Morphology and Thermal Properties of a PLA/PHB/Clay Blend Nanocomposite: The Influence of Process Parameters. *J. Polym. Environ.* **2021**, 1–12. [CrossRef]
36. Wu, N.; Zhang, H. Mechanical properties and phase morphology of super-tough PLA/PBAT/EMA-GMA multicomponent blends. *Mater. Lett.* **2017**, *192*, 17–20. [CrossRef]
37. Oguz, H.; Dogan, C.; Kara, D.; Ozen, Z.T.; Ovali, D.; Nofar, M. Development of PLA-PBAT and PLA-PBSA bio-blends: Effects of processing type and PLA crystallinity on morphology and mechanical properties. *AIP Conf. Proc.* **2019**, *2055*, 030003. [CrossRef]
38. Ojijo, V.; Sinha Ray, S.; Sadiku, R. Role of specific interfacial area in controlling properties of immiscible blends of biodegradable polylactide and poly[(butylene succinate)-co-adipate]. *ACS Appl. Mater. Interfaces* **2012**, *4*, 6690–6701. [CrossRef] [PubMed]
39. Nofar, M.; Oguz, H.; Ovalı, D. Effects of the matrix crystallinity, dispersed phase, and processing type on the morphological, thermal, and mechanical properties of polylactide-based binary blends with poly[(butylene adipate)-co-terephthalate] and poly[(butylene succinate)-co-adipate]. *J. Appl. Polym. Sci.* **2019**, *136*, 47636. [CrossRef]
40. Mehrabi Mazidi, M.; Edalat, A.; Berahman, R.; Hosseini, F.S. Highly-Toughened Polylactide- (PLA-) Based Ternary Blends with Significantly Enhanced Glass Transition and Melt Strength: Tailoring the Interfacial Interactions, Phase Morphology, and Performance. *Macromolecules* **2018**, *51*, 4298–4314. [CrossRef]
41. Ye, J.; Yao, T.; Deng, Z.; Zhang, K.; Dai, S.; Liu, X. A modified creep model of polylactic acid (PLA-max) materials with different printing angles processed by fused filament fabrication. *J. Appl. Polym. Sci.* **2021**, *138*, 50270. [CrossRef]
42. Niaza, K.V.; Senatov, F.S.; Stepashkin, A.; Anisimova, N.Y.; Kiselevsky, M. V Long-Term Creep and Impact Strength of Biocompatible 3D-Printed PLA-Based Scaffolds. *Nano Hybrids Compos.* **2017**, *13*, 15–20. [CrossRef]
43. Morreale, M.; Mistretta, M.C.; Fiore, V. Creep behavior of poly(lactic acid) based biocomposites. *Materials* **2017**, *10*, 395. [CrossRef]
44. Spathis, G.; Kontou, E. Creep failure time prediction of polymers and polymer composites. *Compos. Sci. Technol.* **2012**, *72*, 959–964. [CrossRef]
45. Yang, J.L.; Zhang, Z.; Schlarb, A.K.; Friedrich, K. On the characterization of tensile creep resistance of polyamide 66 nanocomposites. Part II: Modeling and prediction of long-term performance. *Polymer* **2006**, *47*, 6745–6758. [CrossRef]
46. Louchet, F.; Duval, P. Andrade creep revisited. *Int. J. Mater. Res.* **2009**, *100*, 1433–1439. [CrossRef]
47. Plazek, D.J. Dynamic mechanical and creep properties of a 23% cellulose nitrate solution; Andrade creep in polymeric systems. *J. Colloid Sci.* **1960**, *15*, 50–75. [CrossRef]
48. Bucknall, C.B.; Heather, P.S.; Lazzeri, A. Rubber toughening of plastics. *J. Mater. Sci.* **1989**, *24*, 2255–2261. [CrossRef]
49. Gurson, A.L. Continuum theory of ductile rupture by void nucleation and growth: Part I—Yield criteria and flow rules for porous ductile media. *J. Eng. Mater. Technol.* **1977**, *99*, 2–15. [CrossRef]
50. Bucknall, C.B. *Toughened Plastics*; Springer: Berlin/Heidelberg, Germany, 1977; ISBN 085334695X.
51. Xu, S.A.; Tjong, S.C. Deformation mechanisms and fracture toughness of polystyrene/high-density polyethylene blends compatibilized by triblock copolymer. *J. Appl. Polym. Sci.* **2000**, *77*, 2024–2033. [CrossRef]
52. Meaurio, E.; López-Rodríguez, N.; Sarasua, J.R. Infrared spectrum of poly (L-lactide): Application to crystallinity studies. *Macromolecules* **2006**, *39*, 9291–9301. [CrossRef]
53. Braun, B.; Dorgan, J.R.; Dec, S.F. Infrared spectroscopic determination of lactide concentration in polylactide: An improved methodology. *Macromolecules* **2006**, *39*, 9302–9310. [CrossRef]
54. Yao, S.-F.; Chen, X.-T.; Ye, H.-M. Investigation of structure and crystallization behavior of Poly (butylene succinate) by fourier transform infrared spectroscopy. *J. Phys. Chem. B* **2017**, *121*, 9476–9485. [CrossRef]
55. Phua, Y.J.; Chow, W.S.; Mohd Ishak, Z.A. Reactive processing of maleic anhydride-grafted poly (butylene succinate) and the compatibilizing effect on poly (butylene succinate) nanocomposites. *Express Polym. Lett.* **2013**, *7*, 340–354. [CrossRef]

56. Levchik, S.V.; Weil, E.D. A review on thermal decomposition and combustion of thermoplastic polyesters. *Polym. Adv. Technol.* **2004**, *15*, 691–700. [CrossRef]
57. McManis, G.E. Infrared absorption spectra of vinyl esters of carboxylic acids. *Appl. Spectrosc.* **1970**, *24*, 495–498. [CrossRef]
58. Coltelli, M.-B.; Toncelli, C.; Ciardelli, F.; Bronco, S. Compatible blends of biorelated polyesters through catalytic transesterification in the melt. *Polym. Degrad. Stab.* **2011**, *96*, 982–990. [CrossRef]
59. Ding, Y.; Feng, W.; Huang, D.; Lu, B.; Wang, P.; Wang, G.; Ji, J. Compatibilization of immiscible PLA-based biodegradable polymer blends using amphiphilic di-block copolymers. *Eur. Polym. J.* **2019**, *118*, 45–52. [CrossRef]
60. Qi, Z.; Ye, H.; Xu, J.; Chen, J.; Guo, B. Improved the thermal and mechanical properties of poly (butylene succinate-co-butylene adipate) by forming nanocomposites with attapulgite. *Colloids Surfaces A Physicochem. Eng. Asp.* **2013**, *421*, 109–117. [CrossRef]
61. Zhou, W.; Wang, X.; Wang, P.; Zhang, W.; Ji, J. Enhanced mechanical and thermal properties of biodegradable poly (butylene succinate-co-adipate)/graphene oxide nanocomposites via in situ polymerization. *J. Appl. Polym. Sci.* **2013**, *130*, 4075–4080.
62. Lascano, D.; Quiles-Carrillo, L.; Balart, R.; Boronat, T.; Montanes, N. Toughened poly (lactic acid)-PLA formulations by binary blends with poly(butylene succinate-co-adipate)-PBSA and their shape memory behaviour. *Materials* **2019**, *12*, 622. [CrossRef]
63. Nofar, M.; Tabatabaei, A.; Sojoudiasli, H.; Park, C.B.; Carreau, P.J.; Heuzey, M.C.; Kamal, M.R. Mechanical and bead foaming behavior of PLA-PBAT and PLA-PBSA blends with different morphologies. *Eur. Polym. J.* **2017**, *90*, 231–244. [CrossRef]

Article

Essential Work of Fracture and Evaluation of the Interfacial Adhesion of Plasticized PLA/PBSA Blends with the Addition of Wheat Bran By-Product

Laura Aliotta [1,2,*], Alessandro Vannozzi [1,2], Patrizia Cinelli [1,2,3], Maria-Beatrice Coltelli [1,2,*] and Andrea Lazzeri [1,2,3]

1. Department of Civil and Industrial Engineering, University of Pisa, 56122 Pisa, Italy; alessandrovannozzi91@hotmail.it (A.V.); patrizia.cinelli@unipi.it (P.C.); andrea.lazzeri@unipi.it (A.L.)
2. National Interuniversity Consortium of Materials Science and Technology (INSTM), 50121 Florence, Italy
3. Planet Bioplastics s.r.l., Via San Giovanni Bosco 23, 56127 Pisa, Italy
* Correspondence: laura.aliotta@dici.unipi.it (L.A.); maria.beatrice.coltelli@unipi.it (M.-B.C.)

Abstract: In this work biocomposites based on plasticized poly(lactic acid) (PLA)–poly(butylene succinate-*co*-adipate) (PBSA) matrix containing wheat bran fiber (a low value by-product of food industry) were investigated. The effect of the bran addition on the mechanical properties is strictly correlated to the fiber-matrix adhesion and several analytical models, based on static and dynamic tests, were applied in order to estimate the interfacial shear strength of the biocomposites. Finally, the essential work of fracture approach was carried out to investigate the effect of the bran addition on composite fracture toughness.

Keywords: biocomposites; natural fibers; poly(lactic acid) (PLA); fracture mechanics

Citation: Aliotta, L.; Vannozzi, A.; Cinelli, P.; Coltelli, M.-B.; Lazzeri, A. Essential Work of Fracture and Evaluation of the Interfacial Adhesion of Plasticized PLA/PBSA Blends with the Addition of Wheat Bran By-Product. *Polymers* 2022, *14*, 615. https://doi.org/10.3390/polym14030615

Academic Editor: Jose-Ramon Sarasua

Received: 18 January 2022
Accepted: 1 February 2022
Published: 4 February 2022

Publisher's Note: MDPI stays neutral with regard to jurisdictional claims in published maps and institutional affiliations.

Copyright: © 2022 by the authors. Licensee MDPI, Basel, Switzerland. This article is an open access article distributed under the terms and conditions of the Creative Commons Attribution (CC BY) license (https://creativecommons.org/licenses/by/4.0/).

1. Introduction

In the last years a growing interest towards the development of new advanced biobased polymeric products that are sustainable, eco-friendly, eco-efficient and biodegradable has arisen for proposing valid alternatives in the global market, that is at present mainly dominated by petroleum-derived products. Thanks to new governments regulations not only researchers but also industries are seeking for more ecologically and friendly materials [1–3]. There are several market branches extremely valuable for biodegradable plastic materials such as single use items and those applications where collection and recycling are difficultly achieved, and then for a correct waste management compostability becomes an advantage; for this reason, it is expected that the biodegradable and biobased polymer market will increase in the coming years [4,5].

In this context, among biobased and biodegradable matrices which are industrially compostable and commercially available on the market, poly(lactic acid) (PLA) is one of the most attractive due to its relatively low production cost compared to other biobased polymers [6,7]. PLA exhibits very interesting mechanical properties (about 3 GPa as Young's modulus, 60 MPa as tensile strength and an elongation at break between 3–6%) [8,9]; however, it also presents some drawbacks to be overcome to reach end-users demands. In particular, PLA is very stiff and brittle and for this reason the addition of plasticizers and/or rubber particles is often required to enhance its elongation at break and tensile toughness [10–13].

An economic and assessed method to tune polymeric matrix properties is the polymers blending [14]. In literature many flexible biobased polymers have been successfully blended with PLA such as: poly(butylene succinate) (PBS) [15–17], poly(butylene succinate-*co*-adipate) (PBSA) [18–20], polycaprolactone (PCL) [21–23] and poly(butylene adipate-*co*-terephthalate) (PBAT) [24,25].

In this article, on the basis of a previous study [18], a PLA/PBSA blend containing 60 wt.% of PLA and 40 wt.% of PBSA was selected as matrix due to its good flexibility and fracture properties. To this binary blend a by-product natural filler has been added. It is generally expected that the main address for a composite material is to exhibit enhanced physical and mechanical properties when it is compared to its individual components [26]. Nevertheless, in the case of very short fiber or particulate composites, the fibers cannot bear efficiently the load and their randomly orientation does not allow specific reinforcements along the fibers direction. In this case the benefit is related to the cost savings, lighter product weight, valorization of waste products and degradability promotion (in particular the final product disintegration) [27–31].

In response to the demand for extending biobased polymers applications while reducing the final materials cost, different studies are present in literature where the incorporation of low-cost and highly-available natural fillers and short fibers (derived from agricultural and or industrial waste) into a biobased polymeric matrices has been investigated [32–35]. More research is ongoing to optimize the valorization of agriculture residues as fillers in bio-composites. At this purpose, due to the large amount of grain by-products generated by the main food production, bran is very interesting due to its low cost and wide availability and it is a relevant cellulosic based filler for bioplastic production [36–39].

To improve the filler dispersion in the polymeric matrix and at the same time to counterbalance the stiffening effect caused by the bran addition, also improving the processability, a plasticizer was added. In particular, Triacetin was added considering the very interesting results found in literature in which Triacetin was added as plasticizer in natural fiber composites [40–42]. Ibrahim et al. [1], for example, demonstrated that PLA biocomposites with good mechanical and thermomechanical properties can be obtained using kenaf bast fiber as reinforcement and Triacetin as plasticizer. Furthermore, Pelegrini et al. [43] observed that Triacetin is a good processing aid and coupling agent that does not influence negatively the PLA biocomposites degradation.

In this study, after a preliminary characterization of wheat bran in which its granulometry and aspect ratio was determined; the effect due to the addition of different bran amount (from 10 to 30 wt.%) in a PLA/PBSA based polymeric matrix (plasticized with a fixed amount of Triacetin) was studied. Analytical models were applied in order to predict the mechanical characteristics and the interface adhesion between fibers and matrix that is fundamental to understand the interactions existing between the bran particles and the surrounding matrix. The mechanical properties, in fact, are connected to the fiber and matrix strengths, to the fibers distribution and to the interfacial shear strength (IFSS or τ) [6,27,44,45]. The last parameter, IFSS, is strictly related to different factors such as the interface thickness, the adhesion strength and the energy surface filler [27,46,47]. An IFSS good evaluation was obtained by the application of analytical models based on static and dynamical mechanical tests. Furthermore, for the first time for this biocomposites typology, the fracture toughness and crack resistance were investigated by the essential work of fracture (EWF) approach [48].

2. Theoretical Analysis

Two analytical approaches were followed: one based on the application of analytical models taking the data from the static tensile tests, and one taking the data from dynamical mechanical tests.

2.1. Analytical Predictive Model Based on Static Tests

For rigid fillers and for ultra-short fibers composites the effect of the fiber content correlated to the fiber/matrix adhesion on the composite stress at break lies among two limits called upper and lower bound [49]. Where low adhesion exists, the load is sustained only by the polymeric matrix and a simple expression (derived by Nicolais and Nicodemo [50]) can be written (Equation (1)):

$$\sigma_c = \sigma_m \left(1 - 1.21 \, V_f^{\frac{2}{3}}\right) \quad (1)$$

In Equation (1) σ_c and σ_m are, respectively, the composite strength and the matrix strength while V_f is the filler volume fraction. This equation represents the lower bound for the prediction of the tensile strength of not well bonded particulate filled composites. On the other hand, the upper bound can be obtained considering that the filler decreases the load bearing capacity of the matrix (Equation (2)) [49,51].

$$\sigma_c = \sigma_m \left(1 - V_f\right) \quad (2)$$

Pukanszky et al. [52,53] provided an empirical relationship for composites strength prediction (Equation (3)) where an empirical constant, named B, was correlated to: the particles surface area, particles density and interfacial bonding energy. However, the B parameter does not provide a numerical value of the interfacial shear stress (MPa), it is able to give information about the filler/matrix adhesion. In fact, a value of B equal or very close to zero corresponds to poor interfacial bonding where the fillers will not carry any load.

$$\sigma_c = \sigma_m \frac{1 - V_f}{1 + 2.5 V_f} \exp\left(B V_f\right) \quad (3)$$

By writing Equation (3) in a linear form (Equation (4)), is obtained a linear correlation where the interaction parameter, B, corresponds to the slope of the Pukánszky's plot (obtained plotting the natural logarithm of Pukánszky's reduced strength, σ_{red}, against the volume filler fraction).

$$\ln \sigma_{red} = \ln \frac{\sigma_c \left(1 + 2.5 V_f\right)}{\sigma_m \left(1 - V_f\right)} = B V_f \quad (4)$$

Lazzeri and Phuong [54] correlated the Pukánszky's interaction B parameter with the interfacial shear stress (IFFS). In particular, for composites having a length below the critical length (L_c) the failure of the composite occurs by plastic flow of the matrix and the interfacial shear strength (IFSS) can be calculated with the following expression:

$$\tau \, (\text{or } IFSS) = \frac{2\sigma_m (B - 2.04)}{\eta_0 a_r} \quad (5)$$

where η_0 is the fiber orientation factor, a_r is the fibers aspect ratio. In the case of particulate fillers the fibers orientation factor and the fibers aspect ratio can be taken equal to one; for very short aspect ratio fibers (such as those used in this work), a randomly fiber orientation can be considered and a value $\eta_0 = 3/8$ can be taken [27,55]. It is evident, from Equation (5), that the IFSS is directly proportional to B and 2.04 is the lower limit for the IFSS estimation.

2.2. Analytical Predicitve Model Based on Dynamic Tests

The interfacial strength between the polymeric matrix and the different volume content of wheat bran added, can be tracked using the so called "adhesion factor" (A). The damping factor ($tan\delta$), obtainable from DMTA tests, is an indicator of the material molecular motions. Thanks to its estimation it is possible to have an idea of the fiber-matrix interfacial bonding. The adhesion parameter was introduced by Kubat et al. [56] starting on the assumption that the mechanical loss factor ($tan\delta_c$) of the composite can be given as follows (Equation (6)):

$$tan\delta_c = V_f tan\delta_f + V_i tan\delta_i + V_m tan\delta_m \quad (6)$$

where the subscripts f, i, and m subscripts denote the filler, the interphase, and matrix; while V_f is the volume filler fraction. Making the assumption that the volume fraction of the interphase is rather small ($\delta_f \approx 0$), its contribution can be neglected and Equation (6) can be rearranged as follows [57] (Equation (7)):

$$\frac{tan\delta_c}{tan\delta_m} \approx \left(1 - V_f\right)(1 + A) \tag{7}$$

with the "A" term, called adhesion factor. Equation (7) can be rewritten in order to explicit the adhesion factor as follows (Equation (8)):

$$A = \frac{1}{1 - V_f} \frac{tan\delta_c}{tan\delta_m} - 1 \tag{8}$$

Calculating the adhesion factor from DMTA experiments, it is possible to interpret the interactions between the fillers and the polymeric matrix. Generally, a low value of the adhesion factor is an indication of good adhesion (or high degree of interaction between the two phases) due to the reduction of the macromolecular mobility induced by the presence of the filler [58].

2.3. Essential Work of Fracture (EWF) Approach

In order to characterize the polymers fracture, the linear elastic fracture mechanics (LEFM) approach is generally adopted. However, the characterization of fracture toughness by LEFM theory is difficult when relatively ductile polymers are considered due to the formation of a large plastic zone prior to crack initiation that violates the validity of LEFM approach. The fracture event of a relatively ductile polymer, having a marked plastic deformation zone at the crack tip, can be investigated by the essential work of fracture approach (EWF) originally suggested by Broberg [59] and then developed by Cotterell, Mai and co-workers [48,60–62]. According to this methodology, the fracture process zone is divided into two regions: an inner region (where the fracture process occurs) and an outer region (where the plastic deformation occurs). The total work of fracture follows this regions division and it can be separated in two contributions: the first contribution is related to the work spent in the inner fracture zone (work of fracture) and the work spent in the plastic deformation zone (non-essential work of fracture) [48].

For the application of the EWF theory, two different kinds of specimen can be used: the single edge notched (SENT) specimen and the double edge notched (DENT) specimen. Generally, the EWF approach is applied to thin sheets having a thickness between 1 mm and 0.2 mm [63]; the thickness of the specimen used (1.5 mm), although slightly higher than the standard range, is very low and consequently an attempt of applying the EWF approach was maintained.

The SENT specimen configuration illustrated in Figure 1 was adopted.

The total work of fracture, W_f, is defined as:

$$W_f = W_e + W_p \tag{9}$$

where W_e is the work spent for the formation of two new fracture surfaces and it is spent during the fracture process and corresponds to the resistance to crack initiation. For specimens having a given thickness, W_e is proportional to the ligament length, l (where $l = W - a$). In the Equation (9), W_p is a volume energy, it is proportional to l^2, and corresponds to the energy for activating the plastic deformation mechanism antagonists to the crack propagation. Consequently, Equation (9) can be rewritten in the following form (Equation (10)):

$$W_f = w_e t l + \beta w_p t l^2 \tag{10}$$

The specific total work of fracture can be written in the following form:

$$w_f = \left(\frac{W_f}{tl}\right) = w_e + \beta w_p l \qquad (11)$$

where w_e and w_p are the specific essential work of fracture and the specific non-essential work of fracture, respectively; β is the plastic zone shape factor while t, and l are related to the specimen geometry and correspond to the thickness and ligament length of the SENT specimen, respectively.

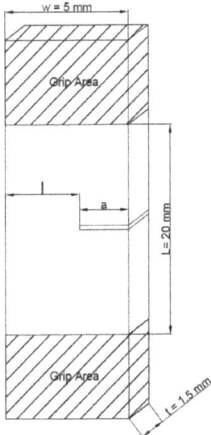

Figure 1. Schematic diagrams showing the single-edge-notched-tension (SENT).

With the assumption that w_e is a material constant and w_p and β are independent from the ligament length, it is possible to plot Equation (11) as a straight line in a graph w_f vs. l. Consequently, w_e can be determined from the intercept of the Y-axis of the w_f versus l plot, while βw_p is the slope of the straight line. It must be pointed out that β depends on the specimen geometry and on the initial crack length so the straight line relationship can be obtained only if the geometric similarity is retained for all ligaments lengths.

3. Materials and Methods

3.1. Materials

The materials used in this work are listed below:

- Poly(lactic acid) (PLA), trade name Luminy LX175, provided by Total Corbion PLA (Rayong, Thailand). It is a biodegradable PLA derived from natural resources that appears as white pellets. This extrusion grade PLA can be used alone or mixed with other polymers/additives to produce blends/composites on conventional equipment for film extrusion, thermoforming or fiber spinning (D-lactic acid unit content about 4%, density: 1.24 g/cm^3; melt flow index (MFI) (210 °C/2.16 kg): 6 g/10 min).
- Poly(butylene succinate-*co*-adipate) (PBSA), trade name BioPBS FD92PM, was purchased from Mitsubishi Chemical Corporation (Tokyo, Japan) and it is a copolymer of succinic acid, adipic acid and butandiol. It is a soft and flexible semicrystalline polyester that appears as matt white pellets (density:1.24 g/cm^3; MFI (190 °C, 2.16 kg): 4 g/10 min).
- Triacetin (TA), or glycerol triacetate, was purchased from Sigma-Aldrich (St. Louis, MO, USA). It appears as an oily transparent and odorous liquid and it was used as biobased and biodegradable food contact plasticizer (CAS number: 102-76-1; density:1.16 g/cm^3; boiling point: 258 °C; Mw = 218.20 g/mol).

- Wheat bran (WB) was provided by WeAreBio organic food; it appears as light brown powder and was used as filler for the PLA/PBSA plasticized blends formulations (CAS number: 130498-22-5; apparent density: 0.51 g/cm^3; dietetic soluble fibre: 0.93% (p/p); dietetic insoluble fibre: 19.70% (p/p)).

3.2. SFT-IR Characterization

Infrared spectrum of bran was recorded in the 550–4000 cm^{-1} range with a Nicolet 380 Thermo Corporation Fourier Transform Infrared (FTIR) Spectrometer (Thermo Fisher Scientific, Waltham, MA, USA) equipped with Smart Itx ATR (Attenuated Total Reflection) accessory with a diamond plate, collecting 256 scans at 2 cm^{-1} resolutions.

3.3. SEM Analysis

Wheat bran morphology was investigated by scanning electron microscopy (SEM) with a FEI Quanta 450 FEG instrument (Thermo Fisher Scientific, Waltham, MA, USA). The bran powder was prior sputtered with a layer of platinum and then observed by SEM. The metallic layer makes the surface electrically conductive, allowing the electrons to generate the images. From the SEM images obtained, the dimensional distribution of the wheat bran powder was also evaluated. More than 200 particles were examined by using ImageJ® software (National Institutes of Health and the Laboratory for Optical and Computational Instrumentation, Madison, WI, USA).

In addition, the composites morphologies were investigated by SEM, on a cryogenic fractured cross-sections of tensile specimens, to evaluate the fiber-matrix adhesion and the fibers dispersion inside the matrix.

3.4. Samples Preparation

PLA/PBSA blends were prepared using an Haake Minilab II (Thermo Scientific Haake GmbH, Karlsruhe, Germany) co-rotating conical twin-screw mini-extruder. Before processing the materials were dried in an air circulated oven at 60 °C for 1 day. For each extrusion cycle about 6 g of material, manually mixed with the additives, were fed into the mini extruder at 190 °C and 110 rpm. Subsequently the molten material was transferred, through a preheated cylinder, to a Thermo Scientific Haake MiniJet Mini-Injection Molding System for preparation of tensile dog-bone specimens (Haake bar type 3 with gauge dimensions: 25 × 5 × 1.5 mm^3). The cylinder temperature was set equal to the extrusion temperature (190 °C) while the mold temperature was set at 50 °C. In order to completely fill the mold, 300 bar of pressure for 15 sec was set followed by a post pressure of 200 bar for 4 sec.

The blends compositions are listed in Table 1. On the basis of a previous work [18] a PLA/PBSA matrix containing 60 wt.% of PLA and 40 wt.% of PBSA was selected. The matrix was plasticized with Triacetin maintaining fixed the ratio between the matrix and the plasticizer (equal to 9), for all the formulations. The ratio between the matrix and plasticizer was set on the basis of literature work [40–42,64] in order to ensure a good processability and a high flexibility to counterbalance the effect of fiber addition that tend to embrittle the matrix.

Table 1. Blends name and compositions.

Blend Name	Matrix (PLA 60 wt.%-PBSA 40 wt.%)	Triacetin wt.%	Wheat Bran wt.%
PLA_PBSA_TA	90	10	-
PLA_PBSA_TA_10	81	9	10
PLA_PBSA_TA_20	72	8	20
PLA_PBSA_TA_30	63	7	30

3.5. Mechanical Characterization

Tensile tests were carried out, at room temperature, on Haake type 3 dog-bone specimens by an MTS Criterion testing machine model 43 (MTS System Corporation, Eden Praire, MN, USA) equipped with a load cell of 10 kN. The instrument is interfaced to a computer running MTS Elite Software in order to record the results of the tensile tests. The gauge separation was set at 25 mm and the crosshead speed was 10 mm/min. The main mechanical properties were collected. Tests were conducted after 3 days from the samples injection moulding and during this time the specimens were stored in a dry keeper (SANPLATEC Corp., Osaka, Japan) in controlled atmosphere (room temperature and 50% humidity). At least 10 specimens were tested for each blend composition and the average values of the main mechanical properties were reported.

For the EWF characterization the SENT geometry was used. The SENT specimens were obtained from the injection molded Haake type 3 specimens appropriately cut (size: $20 \times 5 \times 1.5$ mm^3). SENT specimens with different crack length (from 1.5 mm up to 3.5 mm) were produced by sharp notch and during the cutting process, compressed air was used in order to avoid the "notch closing" phenomenon caused by excessive overheating generated by the cutter. The ligament length was checked by optical microscope. The tensile load to the SENT specimen was applied using the previous mentioned MTS Criterion testing machine at a crosshead speed of 0.5 mm/min. At least five samples for each ligament lengths were tested and the average values of the area under the load displacement curves were reported.

Dynamic mechanical thermal analysis (DMTA) was carried out using a Gabo Eplexor® DMTA (Gabo Qualimeter, Ahiden, Germany) with a 100 N load cell. At least three specimens were tested for each composition. The test bars (size: $20 \times 5 \times 1.5$ mm^3) were obtained by cutting the tensile specimens' dog-bone specimens. The samples bars were mounted on the machine in tensile configuration. The temperature range adopted for the test varied from -80 to $120\ °C$ with heating rate of $2\ °C$/min and at a constant frequency of 1 Hz. The relaxation temperature, associated with the glass transition, was taken at the maximum of the peak of the damping factor ($tan\delta$).

4. Results and Discussion

4.1. Wheat Bran Characterization

The micrographs of wheat bran fibers (Figure 2a) show that bran particles have various dimension. Mainly the bran powder consists of platelets having low aspect ratio that forms agglomerates having an average size around 250–500 micrometers. However smaller size fraction in large quantity can also be observed. In the SEM micrographs the wheat bran fibers showed a low aspect ratio and their morphology is similar to flakes.

Consequently, for the fibers aspect ratio distribution a flake geometry was assumed where the filler aspect ratio, a_r, can be defined as the ratio between the average fiber diameter (d) (calculated on the larger platelets surface) and the mean thickness of the platelets (h) according to Equation (12) [54]

$$a_r, platelets = \frac{d}{h} \qquad (12)$$

In Figure 2b-c a sharp peak can be observed in the range of 0–200 μm and 0–40 μm for the mean fiber diameter and fiber thickness distributions, respectively. Thanks to the fibers distribution the mean weighted fibers lengths for the calculation of the mean fibers aspect ratio was carried out. In particular, a mean weighted fiber diameter of 219.76 μm and a mean weighted fiber thickness of 51.40 μm were obtained that correspond to a mean fiber aspect ratio of 4.27. Consequently, the composites obtained belong to the family of very short fiber composites [27].

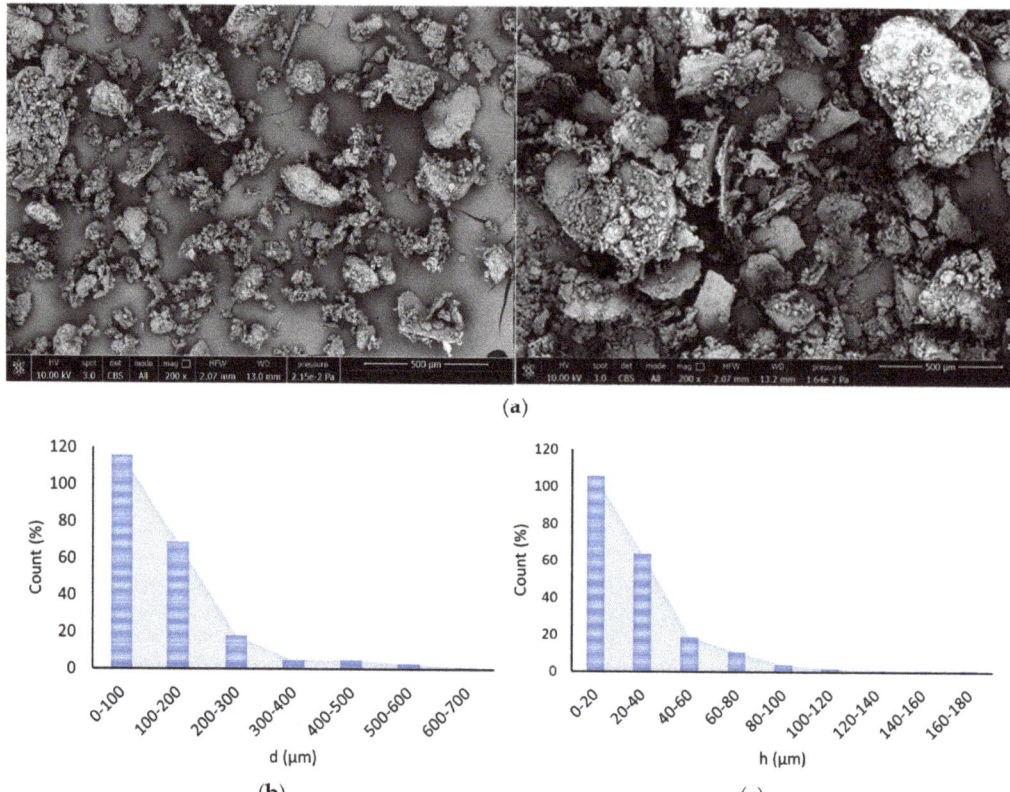

Figure 2. (a) SEM images at 200× of wheat bran fibers (b) average diameter distribution and (c) average thickness distribution.

In the infrared spectra (reported in Figure 3) the hydrophilicity of the wheat fibers is highlighted by the very sharp –OH groups stretching in the region of 3650–3000 cm^{-1}. The peak at 3294 cm^{-1} corresponding to –OH stretching was attributed to specific intramolecular hydrogen bonds of cellulose II [65,66]. The infrared spectra displayed a small adsorption band at 2925 cm^{-1}, typical of the stretching vibrations of the C–H bonds in hemicelluloses and cellulose. The bands at 1639 cm^{-1} and 1540 cm^{-1}, attributable to amide I and amide II vibrations, can be ascribed at the presence of the protein fraction in bran. Moreover, the shoulder at 1639 cm^{-1} is attributable to the acetyl and uronic ester groups of hemicelluloses or to the ester linkage of carboxylic group of the ferulic and p-coumaric acids of lignin [67]. Adsorption bands can also be observed near 1639 cm^{-1} and they are attributed to deformation of the C=O groups of xylan [68] that is the main component of hemicellulose.

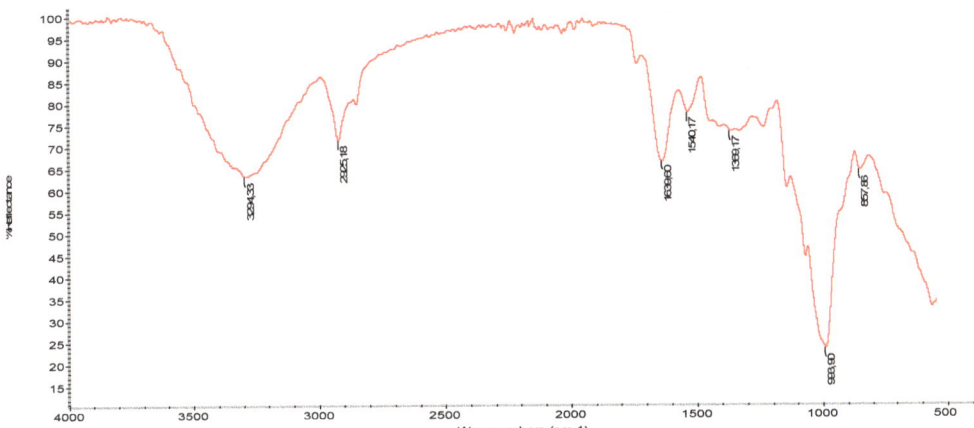

Figure 3. ATR spectrum of wheat bran fibers.

4.2. Composites Results

Tensile tests results reported in Table 2 and Figure 4 evidenced that by increasing the bran content, a decrease in yield stress, stress at break and elongation at break was observed. As it can be expected, the bran addition caused a matrix embrittlement. This behavior is in line with literature papers on similar biocomposites [37,69,70]. The Triacetin addition was revealed a valuable approach to contrast the embrittlement caused by the bran fibers addition. At this purpose, it can be observed that up to 20 wt.% of bran content the elongation at break is still high (about 50%); also with 30 wt.% of Bran, despite the sharp drop down of elongation at break due to the high bran amount, the final elongation at break is still higher (17.3%) if compared to pure PLA (about 4% [6]).

Table 2. Tensile data of PLA/PBSA plasticized blends.

Blend Name	Yield Stress (MPa)	Stress at Break (MPa)	Elongation at Break (%)	Young's Modulus (GPa)
PLA_PBSA_TA	15.7 ± 1.8	20.4 ± 0.9	325.0 ± 19.9	1.9 ± 0.1
PLA_PBSA_TA_10	12.9 ± 0.8	13.1 ± 0.8	229.0 ± 32.6	1.61 ± 0.1
PLA_PBSA_TA_20	11.1 ± 1.3	9.6 ± 1.0	57.7 ± 10.8	1.6 ± 0.1
PLA_PBSA_TA_30	10.9 ± 1.0	8.9 ± 0.8	17.3 ± 2.7	1.6 ± 0.1

The decrease in the stress at break with bran content is ascribable to a low adhesion between the matrix and the fibers that causes an ineffective transmission of the load from the matrix to the fibers.

The application of the analytical models for stress at break (Figure 5a) confirms that low adhesion exists between the wheat bran and the polymeric matrix. The experimental data lies exactly between the lower and upper bound confirming the good applicability of the analytical models to the analyzed system.

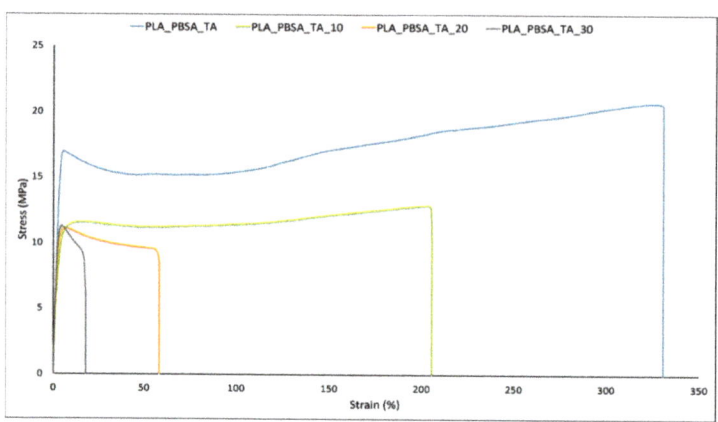

Figure 4. Stress-strain representative curves for plasticized PLA/PBSA composites.

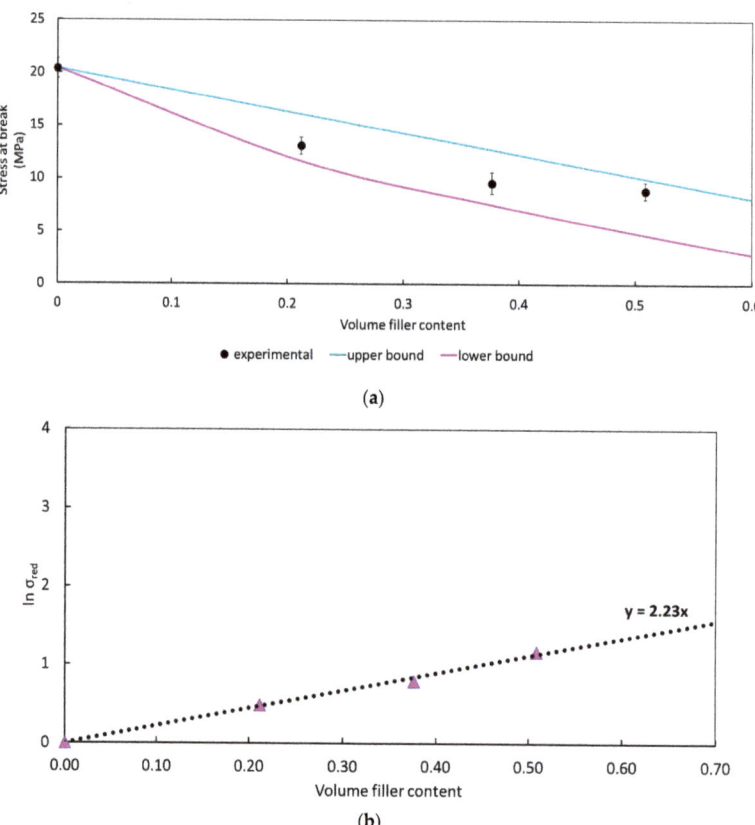

Figure 5. (**a**) Comparison between the experimental composites strength and the values predicted according to the analytical models illustrated in Section 2.1. Volumetric fractions are derived from the weight fractions calculated by using the density of each component taken from data sheets; (**b**) Pukánszky's plot for PLA/PBSA plasticized composites.

The poor adhesion is also confirmed by the B value obtained from the Pukánszky's plot (Figure 5b) where a *B* value equals to 2.23 was obtained. Being the B value obtained greater than 2.04, the Equation (5) was used for the *IFSS* estimation and the value, reported in Table 3, was also compared to other similar composites systems containing natural fibers. The IFSS value obtained can be retained acceptable because it is inferior to its maximum theoretical threshold value achievable calculated by the Von Mises relationship [71] ($\tau = \sigma_m/\sqrt{3}$) equal to 11.78 MPa; furthermore the IFSS value obtained is comparable to similar bio composites described in literature.

Table 3. IFSS obtained for the composites system analyzed compared with literature data of similar systems.

Reference	Matrix	Natural Fibres	IFSS (MPa)
Experimental	PLA/PBSA blend	Wheat bran	4.84
Aliotta et al. [6]	PLA	Cellulose	8.20
Li et al. [72]	PP	Hemp	5.84
Lopez et al. [73]	PP	Softwood fibers	3.85
Nam et al. [74]	PLA	Coir	4.56

The addition of bran fibers causes a decrement of the storage modulus (Figure 6) that is similar to the decrement of the elastic modulus observable in Table 2. This decrement can be ascribed to the slight PLA chain scission induced by the bran addition that, as all natural fibers is sensitive to moisture, and caused the slight elastic modulus decrement and the shift, towards a lower temperature, of the PLA glass transition temperature [75,76]. Independently from the bran content, a shift of the PLA *tanδ* peak (corresponding to the PLA glass transition temperature) of about 5 °C is observed; while for PBSA no deviation in its T_g (registered at around −44 °C) has been detected.

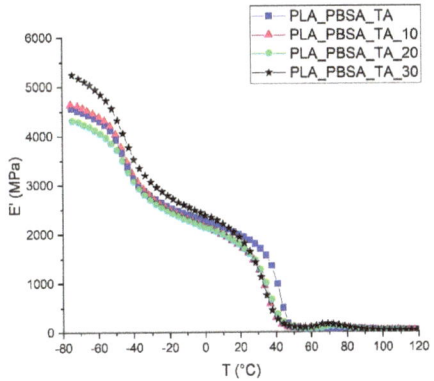 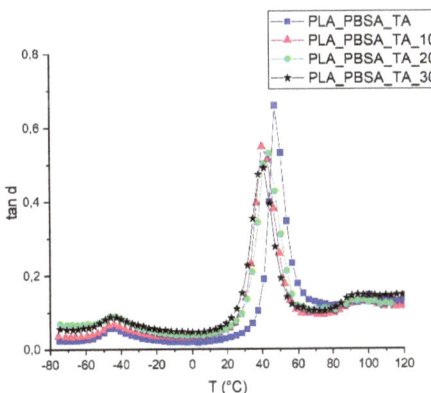

Figure 6. Variation of the storage moduli (E') (**left side**) and damping factors (tan δ) (**right side**) with temperature for PLA/PBSA plasticized bran composites.

The height of the *tanδ* peak is similar for all the composites compositions. However, this height is lower if compared to the *tanδ* height of the matrix. This decrement is due to the introduction of the bran fibers that reduces the molecular mobility of the polymeric matrix [77]. This reduction of the molecular mobility is strictly correlated to the fiber-matrix adhesion: a good adhesion in fact is responsible of a major molecular mobility reduction. The adhesion factor (reported in Table 4) increases with the bran content indicating that the adhesion worsens increasing the fibers content. This result is in accordance to what was found in literature for other similar systems where a decrement of the interfacial shear stress with the increase of the volume fiber content was registered [27,78,79]. The difference in the high hydrophilic nature of the bran and the mainly hydrophobic nature of the matrix can be the main reason of the observed poor adhesion.

Table 4. Bran composites adhesion factor calculated according Equation (8) at room temperature (25 °C).

Blend Name	A
PLA_PBSA_TA_10	1.5
PLA_PBSA_TA_20	2.4
PLA_PBSA_TA_30	5.2

It is also worthy to notice, from the adhesion factor versus temperature illustrated in Figure 7, that the adhesion factor seems to be very sensitive to T_g. Throughout the temperature range analyzed, the adhesion factor increases with the bran amount. This trend is a further confirmation that, at 30 wt.% of bran content, a net decrement of the mechanical properties of the composite occurs due to the slows down of the fiber-matrix interfacial adhesion. Moreover, the adhesion factor in correspondence of the PLA glass transition temperature suddenly increases reaching its maximum value. A similar behavior can be found in literature for other composite systems [57,80] and is ascribed to the higher mobility of the polymeric chains. Above the glass transition the difference in A value between the composites with different bran content are less evident than below the glass transition. Another increment of the adhesion factor, observable by another slight peak, is registered in correspondence of the PLA cold crystallization temperature (at about 90–110 °C [81]) for which the reorganization of the PLA polymeric chains caused another change in polymeric chains mobility, that is decreased due to the occurrence of crystals formation.

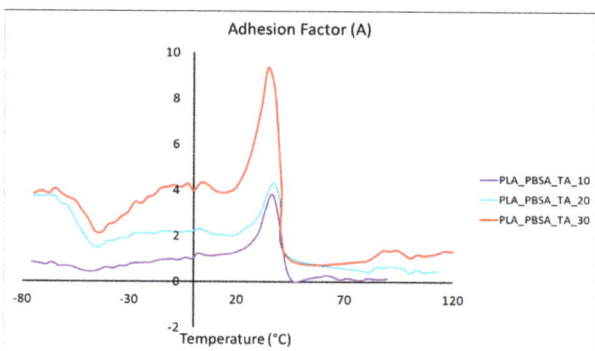

Figure 7. Adhesion factor vs. temperature for plasticized PLA/PBSA bran composites.

The influence of the presence of the wheat bran filler on the fracture of the material was also investigated with the aim of integrating it with the studies of interactions and adhesion.

The load displacement curves Figure 8a–d show a common behavior in the range of the ligament lengths examined (from 1.5 to 3.5 mm). As ligament lengths decreases, the maximum load decreases. Once the maximum has been reached, the materials undergo to a drop down of the load. By increasing the bran content, the load dropping is more marked and smaller displacement values were recorded. The fracture occurs suddenly after that the maximum load is reached and the introduction of the bran particles decreased the material ductility in accordance to the tensile test results.

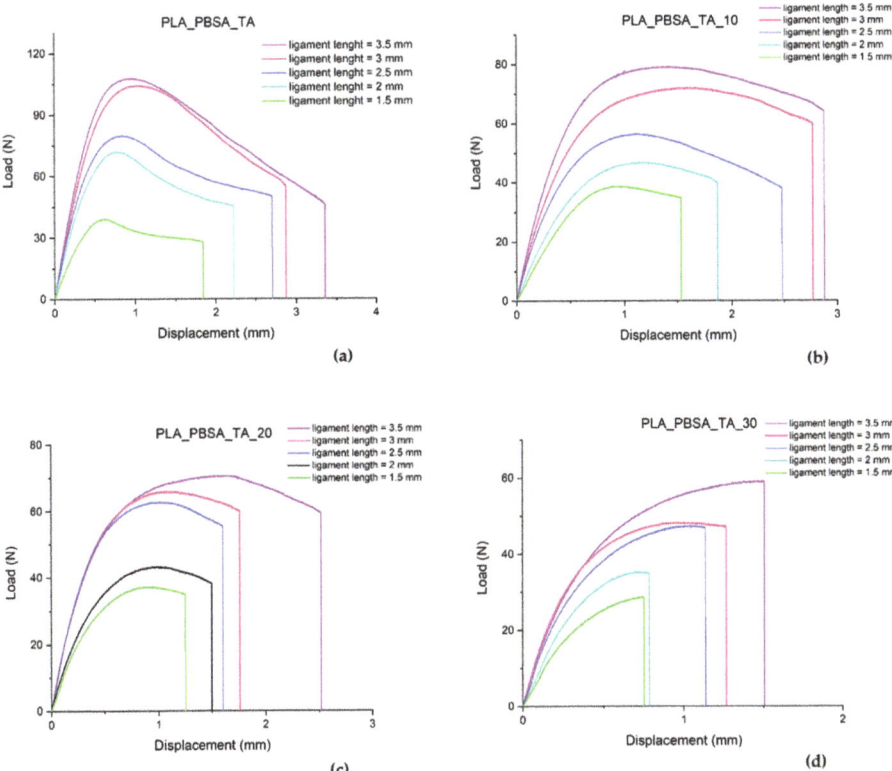

Figure 8. Load displacement curves of plasticized PLA/PBSA blend with and without bran at different ligament lengths.

The plot of the specific total work of fracture versus ligament length for the different blend compositions is shown in Figure 9 while the results of the EWF parameters are reported in Table 5. A good linear relationship between w_f and l can be observed from the regression coefficients that lies in the range of 0.81 and 0.97 allowing the use of the calculated w_e and βw_p parameters. Comparing the data of fracture parameters emerges that both the essential work of fracture and non-essential work of fracture decreases with the bran amount. Since w_e involves both the plastic deformation process of the necking ligament section and the work needed for the cracks to start growing [82], the maximum w_e value will be reach for the material having the highest yield stress. This trend is confirmed, in fact increasing the bran content, a decrease of the yield stress was registered at which corresponds also a decrease of the essential work of fracture. The βw_p increases with the material ductility and it decreases with the yield stress similarly to what was found by Arkhireyaya et al. [83] and in accordance to tensile test results. The lowest value of the essential work of fracture is registered for PLA_PBSA_TA_30 composite where the poor adhesion and the probable presence of agglomerates lowers abruptly the fracture energy.

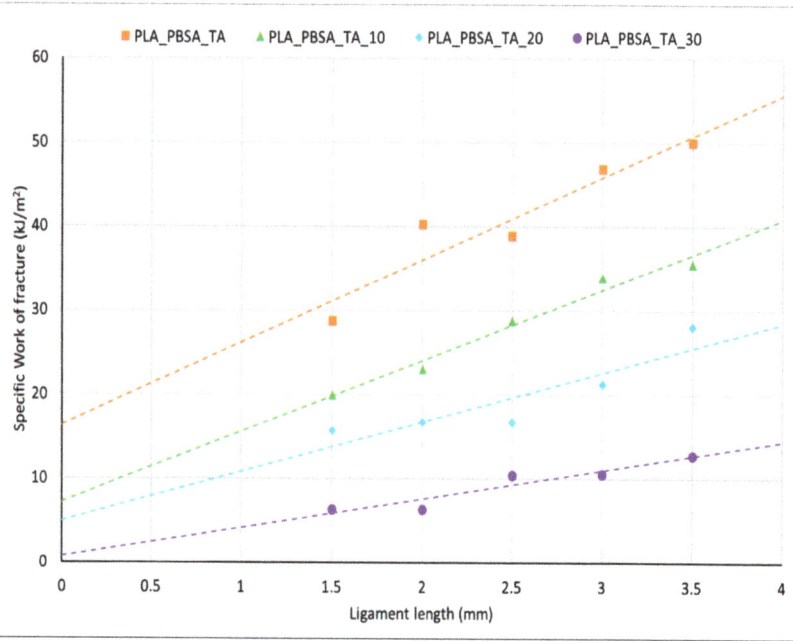

Figure 9. Specific total work of fracture w_f versus ligament length l for plasticized PLA/PBSA blend with and without bran.

Table 5. EWF fracture parameters for PLA/PBSA blends with and without bran.

Blend Name	w_e (kJ/m^2)	βw_p (MJ/m^3)	R^2
PLA_PBSA_TA	16.35	9.82	0.89
PLA_PBSA_TA_10	7.15	8.42	0.97
PLA_PBSA_TA_20	4.97	5.87	0.81
PLA_PBSA_TA_30	0.74	3.42	0.89

The bran particles act as a stress intensification factor, reducing the cross section of the material and leading to a decrease of the tensile properties and fracture resistance. Reasonably because of the high dimension and low aspect ratio, the mechanism of pull—out (that generally leads to a toughness improvement even in the case of natural fibers reinforcement [84]) is not beneficially impacting the fracture resistance of these composites. Similar results were observed by Anuar et al. for composites with Kenaf fibers [85]. Thus, data about essential work of fracture are not largely available in literature, a systematic comparison with different natural fibers is currently difficult.

SEM images (Figure 10) confirm the results. At 30 wt.% of bran (Figure 10C) the detachment from the matrix around the bran fibers becomes more marked. There is some fiber/matrix adhesion (although is not optimal) as evidenced by the magnification at 10 000X, for the composites containing 10 wt.% of bran (Figure 10A1). At 20 wt.% of bran the adhesion starts to worsening (Figure 10B1) the detachment of the bran fiber from the matrix, is present only in some regions of the fibers contours while in another there is continuity between the fiber contour and the matrix. At 30 wt.% of bran (Figure 10C1) an almost complete fiber detachment can be observed. Increasing the fiber content also increases the likelihood of finding fibers agglomerates that contribute to the decrement of the stress at break and to the essential work of fracture.

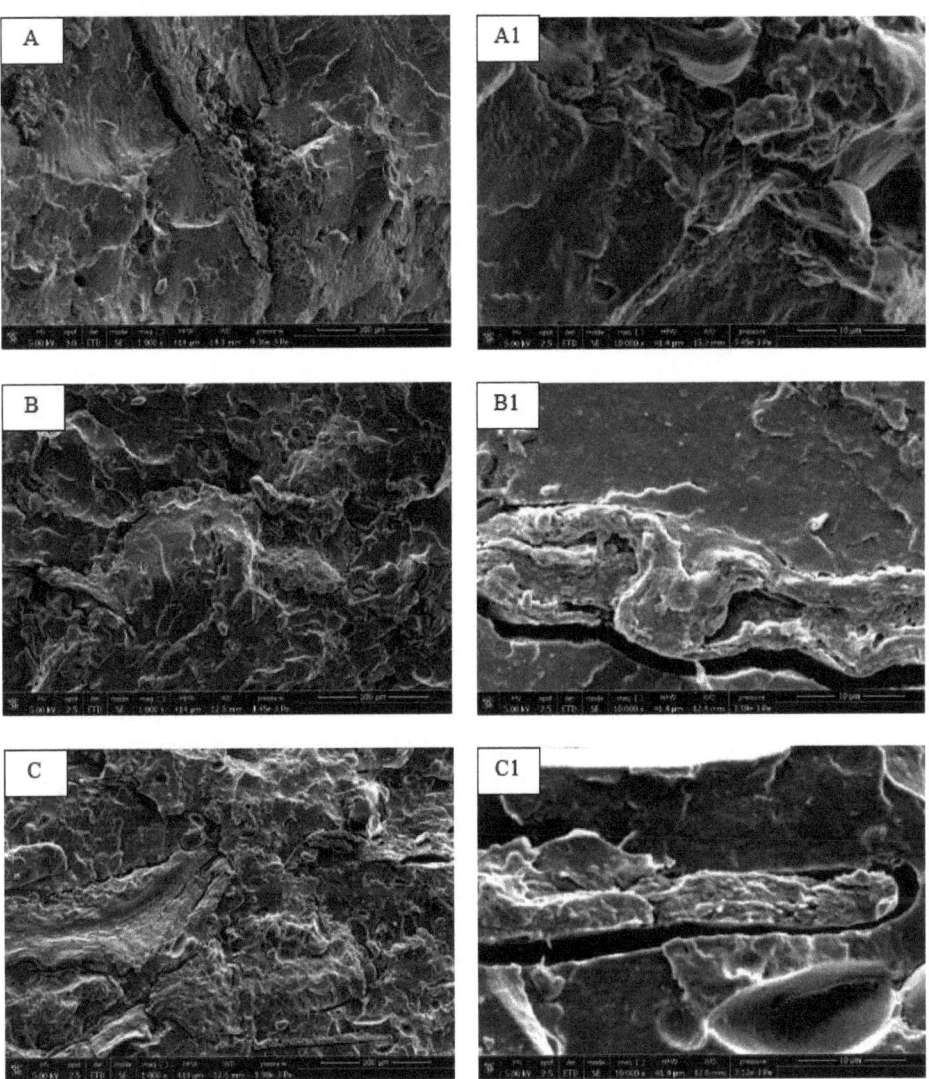

Figure 10. SEM micrographs of cryogenic fractured cross-section for: (**A,A1**) PLA_PBSA_TA_10; (**B,B1**) PLA_PBSA_TA_20; (**C,C1**) PLA_PBSA_TA_30.

5. Conclusions

In order to extend the biobased polymers application and at the same time to reduce the final costs of the material, the design of suitable biocomposites containing natural fibers coming from industrial and/or agricultural waste is a feasible solution. In this work the addition of wheat bran fibers (from 10 to 30 wt.%) was investigated. In order to limit the excessive embrittlement caused by the bran addition a plasticized ductile matrix was chosen (PLA/PBSA matrix containing 60 wt.% of PLA and 40 wt.% plasticized with Triacetin). The evaluation of the interfacial adhesion (interfacial shear stress, *IFSS*) between the fiber and the matrix was carried out adopting analytical models based on static and dynamic tests. For the first time the essential work of fracture (EWF) approach was adopted on this type of biocomposites. The bran addition caused a decrement of the mechanical properties in parallel with the EWF reduction. In particular, the essential work of fracture is highly influenced by the presence and content of bran and this is ascribable to the high irregularity of filler geometry and presence of agglomerates especially where the content of filler is high. The combination of reduced adhesion between matrix- filler and increased intensification of stresses is responsible of the observed mechanical behavior as a function of the bran content.

Author Contributions: Conceptualization, L.A. and M.-B.C.; methodology, L.A. and A.L.; validation, L.A. and A.V.; formal analysis, L.A. and A.V.; investigation, L.A. and A.V.; resources, M.-B.C. and A.L.; writing—original draft preparation, L.A. and A.V.; writing—review and editing, M.-B.C. and P.C.; supervision, M.-B.C., P.C. and A.L.; funding acquisition, M.-B.C. and P.C. All authors have read and agreed to the published version of the manuscript.

Funding: This project has received funding from the Bio-based Industries Joint Undertaking (JU) under the European Union's Horizon 2020 research and innovation programme under grant agreement No 837761 (BIONTOP project, Novel packaging films and textiles with tailored end of life and performance based on bio-based copolymers and coatings). The JU receives support from the European Union's Horizon 2020 research and innovation programme and the Bio-based Industries Consortium.

Institutional Review Board Statement: Not applicable.

Informed Consent Statement: Not applicable.

Data Availability Statement: The data presented in this study are available upon request from the corresponding author.

Acknowledgments: Centre for Instrumentation Sharing–University of Pisa (CISUP) is thanked for its support in the use of FEI Quanta 450 FEG scanning electron microscope.

Conflicts of Interest: The authors declare no conflict of interest.

References

1. Ibrahim, N.A.; Yunus, W.M.Z.W.; Othman, M.; Abdan, K.; Hadithon, K.A. Poly(Lactic Acid) (PLA)-reinforced kenaf bast fiber composites: The effect of triacetin. *J. Reinf. Plast. Compos.* **2010**, *29*, 1099–1111. [CrossRef]
2. Wambua, P.; Ivens, J.; Verpoest, I. Natural fibres: Can they replace glass in fibre reinforced plastics? *Compos. Sci. Technol.* **2003**, *63*, 1259–1264. [CrossRef]
3. Mohanty, A.K.; Misra, M.; Drzal, L.T. Sustainable bio-composites from renewable resources: Opportunities and challenges in the green materials world. *J. Polym. Environ.* **2002**, *10*, 19–26. [CrossRef]
4. Haider, T.P.; Völker, C.; Kramm, J.; Landfester, K.; Wurm, F.R. Plastics of the future? The impact of biodegradable polymers on the environment and on society. *Angew. Chemie Int. Ed.* **2019**, *58*, 50–62. [CrossRef] [PubMed]
5. Garrison, T.F.; Murawski, A.; Quirino, R.L. Bio-Based Polymers with Potential for Biodegradability. *Polymers* **2016**, *8*, 262. [CrossRef] [PubMed]
6. Aliotta, L.; Gigante, V.; Cinelli, P.; Coltelli, M.-B.; Lazzeri, A. Effect of a Bio-Based Dispersing Aid (Einar ® 101) on PLA-Arbocel ® Biocomposites: Evaluation of the Interfacial Shear Stress on the Final Mechanical Properties. *Biomolecules* **2020**, *10*, 1549. [CrossRef]
7. Gross, R.A.; Kalra, B. Biodegradable Polymers for the Environment. *Science* **2002**, *297*, 803LP–807LP. [CrossRef]
8. Aliotta, L.; Cinelli, P.; Coltelli, M.B.; Righetti, M.C.; Gazzano, M.; Lazzeri, A. Effect of nucleating agents on crystallinity and properties of poly (lactic acid) (PLA). *Eur. Polym. J.* **2017**, *93*, 822–832. [CrossRef]
9. Raquez, J.M.; Habibi, Y.; Murariu, M.; Dubois, P. Polylactide (PLA)-based nanocomposites. *Prog. Polym. Sci.* **2013**, *38*, 1504–1542. [CrossRef]
10. Gigante, V.; Canesi, I.; Cinelli, P.; Coltelli, M.B.; Lazzeri, A. Rubber Toughening of Polylactic Acid (PLA) with Poly(butylene adipate-*co*-terephthalate) (PBAT): Mechanical Properties, Fracture Mechanics and Analysis of Ductile-to-Brittle Behavior while Varying Temperature and Test Speed. *Eur. Polym. J.* **2019**, *115*, 125–137. [CrossRef]
11. Byun, Y.; Rodriguez, K.; Han, J.H.; Kim, Y.T. Improved thermal stability of polylactic acid (PLA) composite film via PLA-β-cyclodextrin-inclusion complex systems. *Int. J. Biol. Macromol.* **2015**, *81*, 591–598. [CrossRef]
12. Argon, A.S.; Bartczak, Z.; Cohen, R.E.; Muratoglu, O.K. *Novel Mechanisms of Toughening Semi-Crystalline Polymers*; ACS Publications: Washington, DC, USA, 2000; ISBN 1947-5918.
13. Aliotta, L.; Gigante, V.; Coltelli, M.; Cinelli, P.; Lazzeri, A.; Seggiani, M. Thermo-Mechanical Properties of PLA/Short Flax Fiber Biocomposites. *Appl. Sci.* **2019**, *9*, 3797. [CrossRef]

14. Zhao, X.; Hu, H.; Wang, X.; Yu, X.; Zhou, W.; Peng, S. Super tough poly (lactic acid) blends: A comprehensive review. *RSC Adv.* **2020**, *10*, 13316–13368. [CrossRef]
15. Nofar, M.; Tabatabaei, A.; Sojoudiasli, H.; Park, C.B.; Carreau, P.J.; Heuzey, M.C.; Kamal, M.R. Mechanical and bead foaming behavior of PLA-PBAT and PLA-PBSA blends with different morphologies. *Eur. Polym. J.* **2017**, *90*, 231–244. [CrossRef]
16. Yokohara, T.; Yamaguchi, M. Structure and properties for biomass-based polyester blends of PLA and PBS. *Eur. Polym. J.* **2008**, *44*, 677–685. [CrossRef]
17. Sun, S.; Kopitzky, R.; Tolga, S.; Kabasci, S. Polylactide (PLA) and Its Blends with Poly (butylene succinate) (PBS): A Brief Review. *Polymers* **2019**, *11*, 1193.
18. Aliotta, L.; Vannozzi, A.; Canesi, I.; Cinelli, P.; Coltelli, M.; Lazzeri, A. Poly(lactic acid) (PLA)/Poly(butylene succinate-co-adipate) (PBSA) Compatibilized Binary Biobased Blends: Melt Fluidity, Morphological, Thermo-Mechanical and Micromechanical Analysis. *Polymers* **2021**, *13*, 218. [CrossRef]
19. Pradeep, S.A.; Kharbas, H.; Turng, L.S.; Avalos, A.; Lawrence, J.G.; Pilla, S. Investigation of thermal and thermomechanical properties of biodegradable PLA/PBSA composites processed via supercritical fluid-assisted foam injection molding. *Polymers* **2017**, *9*, 22. [CrossRef] [PubMed]
20. Oguz, H.; Dogan, C.; Kara, D.; Ozen, Z.T.; Ovali, D.; Nofar, M. Development of PLA-PBAT and PLA-PBSA bio-blends: Effects of processing type and PLA crystallinity on morphology and mechanical properties. *AIP Conf. Proc.* **2019**, *2055*, 030003. [CrossRef]
21. Urquijo, J.; Guerrica-Echevarría, G.; Eguiazábal, J.I. Melt processed PLA/PCL blends: Effect of processing method on phase structure, morphology, and mechanical properties. *J. Appl. Polym. Sci.* **2015**, *132*, 42641. [CrossRef]
22. Ostafinska, A.; Fortelný, I.; Hodan, J.; Krejčíková, S.; Nevoralová, M.; Kredatusová, J.; Kruliš, Z.; Kotek, J.; Šlouf, M. Strong synergistic effects in PLA/PCL blends: Impact of PLA matrix viscosity. *J. Mech. Behav. Biomed. Mater.* **2017**, *69*, 229–241. [CrossRef] [PubMed]
23. Takayama, T.; Todo, M.; Tsuji, H.; Arakawa, K. Improvement of fracture properties of PLA/PCL polymer blends by control of phase structures. *Kobunshi Ronbunshu* **2006**, *63*, 626–632. [CrossRef]
24. Nakayama, D.; Wu, F.; Mohanty, A.K.; Hirai, S.; Misra, M. Biodegradable Composites Developed from PBAT/PLA Binary Blends and Silk Powder: Compatibilization and Performance Evaluation. *ACS Omega* **2018**, *3*, 12412–12421. [CrossRef] [PubMed]
25. Kumar, M.; Mohanty, S.; Nayak, S.K.; Rahail Parvaiz, M. Effect of glycidyl methacrylate (GMA) on the thermal, mechanical and morphological property of biodegradable PLA/PBAT blend and its nanocomposites. *Bioresour. Technol.* **2010**, *101*, 8406–8415. [CrossRef] [PubMed]
26. Yu, L.; Dean, K.; Li, L. Polymer blends and composites from renewable resources. *Prog. Polym. Sci.* **2006**, *31*, 576–602. [CrossRef]
27. Aliotta, L.; Lazzeri, A. A proposal to modify the Kelly-Tyson equation to calculate the interfacial shear strength (IFSS) of composites with low aspect ratio fibers. *Compos. Sci. Technol.* **2020**, *186*, 107920. [CrossRef]
28. Aliotta, L.; Gigante, V.; Coltelli, M.B.; Cinelli, P.; Lazzeri, A. Evaluation of Mechanical and Interfacial Properties of Bio-Composites Based on Poly (Lactic Acid) with Natural Cellulose Fibers. *Int. J. Mol. Sci.* **2019**, *20*, 960. [CrossRef]
29. Azwa, Z.N.; Yousif, B.F.; Manalo, A.C.; Karunasena, W. A review on the degradability of polymeric composites based on natural fibres. *Mater. Des.* **2013**, *47*, 424–442. [CrossRef]
30. Brebu, M. Environmental degradation of plastic composites with natural fillers-a review. *Polymers* **2020**, *12*, 166. [CrossRef]
31. Nourbakhsh, A.; Ashori, A.; Tabrizi, A.K. Characterization and biodegradability of polypropylene composites using agricultural residues and waste fish. *Compos. Part B Eng.* **2014**, *56*, 279–283. [CrossRef]
32. Guna, V.; Ilangovan, M.; Hu, C.; Venkatesh, K.; Reddy, N. Valorization of sugarcane bagasse by developing completely biodegradable composites for industrial applications. *Ind. Crops Prod.* **2019**, *131*, 25–31. [CrossRef]
33. Quiles-Carrillo, L.; Montanes, N.; Sammon, C.; Balart, R.; Torres-Giner, S. Compatibilization of highly sustainable polylactide/almond shell flour composites by reactive extrusion with maleinized linseed oil. *Ind. Crops Prod.* **2018**, *111*, 878–888. [CrossRef]
34. Carbonell-Verdu, A.; Boronat, T.; Quiles-Carrillo, L.; Fenollar, O.; Dominici, F.; Torre, L. Valorization of Cotton Industry Byproducts in Green Composites with Polylactide. *J. Polym. Environ.* **2020**, *28*, 2039–2053. [CrossRef]
35. Gowman, A.C.; Picard, M.C.; Lim, L.-T.; Misra, M.; Mohanty, A.K. Fruit waste valorization for biodegradable biocomposite applications: A review. *Bioresources* **2019**, *14*, 10047–10092. [CrossRef]
36. Shankar, R.S.; Srinivasan, S.A.; Shankar, S.; Rajasekar, R.; Kumar, R.N.; Kumar, P.S. Review article on wheat flour/wheat bran/wheat husk based bio composites. *Int. J. Sci. Res. Publ.* **2014**, *4*, 1–69.
37. Gigante, V.; Cinelli, P.; Righetti, M.C.; Sandroni, M.; Polacco, G.; Seggiani, M.; Lazzeri, A. On the Use of Biobased Waxes to Tune Thermal and Mechanical Properties of Polyhydroxyalkanoates–Bran Biocomposites. *Polymers* **2020**, *12*, 2615. [CrossRef]
38. Rahman, M.; Ulven, C.A.; Johnson, M.A.; Durant, C.; Hossain, K.G. Pretreatment of wheat bran for suitable reinforcement in biocomposites. *J. Renew. Mater.* **2017**, *5*, 62. [CrossRef]
39. Onipe, O.O.; Beswa, D.; Jideani, A.I.O. Effect of size reduction on colour, hydration and rheological properties of wheat bran. *Food Sci. Technol.* **2017**, *37*, 389–396. [CrossRef]
40. Kumar, R.; Ofosu, O.; Anandjiwala, R.D. Macadamia nutshell powder filled poly lactic acid composites with triacetin as a plasticizer. *J. Biobased Mater. Bioenergy* **2013**, *7*, 541–548. [CrossRef]
41. Harmaen, A.S.; Khalina, A.; Faizal, A.R.; Jawaid, M. Effect of Triacetin on Tensile Properties of Oil Palm Empty Fruit Bunch Fiber-Reinforced Polylactic Acid Composites. *Polym.-Plast. Technol. Eng.* **2013**, *52*, 400–406. [CrossRef]

42. Phuong, V.T.; Lazzeri, A. "Green" biocomposites based on cellulose diacetate and regenerated cellulose microfibers: Effect of plasticizer content on morphology and mechanical properties. *Compos. Part A Appl. Sci. Manuf.* **2012**, *43*, 2256–2268. [CrossRef]
43. Pelegrini, K.; Donazzolo, I.; Brambilla, V.; Coulon Grisa, A.M.; Piazza, D.; Zattera, A.J.; Brandalise, R.N. Degradation of PLA and PLA in composites with triacetin and buriti fiber after 600 days in a simulated marine environment. *J. Appl. Polym. Sci.* **2016**, *133*, 43290. [CrossRef]
44. Prashanth, S.; Subbaya, K.M.; Nithin, K.; Sachhidananda, S. Fiber Reinforced Composites—A Review. *J. Mater. Sci. Eng.* **2017**, *6*. [CrossRef]
45. Bigg, D.M. Mechanical properties of particulate filled polymers. *Polym. Compos.* **1987**, *8*, 115–122. [CrossRef]
46. Lauke, B. Determination of adhesion strength between a coated particle and polymer matrix. *Compos. Sci. Technol.* **2006**, *66*, 3153–3160. [CrossRef]
47. Zappalorto, M.; Salviato, M.; Quaresimin, M. Influence of the interphase zone on the nanoparticle debonding stress. *Compos. Sci. Technol.* **2011**, *72*, 49–55. [CrossRef]
48. Wu, J.; Mai, Y.W. The essential fracture work concept for toughness measurement of ductile polymers. *Polym. Eng. Sci.* **1996**, *36*, 2275–2288. [CrossRef]
49. Fu, S.Y.; Feng, X.Q.; Lauke, B.; Mai, Y.W. Effects of particle size, particle/matrix interface adhesion and particle loading on mechanical properties of particulate-polymer composites. *Compos. Part B Eng.* **2008**, *39*, 933–961. [CrossRef]
50. Nicolais, L.; Nicodemo, L. Strength of particulate composite. *Polym. Eng. Sci.* **1973**, *13*, 469. [CrossRef]
51. Nicolais, L.; Nicodemo, L. The effect of particles shape on tensile properties of glassy thermoplastic composites. *Int. J. Polym. Mater. Polym. Biomater.* **1974**, *3*, 229–243. [CrossRef]
52. Turcsányi, B.; Pukánszky, B.; Tüdős, F. Composition dependence of tensile yield stress in filled polymers. *J. Mater. Sci. Lett.* **1988**, *7*, 160–162. [CrossRef]
53. Pukánszky, B.; Turcsanyi, B.; Tudos, F. Effect of interfacial interaction on the tensile yield stress of polymer composites. In *Interfaces in Polymer, Ceramic and Metal Matrix Composites*; Elsevier: Amsterdam, The Netherlands, 1988; pp. 467–477.
54. Lazzeri, A.; Phuong, V.T. Dependence of the Pukánszky's interaction parameter B on the interface shear strength (IFSS) of nanofiller- and short fiber-reinforced polymer composites. *Compos. Sci. Technol.* **2014**, *93*, 106–113. [CrossRef]
55. Sanadi, A.R.; Young, R.A.; Clemons, C.; Rowell, R.M. Recycled Newspaper Fibers as Reinforcing Fillers in Thermoplastics: Part I-Analysis of Tensile and Impact Properties in Polypropylene. *J. Reinf. Plast. Compos.* **1994**, *13*, 54–67. [CrossRef]
56. Kubát, J.; Rigdahl, M.; Welander, M. Characterization of interfacial interactions in high density polyethylene filled with glass spheres using dynamic-mechanical analysis. *J. Appl. Polym. Sci.* **1990**, *39*, 1527–1539. [CrossRef]
57. Ghasemi, I.; Farsi, M. Interfacial Behaviour of Wood Plastic Composite: Effect of Chemical Treatment on Wood Fibre. *Iran. Polym. J.* **2010**, *19*, 811–818.
58. Wei, L.; McDonald, A.G.; Freitag, C.; Morrell, J.J. Effects of wood fiber esterification on properties, weatherability and biodurability of wood plastic composites. *Polym. Degrad. Stab.* **2013**, *98*, 1348–1361. [CrossRef]
59. Broberg, K.B. On stable crack growth. *J. Mech. Phys. Solids* **1975**, *23*, 215–237. [CrossRef]
60. Mai, Y.W.; Cotterell, B. Effect of specimen geometry on the essential work of plane stress ductile fracture. *Eng. Fract. Mech.* **1985**, *21*, 123–128. [CrossRef]
61. Cotterell, B.; Reddel, J.K. The essential work of plane stress ductile fracture. *Int. J. Fract.* **1977**, *13*, 267–277. [CrossRef]
62. Mai, Y.-W.; Cotterell, B. On the essential work of ductile fracture in polymers. *Int. J. Fract.* **1986**, *32*, 105–125. [CrossRef]
63. Williams, J.G.; Rink, M. The standardisation of the EWF test. *Eng. Fract. Mech.* **2007**, *74*, 1009–1017. [CrossRef]
64. Oksman, K.; Skrifvars, M.; Selin, J.F. Natural fibres as reinforcement in polylactic acid (PLA) composites. *Compos. Sci. Technol.* **2003**, *63*, 1317–1324. [CrossRef]
65. Yang, H.; Yan, R.; Chen, H.; Lee, D.H.; Zheng, C. Characteristics of hemicellulose, cellulose and lignin pyrolysis. *Fuel* **2007**, *86*, 1781–1788. [CrossRef]
66. Zuluaga, R.; Putaux, J.L.; Cruz, J.; Vélez, J.; Mondragon, I.; Gañán, P. Cellulose microfibrils from banana rachis: Effect of alkaline treatments on structural and morphological features. *Carbohydr. Polym.* **2009**, *76*, 51–59. [CrossRef]
67. Cherian, B.M.; Pothan, L.A.; Nguyen-Chung, T.; Mennig, G.; Kottaisamy, M.; Thomas, S. A novel method for the synthesis of cellulose nanofibril whiskers from banana fibers and characterization. *J. Agric. Food Chem.* **2008**, *56*, 5617–5627. [CrossRef]
68. Siqueira, G.; Bras, J.; Dufresne, A. Luffa cylindrica as a lignocellulosic source of fiber, microfibrillated cellulose and cellulose nanocrystals. *BioResources* **2010**, *5*, 727–740.
69. Robin, F.; Dubois, C.; Curti, D.; Schuchmann, H.P.; Palzer, S. Effect of wheat bran on the mechanical properties of extruded starchy foams. *Food Res. Int.* **2011**, *44*, 2880–2888. [CrossRef]
70. Fu, Z.; Wu, H.; Wu, M.; Huang, Z.; Zhang, M. Effect of Wheat Bran Fiber on the Behaviors of Maize Starch Based Films. *Starch-Stärke* **2020**, *72*, 1900319. [CrossRef]
71. Pegoretti, A.; Della Volpe, C.; Detassis, M.; Migliaresi, C.; Wagner, H.D. Thermomechanical behaviour of interfacial region in carbon fibre/epoxy composites. *Compos. Part A Appl. Sci. Manuf.* **1996**, *27*, 1067–1074. [CrossRef]
72. Li, Y.; Pickering, K.L.; Farrell, R.L. Determination of interfacial shear strength of white rot fungi treated hemp fibre reinforced polypropylene. *Compos. Sci. Technol.* **2009**, *69*, 1165–1171. [CrossRef]

73. Serrano, A.; Espinach, F.X.; Julian, F.; Del Rey, R.; Mendez, J.A.; Mutje, P. Estimation of the interfacial shears strength, orientation factor and mean equivalent intrinsic tensile strength in old newspaper fiber/polypropylene composites. *Compos. Part B Eng.* **2013**, *50*, 232–238. [CrossRef]
74. Nam, T.H.; Ogihara, S.; Kobayashi, S. Interfacial, mechanical and thermal properties of coir fiber-reinforced poly (lactic acid) biodegradable composites. *Adv. Compos. Mater.* **2012**, *21*, 103–122. [CrossRef]
75. Mohammed, L.; Ansari, M.N.M.; Pua, G.; Jawaid, M.; Islam, M.S. A review on natural fiber reinforced polymer composite and its applications. *Int. J. Polym. Sci.* **2015**, *2015*, 1–15. [CrossRef]
76. Coltelli, M.-B.; Bertolini, A.; Aliotta, L.; Gigante, V.; Vannozzi, A.; Lazzeri, A. Chain Extension of Poly(Lactic Acid) (PLA)—Based Blends and Composites Containing Bran with Biobased Compounds for Controlling Their Processability and Recyclability. *Polymers* **2021**, *13*, 3050. [CrossRef] [PubMed]
77. Jacob, M.; Francis, B.; Thomas, S. Dynamical Mechanical Analysis of Sisal/Oil Palm Hybrid Fiber-Reinforced Natural Rubber Composites. *Polymer Composites* **2006**, *38*, 1504–1518. [CrossRef]
78. Thomason, J.L. The influence of fibre length and concentration on the properties of glass fibre reinforced polypropylene: 7. Interface strength and fibre strain in injection moulded long fibre PP at high fibre content. *Compos. Part A Appl. Sci. Manuf.* **2007**, *38*, 210–216. [CrossRef]
79. Gigante, V.; Seggiani, M.; Cinelli, P.; Signori, F.; Vania, A.; Navarini, L.; Amato, G.; Lazzeri, A. Utilization of coffee silverskin in the production of Poly (3-hydroxybutyrate-*co*-3-hydroxyvalerate) biopolymer-based thermoplastic biocomposites for food contact applications. *Compos. Part A Appl. Sci. Manuf.* **2021**, *140*, 106172. [CrossRef]
80. Correa, C.A.; Razzino, C.A.; Hage, E. Role of maleated coupling agents on the interface adhesion of polypropylene-wood composites. *J. Thermoplast. Compos. Mater.* **2007**, *20*, 323–339. [CrossRef]
81. Aliotta, L.; Gazzano, M.; Lazzeri, A.; Righetti, M.C. Constrained Amorphous Interphase in Poly (L -lactic acid): Estimation of the Tensile Elastic Modulus. *ACS Omega* **2020**, *5*, 20890–20902. [CrossRef]
82. Mai, Y.; Cotterell, B.; Horlyck, R.; Vigna, G. The essential work of plane stress ductile fracture of linear polyethylenes. *Polym. Eng. Sci.* **1987**, *27*, 804–809. [CrossRef]
83. Arkhireyeva, A.; Hashemi, S. Fracture behaviour of polyethylene naphthalate (PEN). *Polymer* **2002**, *43*, 289–300. [CrossRef]
84. Alvarez, V.; Vazquez, A.; Bernal, C. Effect of microstructure on the tensile and fracture properties of sisal fiber/starch-based composites. *J. Compos. Mater.* **2006**, *40*, 21–35. [CrossRef]
85. Anuar, H.; Ahmad, S.H.; Rasid, R.; Surip, S.N.; Czigany, T.; Romhany, G. Essential work of fracture and acoustic emission study on TPNR composites reinforced by kenaf fiber. *J. Compos. Mater.* **2007**, *41*, 3035–3049. [CrossRef]

Review

Liquid and Solid Functional Bio-Based Coatings

Vito Gigante [1,2], Luca Panariello [1,2], Maria-Beatrice Coltelli [1,2,*], Serena Danti [1,2], Kudirat Abidemi Obisesan [3], Ahdi Hadrich [4], Andreas Staebler [5], Serena Chierici [6], Ilaria Canesi [7], Andrea Lazzeri [1,2,7] and Patrizia Cinelli [1,2,7,*]

1. Department of Civil and Industrial Engineering, University of Pisa, 56122 Pisa, Italy; vito.gigante@dici.unipi.it (V.G.); luca.panariello@ing.unipi.it (L.P.); serena.danti@unipi.it (S.D.); andrea.lazzeri@unipi.it (A.L.)
2. Interuniversity Consortium of Materials Science and Technology (INSTM), 50121 Florence, Italy
3. IRIS Technology Solutions (IRIS), 08940 Cornellà de Llobregat, Spain; kabidemi@iris-eng.com
4. Biomass Valorization Platform-Materials, CELABOR s.c.r.l., 4650 Chaineux, Belgium; ahdi.hadrich@celabor.be
5. Fraunhofer-Institute for Process Engineering and Packaging, 85354 Freising, Germany; andreas.staebler@ivv.fraunhofer.de
6. Stazione Sperimentale per l'Industria delle Conserve Alimentari (SSICA), 43121 Parma, Italy; serena.chierici@ssica.it
7. Planet Bioplastics s.r.l., 56017 Pisa, Italy; ilariacanesi@planetbioplastics.it
* Correspondence: maria.beatrice.coltelli@unipi.it (M.-B.C.); patrizia.cinelli@unipi.it (P.C.); Tel.: +39-0502217856 (M.-B.C.); +39-0502217869 (P.C.)

Citation: Gigante, V.; Panariello, L.; Coltelli, M.-B.; Danti, S.; Obisesan, K.A.; Hadrich, A.; Staebler, A.; Chierici, S.; Canesi, I.; Lazzeri, A.; et al. Liquid and Solid Functional Bio-Based Coatings. *Polymers* **2021**, *13*, 3640. https://doi.org/10.3390/polym13213640

Academic Editor: Begoña Ferrari

Received: 20 September 2021
Accepted: 20 October 2021
Published: 22 October 2021

Publisher's Note: MDPI stays neutral with regard to jurisdictional claims in published maps and institutional affiliations.

Copyright: © 2021 by the authors. Licensee MDPI, Basel, Switzerland. This article is an open access article distributed under the terms and conditions of the Creative Commons Attribution (CC BY) license (https://creativecommons.org/licenses/by/4.0/).

Abstract: The development of new bio-based coating materials to be applied on cellulosic and plastic based substrates, with improved performances compared to currently available products and at the same time with improved sustainable end of life options, is a challenge of our times. Enabling cellulose or bioplastics with proper functional coatings, based on biopolymer and functional materials deriving from agro-food waste streams, will improve their performance, allowing them to effectively replace fossil products in the personal care, tableware and food packaging sectors. To achieve these challenging objectives some molecules can be used in wet or solid coating formulations, e.g., cutin as a hydrophobic water- and grease-repellent coating, polysaccharides such as chitosan-chitin as an antimicrobial coating, and proteins as a gas barrier. This review collects the available knowledge on functional coatings with a focus on the raw materials used and methods of dispersion/application. It considers, in addition, the correlation with the desired final properties of the applied coatings, thus discussing their potential.

Keywords: coatings; active molecules; barrier properties

1. Introduction

The production of items derived from sustainable and renewable resources, not toxic for humans and the environment, is a pressing challenge facing our society [1]. In this context, the production of sustainable coatings with improved and multifunctional performances is necessary [2]. As such, the search for coatings that have to be bio-based, with good barrier, water resistance and antimicrobial features is underway [3]. Nowadays, extensively used materials, with excellent moisture barrier properties for the production of coatings, are fundamentally petro-based. This must be the barrier to break down in research in the coming years [4].

Before going into the detail of the review subject, it is necessary to underline and to clarify the definition of biodegradable and bio-based polymers used for coatings formulation. The concept of biodegradation refers to biodegradable polymers that can be disintegrated and catabolized to CO_2 and H_2O by bacteria and/or enzymes [5,6]. Instead, bio-based polymers can be categorized based on their source, process technique, and formulation following the classification shown in Figure 1 [7]. In detail, bio-based polymers can be derived from biomass (like polysaccharides, lipids and proteins), from bio-derived monomers achieved by fermentation, such as lactic acid oligomers (OLAs); finally it is possible to find polymers developed from microorganisms (e.g., polyhydroxyalkanoates, PHAs) [8].

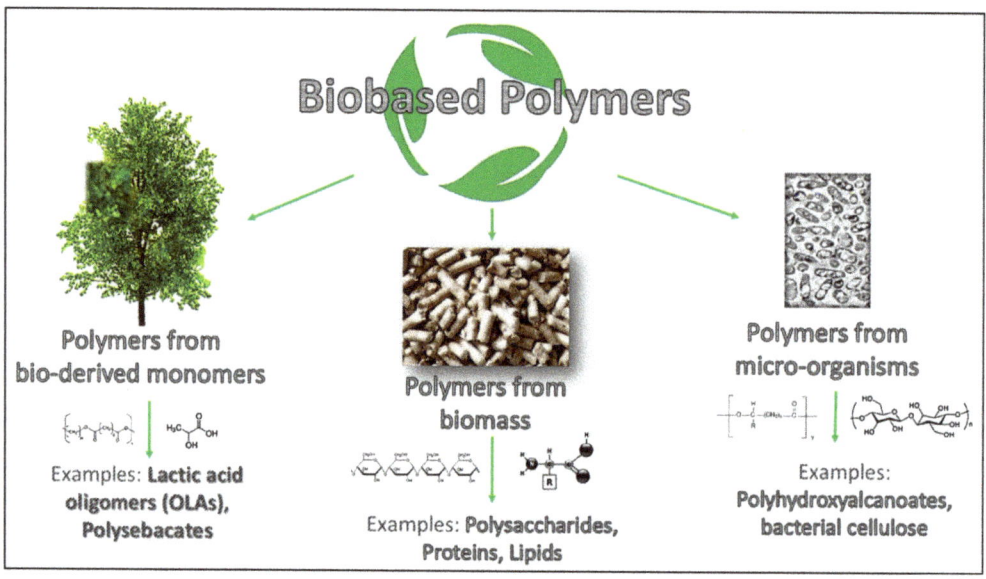

Figure 1. Schematic overview of bio-based polymers' differences.

In the field of coatings, these bio-based polymers represent the "new pathway to follow" because they can specifically provide to substrates multiple functionalities, also in relation to their processing conditions, without being petro-based [9].

Generally, functional bio-based coatings can be applied with the aim to improve the surface characteristics of a substrate (adhesion, wettability, water repellence, anti-corrosion properties and gas barrier. In other cases the coating can guarantee new properties in the final product, being an essential part of it [10].

This review will, therefore, be focused on the state of the art of bio-based and sustainable coatings production, with a detailed analysis of their application on cellulosic and plastic substrates. Moreover, the involvement of biomolecules in the coating formulations, but also the main technological innovations and the difference among liquid and solid preparation of bio-based sustainable coatings will be described in the following sections.

Indeed, to develop sustainable coatings for cellulose or bioplastic substrates is a technological goal of huge importance and it has become mandatory in the bioeconomy and circular economy context, aimed at imparting proper functional characteristics, based on biopolymer and functional materials coming from agro-food waste streams [11].

Coatings based on polymers, polymeric composites, and nanocomposites are used in several applications and sectors: (aerospace, automotive, marine structures, biomedical devices, decorative stuff, energy items, packaging). High-quality material is usually attained by thoroughly modulating layer/substrate.

Starting from paper substrates, it is well known that they are made of the most available bio-based material: the cellulose [12]. The use of cellulose-coated materials in personal care and disposable items for food (i.e., tableware) could be helpful for the environment and it is a route that has been followed in recent years [13]. The limits yet to be overcome are the hydrophilicity and low barrier properties typical of a non-woven fibrous system. For this reason, materials based on cellulose combined with poly(ethylene) are still widely present on the market, and these petro-based products are currently preferred, despite their negative environmental impact.

Regarding the application of coatings on plastic substrates, this is becoming increasingly necessary with the development of items with novel bio-based and biodegradable plastics based for example on poly (lactic acid) PLA, poly (butylene succinate) PBS, poly(butylene succinate-adipate) PBSA, as they do not show adequate barrier properties and are not able to withstand the rigours of the market [14–17]. In fact, since they do not present barrier properties comparable to traditional plastics [18], they need a protective layer. Obviously, in order not to affect the renewability and biodegradability of the product, a coating must also be developed with the characteristics of being bio-based and environmentally friendly [19].

Therefore, a considerable research effort "is on the agenda" investigating and formulating new bioplastics and new sustainable coating systems [20]. While their actual impact on the market is growing, it needs to be sharpened in many other applications.

The critical issue to achieving real progress towards sustainable materials is to intercept society's willingness to achieve sustainability; consumers must understand that obtaining sustainable products also means reducing global costs [21].

2. Bio-Based Coatings—Properties, Processing, Testing and Applications

2.1. Key Properties of Bio-Based Coatings

The innovation on bio-based coatings accompanies food packaging novelties and personal care applications. More specifically, the largest part of the bio-based coating research activity is primarily focused upon low-end (i.e., short-lived) bioplastic-based food packaging and paper coating for personal care. In contrast, fewer innovations are dedicated to coating for high-end (i.e., durable) applications [22]. Food products, indeed, endure many chemical, physical, and bacterial modifications when stored [23]. The shielding coating achieved during processing retards the damaging food deterioration, but also its quality is improved. For this reason, modification of the packaging, together with the development of eco-sustainable materials for packaging applications [24], can slow down deterioration rate of the packaged product, and hence, extend the shelf life of food [25].

Regarding personal care products, the goal to achieve is to reach tailored specific functional assets via a proper coating that can widen a large range of application, improving properties and favouring their use [26]. In addition, the production of bio-based films to coat personal care products, able to provide antimicrobial properties through the insertion of active biomolecules into a primer, is an encouraging alternative with respect to the direct application of antimicrobials in the food [27].

Nonetheless, to increase sustainability, the polymers should be bio-based, but green synthesis methods, which favour the use of non-toxic and environmentally friendly solvents, preferably relying on water-based or powder coatings, should be adopted [28]. Furthermore, coating cellulose or bioplastics with proper functional coatings based on biopolymer and functional materials deriving from agro-food waste streams will improve their performances, thus enabling them to replace effectively fossil-based products in personal care, tableware, and food-packaging sectors [29].

Table 1 briefly describes the most employed biopolymers used as a coating on cellulosic or plastic substrates, their preparation, application methods (that will be evidenced more n-depth in Section 4) and their key properties.

Table 1. Brief overview of biomolecules mostly used for coating formulations.

Bio-Based Polymer	Preparation	Application Method	Properties Improved and Main Results	REF
Chitin	0–2 wt.% chitin nanowhiskers dissolved in H_2SO_4 and glycerol.	Casting method on maize-starch films.	Evident antimicrobial resistance vs. Gram-positive *Listeria monocytogenes*.	[30]
Chitin	2 wt.% of water suspension of nanofibrils dispersed in PEG 8000.	Spray dryer on bioplastics films.	Antimicrobial and skin-regenerative improvements.	[31]
Chitosan	Chitosan (2 wt.%) and glycerol (2 wt.%) dissolved in a 1% (vol/vol) aqueous solution of acetic acid.	Chromatography plate coater application onto PP films. corona-treated	Evident antimicrobial resistance vs. *Listeria monocytogenes*, *Staphylococcus aureus*, and *Escherichia coli*.	[32]
Chitosan	Chitosan concentration of 0.02 g/mL in acetic acid mixed in equal volumes with hydroxypropyl methylcellulose.	Thin-layer chromatography plate coater on plastic films.	Excellent long-term antilisterial effect.	[33]
Lignin	Dissolution in acetone of different amounts of softwood kraft lignin and evaporation of the solvent.	Erichsen coater on to a paperboard substrate.	Evident decrease in Oxygen Transmission Rate (OTR) value and a stable contact angle with respect to paperboard alone.	[34]
Lignin	Lignin esterefied with palitic and lauric acid chloride in a mixture 3:1 ethanol/water.	Erichsen coater on a commercial paperboard substrate.	Good barrier properties against O_2 and H_2O	[35]
Cellulose derivates	Cellulose nitrate ester (CMCN) were dissolved in mixed solvents systems in different amounts.	Solvent casting method.	Gas and water barrier optimized.	[36]
Cellulose derivates	Hydroxypropyl methylcellulose acetate succinate plasticized with triethyl citrate and acetylated monoglyceride	Centrifugal granulator for feeding the coating powder and spraying simultaneously the plasticized.	Improved gastric resistance, coating efficiency, and processing stability	[37]
Proteins	Whey proteins with hydrolysed lactose at different contents	"Bird-type" applicator onto paperboard substrates	Good grease resistance and minimization of plasticizer migration	[38]
Proteins	12 g of whey proteins in 6 g of glycerol and 30 g of deionized water	Compression molding onto cellulosic substrates	Gas-barrier properties improvements	[39]

Connecting bio-based coating properties to final applications can be very useful in new product development. Those relations will be more extensively considered in the following sections of this review. In particular, three main properties for coatings will be considered:

- Antimicrobial coatings produced with chitin nano-fibrils and/or chitosan can be useful for cellulose tissues (e.g., personal care), paper and cardboard (e.g., packaging for fresh products like pasta, tableware), woven and non-woven (e.g., sanitary, personal care), plastic substrates (e.g., bio-polyesters) for active packaging.
- Gas barrier improvements for multilayer food packaging (e.g., bio polyester-based), with sustainable end of life options could be achieved by protein-based coatings
- Water-repellent properties for paper cups, but also non-food packaging, could be imparted by including cutin, thanks to its hydro-repellence

Polymeric coatings can be applied on several substrates, using many technologies, and with different approaches that depend on the nature of the coating (i.e., liquid or solid, detailed in the next sections). Although going in depth into the details of such technologies is behind the purpose of this review a marginal description of the main technologies, such as extrusion/dispersion coating and solution application, is necessary to comprehend how to exploit and develop bio-based coatings [40].

Thermoplastic polymers can be applied on bioplastic or cellulosic substrates with the technique of cast extrusion coating. Differently, biopolymers lacking of thermoplastic behaviour—as for example proteins, polysaccharides and fatty acids—can be also coated by polymer dissolution in a suitable solvent, or dispersing it in a solvent via dispersion coating [9].

Anyway, the use of these renewable materials in coatings faces issues and technical challenges due to low adhesion of the bio-based coatings on both plastic and cellulosic substrates [41]. Indeed proteins, chitosan and chitin have shown difficulty in adhering to plastic substrates; coating of cellulosic substrates have to face the challenge of moisture and temperature sensitivity [42].

2.2. Main Physico-Chemical Surface Treatments and Measurement Protocols

There are many physical and chemical processes employed for activating the surface of materials. Plasma-treated wood presented a substantially improved adhesion to the coating, leading to increased durability and a reduced attack by blue stain fungi. In the Durawood project [43], plasma was used as a pre-treatment before wood coating. Plasma-treated wood presented a substantially improved adhesion to the coating, leading to an increased durability, as well as a reduced attack by blue stain fungi. Unlike chemical treatment, plasma treatment does not require the use of chemicals and does not generate by-products. It can be promising for surface decontamination and finally for process intensification as it is expected to speed up the impregnation of the applied liquid.

Moreover, it is envisaged that coatings of several microns thickness will be applied to reach the multifunctional requirements of these applications, possibly in a subsequent step. As such, monitoring of these characteristics is needed. A number of monitoring techniques exist for thin printed coatings in the sub-micro/micro ranges. Most of them are in fact implemented off line and require sample preparation.

Most of them are used offline and require sample preparation. However, according to a recent review article, some combined optical techniques have shown potential for this type of in situ analysis [44]. Spectral reflectance is the most frequently employed technology giving quantifiable data. A white light beam is directed onto the specimen surface and the reflectance is gathered and studied by a spectrometer. Thickness is computed by determining the wavelengths of the interference peaks in the reflectance spectrum, where the thickness of the layer is a function of the wavelength of the peak and the refractive index of the material. [45]. This method is ideal for a thickness between 1 and 50 microns.

As far as the testing methods are concerned, several protocols and procedures have been developed to test antimicrobial properties, gas and water barriers.

An interesting review has shown several methods to evaluate antimicrobial properties [46]. The official standards were published by the Clinical and Laboratory Standards Institute (CLSI) for bacteria and yeasts testing [47], being the agar disk-diffusion test the mainly used technique. In this procedure, microorganism were inoculated by agar plates following standard procedures. Then, filter paper discs are placed on the agar surface. The Petri dishes are protected under suitable conditions. Commonly, the antimicrobial agent diffuses into the agar and inhibits germination and growth of the tested microorganism and then the diameter of growth inhibition zone (i.e., called "halo") is measured [48].

Regarding the barrier properties, the oxygen permeability, according to ASTM D3985-81, is evaluated as oxygen transmission rate (OTR) and demarcated as the oxygen amount passed through the material of a fixed thickness per unit of area and time [49].

The capacity of water vapour to permeate is measured, according to ASTM E96, instead, as water vapour transmission rate (WVTR), i.e., the quantity of water that passes through a substance of fixed thickness per unit of area and time [50]. The wettability or surface hydrophobicity can be evaluated through static or dynamic water contact angles [51]. Moreover, specifically for paper substrates, water absorption can be defined by the Cobb test (ISO 535). The Cobb value describes the water absorption capacity of a carton-board expressed in g/m^2. If the COBB value is high, the substrate shows the ability to absorb and retain moisture, otherwise the substrate can withstand penetration and retention of moisture [52].

3. Innovative Coatings Based on Chitosan-Chitin, Proteins and Cutin

3.1. Innovation on Chitosan- and Chitin-Based Coatings

Coatings with antimicrobial agents are useful because they can protect surfaces to microbial growth and can also be employed as barriers to humidity and oxygen [53].

Among the biomolecules that can be helpful to guarantee antimicrobial properties, a lot of interest is focused on chitin (and its derivate: chitosan), which is also the second most abundant biopolymer on the earth with an annual production of 10^{12}–10^{14} tons [54,55]. Speaking of numbers, the global demand for chitin in 2015 was above 60,000 tons, while its global production was around 28,000 tons [56]. Chitosan market size was valued at €1.5 billion in 2019, and is projected to reach €4 billion by 2027, according to a report by Global Industry Analysis [57]. The necessity of proper use of this waste material may allow the recovery of value-added goods also in the field of bio-based coatings. The amorphous part of chitin is transformed in chitosan by deacetylation. The difference between chitin and chitosan is not strict and it depends from the degree of deacetylation. Chitosan is a fully or partially deacetylated derivative of chitin, with a typical degree of deacetylation not higher than 65% [58]. Moreover, it can have animal (e.g., shells of crustaceans) or vegetal (e.g., fungi, such as *Aspergillus niger*) origin. Chitosan is characterized by non-toxicity, biodegradability, film-forming capacity, antimicrobial and antioxidant properties and good oxygen barrier properties [59]. The main advantage of chitosan application is the possibility to produce films and coatings with intrinsic antimicrobial properties which mainly differentiates chitosan from the other common antimicrobials (e.g., ethanol, sorbic acid, bacteriocins, lysozyme, essential oils) [60].

The properties of chitosan are related to origin and physico-chemical characteristics. Referring to films and coatings, antimicrobial and barrier properties depend on the molecular weight of chitosan, deacetylation degree, concentration, solvent used for its solubilisation, pH and possible plasticizers or other additives added in the formulations. The antimicrobial activity of chitosan relies on its positive charges, which can interact with negatively charged residues of macromolecules on the microbial cell surface, finally causing membrane leakage [61]. It is thus possible to find many examples of coatings, applied by dipping technique, spraying and other methods, as well as films produced by casting technique for fruit and vegetables, meat, cheese and fish, which avail themselves of chitosan. Antimicrobial properties of chitosan have been largely studied, even in combination with other substances, such as essential oils, or with other film-forming materials, such as proteins and gelatine. The use of chitosan for the edible coatings of fresh vegetables was investigated in depth recently by Tampucci et al. [62] who highlighted the possibility of developing a nutraceutical active coating for tomatoes.

Interestingly, chitin nanofibrils (CNs) can be formed by controlling the deacetylation step, thus avoiding the full conversion to chitosan [63]. In fact, the CNs represent the crystalline part of chitin. The amorphous part of chitin is transformed in any case in chitosan by deacetylation.

CNs have attracted significant interest because of their peculiar properties, including exceptional mechanical properties (Elastic Modulus with values up to 140 GPa), thermal stability (around 300 °C), low density ($\approx$1.5 g/cm^3), renewable bio-based biodegradable and biocompatible character, biological properties, high aspect ratio and high surface area with a wide chemical modification capacity [64]. The first studies on CNs focused on their production processes by applying shear forces using mechanical treatment for physical disintegration of the cell wall along the longitudinal axis. The common mechanical treatments for the defibrillation of chitin fibres are based on high-pressure homogenizer and disk mills [65], less conventional ball milling [66], or high intensity ultrasonication [67]. However, thanks to the tough hydrogen bonds between chitin fibers, large quantities of energy are needed to their disintegration into nanofibers via mechanical treatments. To circumvent the problem of high energy consumption during the defibrillation processes, the mechanical treatment was combined with chemical pretreatment such as (2,2,6,6-tetramethylpiperidine-1-oxyl radical)-mediated oxidation (TEMPO) which was used to weaken the bonds that hold the chitin chains together, facilitating their conversion into CNs [68]. Partial deacetylation associated with partial mechanical scission of the fibrils during disintegration was also used to obtain CNs [69]. In addition, the esterification of hydroxyl groups of chitin by carboxylate groups can significantly improve the mechanical disintegration of chitin using a grinder [70]. Furthermore, a simple acidic treatment of chitin fibres coupled with mechanical treatment using grinder can accelerate their conversion into CNs thanks to the repulsive force caused by the cationization of amino groups [71]. Unfortunately, most of these methods require the use of toxic solvents, which significantly reduce the environmental benefits of CN [72].

Regarding the preparation of poly(lactic acid) (PLA)-based nanocomposites containing CNs, a fine dispersion was achieved thanks to the preparation of pre-mixtures, as described by Coltelli et al. [73,74]. This strategy can be considered to uniformly disperse CNs in biopolyester formulations or hot-melt oligopolyesters for producing functional film or coatings. CNs have been demonstrated to be cytocompatible, interestingly showing anti-inflammatory activity, which make them good vectors for the distribution of biomolecules for skin care and cells restoration [75]. All these findings are suggestive for promising applications in the personal care sector, because of the good compatibility of the CNs with the skin [76,77]. Recent studies are also considering CNs coatings and nanocomposites for some biomedical applications, such as eardrum repair [78].

3.2. Innovation on Protein-Based Coatings

As bio-based materials are potentially useful for protective coatings, the proteins play a fundamental role [79–81]. Specific advantages of proteins (easy to make into films and abundance) allow them to be used extensively for preparing biodegradable films [82].

Proteins are natural polymers synthesized by all living organisms for a wide range of reasons. There are twenty different monomeric units, called proteinogenic amino acids, whereas the structure and properties of a specific protein is determined by the number, sequence and types of amino acid. Therefore, different proteins as oxygen barrier layers have received some attention in the literature [83–86]. The excellent barrier properties of protein-based films are due to covalent and non-covalent intermolecular interactions caused by free functional groups of the amino acids in the polypeptide chain. These cause the formation of a protein network, acting as an efficient barrier for oxygen [87–89]. However, as a result of these interactions, protein-based films and coatings are usually brittle and require to be added with plasticisers [90]. Glycerol (GLY), a characteristic polyol, shows high capacity to resist to the water, and it can be added to the solution to increase the ductility of the final film [91]. On the other hand, these plasticisers increase oxygen permeability due to the increased free volume in the protein network [80]. Therefore, developing suitable protein-based formulations combining both good barrier as well as mechanical properties is of utmost importance [92].

Micellar proteins obtained from different sources have been used to develop a lacquering adhesive having the unique property of combining a high adhesive strength with an excellent barrier against oxygen [93]. Unfortunately, the adhesive strength could not be quantified as a rupture of the paper substrate occurred before the protein coating failed. This, however, indicates that the bond strength of the coating was exceeding the cohesion strength of the substrate [94]. Because of the huge capability to act against the oxygen permeation, protein-based polymers are helpful for producing sustainable coatings more than polysaccharides and lipids. For example, the oxygen permeability of soy protein-based films is lower with respect to pectin, starch and even polyethylene (PE) according to Schmid et al. [95]. Clearly, the tremendous gas barrier improvement and the increasing of mechanical resistance make the protein-based biopolymers one of the most useful solutions for the future trends in packaging [96].

3.3. Innovation on Cutin-Based Coatings

Cutin is a crosslinked polyester formed mainly by condensed polyhydroxylated acid [97] and is the main constituent of the cuticles of the plant. The primary role accredited to plant cuticles is to be water repellent, to avoid leakages from internal tissues [98,99]. They also act as gas obstacles, UV inhibitors and thermal controllers [100]. Cutin can be depolymerized by cleaving the ester bonds using alkaline hydrolysis, with NaOH or KOH in water, transesterification with methanol containing BF_3 or $NaOCH_3$, reductive cleavage by exhaustive treatment with $LiAlH_4$ in THF, or with trimethylsilyl iodide in organic solvents [101]. Nevertheless, these methodologies are not satisfactory for large-scale cutin extraction, because of the steps involved and the impact of solvents and chemicals in terms of environmental and economic sustainability. Instead, the method patented by Cigognini et al. [102] is solvent-free and does not require pretreatment for cuticle isolation. This innovation allowed a pilot plant to be designed that extracts cutin from tomato by-products at a semi-industrial scale [103].

The first application of tomato cutin was the development of a bio-lacquer to coat food metal packaging. This application was patented and consists of a solvent based formulation [102]. Insoluble and thermostable coatings have been prepared from aleuritic acid as it is or added to palmitic acid, by melt-condensation polymerization in air without using solvents and catalysts [97,104]. The polyesters formulated can substitute plastic polymers or be applied as a coating. Tomato cutin was used in combination with sodium alginate and beeswax in a green solvent (i.e., water and ethanol) to obtain a hydrophobic free-standing film [105]. The work revealed that the thermal treatment (i.e., 150 °C, 8 h) represents a sustainable route to create structured, composite networks. Manrich et al. described the combination of cutin with pectin for the production of water-resistant plastic wraps [106], or as coating for plastic and bioplastic to confer hydrophobicity. Biodegradable polyester film has been prepared from aleuritic acid by melt-polycondensation in air. The film showed good water barrier properties and biocompatibility [107]. Similarly, films obtained by non-catalyzed melt-polycondensation of three types of tomato pomace by-products demonstrated high hydrophobicity. Furthermore, all these studies indicate that cutin has a valuable potential for packaging applications.

4. Liquid Bio-Based Coatings

One of the main methodologies used in the coating of cellulosic and plastic substrates is represented by the application of a liquid suspension/solution of functional molecules.

Among liquid application techniques, the most used are mentioned in Table 2, summarizing the description and main results of spray drying, electrospray, airbrush spraying, spin coating, dipping, solution casting, flexography and gravure roll coating.

Table 2. Liquid coatings techniques and main results regarding liquid bio-based coatings.

\	LIQUID COATINGS	
Technique	Description	Meaningful Applications in Liquid Bio-Based Coatings
Spray Drying	Transformation of a solution in which are dispersed particles into dried ones, thanks to a gaseous hot drying medium [108].	[109–111]
Electrospray	Liquid atomization through commanding electrical forces on the flow of a liquid injection from a cilindric die. This technique gaurantees uniform droplets generation [112].	[113,114]
Airbrush Spraying	Polymer solutions are sprayed through an airbrush supplied by a nitrogen line and fixed on a mechanic arm over a hot plate [115].	[116,117]
Spin Coating	The material used to coat is present at the centre of the substrate, then it is rotated at high speed until centrifugal force spreads the coating material [118].	[119,120]
Dipping	The solution substrate is immersed in the coating for effective formation of the complete material [121].	[122,123]
Solution Casting	A polymer is dissolved in a solution into which an inner diameter mold is immersed. The solvent is removed to leave a solid cast layer. This layer can be laminated or coated before being stripped from the mold [124].	[125–127]
Flexography	Flexographic assumes the possibility to widespread liquid inks with a low viscosity on paper, cardboard, or plastic films [128].	[129,130]
Gravure Roll Coater	Coating is introduced onto the surface of an engraved roll, then it is partially submersed in or by an enclosed applicator head that holds the coating against the roll [131].	[132–134]

Each method can be considered a valid technique for wet coating application and the specific choice depends on the physico-chemical features of the coating and the surface properties of the chosen substrate. For instance, coatings based on polysaccharides or proteins exhibit a considerably polar component in terms of surface energy, while the cutin, composed of ω-hydroxy acids, forms hydrophobic films [135]. Similarly, the surface energy of fossil-based plastics, such as polyolefins, showed a high dispersive component [136,137], bioplastics, such as polyesters, displayed a progressive increase in the polar component [138], whereas polysaccharides showed a predominance of the polar component [139]. It was reported in the literature that good adhesion between coating and substrate strongly depends on the interfacial surface energy and the topography/geometry of the adherent bodies [140]. As the wet coating was applied through the use of a liquid it was necessary for optimal conditions to be established in the substrate and the coating solution/suspension. Commonly the evaluation of surface energy of a liquid on a solid surface is defined by the contact angle expressed by the Young's equation and the relative work of adhesion expressed by Dupré's equation [141,142]. Surface energy of the liquid depends not only on the selected coating but also on the chosen solvent and the presence of surfactants [143–145]. Instead, factors such as concentration [146,147], viscosity [148,149], and wettability also influence the homogeneity of the coating, the drying speed, and the choice of application method. Instead, factors as the concentration and viscosity, in addition to the wettability, also influence the homogeneity of the coating, the drying speed and the choice of application method. Regarding the morphology, as the liquid coating assumes the shape of the solid, it was important to evaluate the roughness and the absorbency/porosity of the substrate. In literature it was reported that roughness has a strong influence on the wettability of the surface showing lower values of contact angle at higher levels of roughness [150–152]. The presence of porous or high-absorbency substrates highly influences the coating process by increasing the wettability and changing the drying kinetic [153–157]. Although surface roughness and porosity can increase the wettability of a surface, they have a significant influence on the coating morphology and thickness uniformity [158,159]. Other aspects that influence the coating are the process parameters, such as the deposition rate [160], the drying temperature [161] and the use of air or vacuum drying [162,163].

Application of coating with a wet technique had some advantages that were suitable for increasing the development of bio-based coatings. The use of a room temperature application avoids the thermal degradation and hydrolysis of bio-based materials, which are inherently sensitive to these processes [164–166]. Moreover, the use of a liquid medium allows the wettability of this type of coating to be tuned. For instance, a concentrated coating can be more suitable for blade or dipping application than a diluted one, which conversely can be more suitable for spray application. Particular attention must be paid to the choice of solvent/suspending agent, favouring bio-based and non-toxic liquids. The use of non-toxic substances for humans and environment should be deeply investigated because it could interfere with processes such as biodegradation [167–169]. Unfortunately, the preparation of optimal solution and dispersion for coating could not be easy to achieve. Solution guarantees a homogeneous distribution of the coating layer in wet medium, but the coatings are strongly influenced by properties like viscosity and possible formation of gel structures [170]. Dispersion has a weak influence on the physical properties of the coatings but they request a stabilization. In particular, with the increasing availability of nanometric biomolecules, such as the CNs [75,171] or the cellulose nanowhiskers [119], these problems were amplified due to the increase in the surface area. Consequently, high energetic dispersion and homogenization techniques, such as the ultraturrax homogenization [172,173], sonication [174,175] and high pressure homogenization (HPH) [176,177], were increasingly applied. If the operative parameters and the homogenization techniques did not allow an optimal wet coating to be prepared, the use of biosurfactant [178,179] or a bio-based primer [180–182] becomes necessary.

5. Solid Functional Bio-Based Coatings

In recent years, solid coatings have been developed in an exponential way and the necessities of functional coatings have also gradually been fortified [183]. As described in Table 3, among the widely used solid coating application techniques, the most common are: co-extrusion, compression molding, fluidized bed dipping, electrostatic spray and roll coating.

A differentiation can be made between hot melt coatings (HMCs) and powder coatings. HMCs have been in use since the fifties, they relies on thermoplastic solid materials achieved without the use of solvents, which are inherently solid below 80 °C and they become low-viscosity fluids at higher [184,185].

HMC is made of thermoplastic materials that can be easily spread upon heating. When the hot melt is in a fluid state, it flows onto the substrate. When the hot melt is then cooled, the coating solidifies and forms a bond to the substrate [186].

Today, HMCs are involved in the production of items in many manufacturing fields, from packaging to paper industry, and their development is increasing considering the step ahead made in the hot melt coating application methods [187].

Table 3. Description of solid coatings application techniques and main results on solid coatings.

Solid Coatings		
Technique	Description	Meaningful applications with solid coatings
Co-extrusion	Co-extrusion is a process that allows the simultaneous extrusion of two or more materials along the same production line, resulting in a multilayer final product [188].	[189]
Compression molding	A method based on the application of a pressure on a powder or another solid placed on a substrate in the lower plate of the press. The equipment is heated guaranteeing a good adhesion between the layers [190].	[191]
Fluidized bed dipping	A powder is transformed in an entirely consolidated film thanks to electrostatic forces [192].	[193]
Electrostatic Spray	The coating method is characterized by the deposition of the solid coating through electrostatic atomization [194].	[195]
Roll-to-Roll Coating	The coating or printing process is performed spreading a solid coating on a moving substrate, constituted above all by thin and flexible polymers, papers, ot textiles [196].	[197]

As they form a strong bond quickly, simply by cooling, they are compatible with many materials Achanta et al. [198] stated that HMC methods of applications are very attractive in all sectors in which there is a fundamental necessity to develop novel, simple, efficient, precise, and cost-effective coating processes.

The driving force for the employment of HMCs (and their strength compared to water-based film-coating technology) is to avoid the use of hazardous and toxic solvents as described by several literature works [199,200].

On this premise, since there is no necessity for solvent evaporation, the time for the process to be completed is shorter; consequently are eliminated all solvent disposal/treatment associated with organic solvents., making HMC environment-friendly materials [183].

Although water-based coating systems are useful, they are not completely flawless. A difficult problem encountered with waterborne coating systems is the variation in the dispersion of the coating. In fact, it is virtually impossible to control the presence or growth of microbes in coating dispersions without damage [201]. In addition, HMCs offer significant technical advantages, i.e., faster and cheaper coating processes and less risk of dissolution of biomolecules during treatment [202].

However, although this technique has been described well by many review papers, like by Lopes et al. [203], its application is scarce in producing coatings out of the pharmaceutical sector. The motivation is the necessity to mix in the correct way "active molecules", able to guarantee the achievement of the desired HMC properties, with oligomers, which act as primers during a low-temperature extrusion process (to ensure that the hot melt has the right melt strength to be processed) [204].

Improving the solubility of water-insoluble molecules remains a real challenge in the development of HMC formulations, as the bioavailability of active ingredients is controlled by their solubility in water [205]. Improving the solubility of the couple "active molecule–oligomer" is one of the challenges nowadays.

Finally, it is possible to conclude that HMCs represent the best strategy to develop coatings for bioplastics and cellulose with highly diffused industrial technologies, such as extrusion coating, in which the adhesion of the coating to the bioplastic substrate is very critical, as pointed out by Correlo et al. [206].

Another solid coating can be achieved in the form of a powder. In fact, with environmental regulations becoming more stringent, an urgent problem is to reduce the use of volatile organic compounds (VOCs). An approach based on powder coatings, which is inherently solvent free is perfect from this point of view. Such coatings represent the final destination along the road to VOC reduction [207].

Because of their superior application properties and environmental friendliness, the use of powder coatings has grown very rapidly in recent years and the demand for functional powder coatings has gradually intensified. The components of powder coatings are extruded, crushed and screened to obtain powder for coating [208]. Powder coatings are usually operated first by electrostatic spraying and fluidised bed impregnation methods. Then, the powder is heated until it melts and hardens.

The most commonly used method for thermoplastic systems is the fluidised bed process. Here, a hot metal test piece is immersed in the fluidised powder. The powder dissolves, melts and cures, resulting in a smooth polymer surface on the test piece [209]. Due to the partial crystallisation of polyester resin, the effect on the properties of the powder coating film, especially the mechanical properties, cannot be ignored in industrial applications [210].

The production of a polymeric powder coating by extrusion is, actually, a multi-step process. Indeed, it can be labelled a "batch process" because it involves weighing, premixing, extrusion and milling, weighting the "ingredients" in prescribed ratios, pre-mixing them in the solid state, feeding them into an extruder so to obtain a molten homogenous mixture. The molten material, after cooling, is subsequently crushed into flakes of about 10–20 mm and then finally ground by disc or hammer mill to obtain particles with size in the range of 2–100 μm with a distribution peak of about 50 μm [211].

Powder coating formulations exist on the market either as thermosetting or thermoplastic but they are fossil-based. Concerning biopolyester thermoplastic-based powder coatings there are still many steps ahead to reach an industrial application. Interestingly, Van Haverman et al. [212] developed alkyd resins for high-solid powder coatings completely based on commercially available renewable resources.

As interest continues to focus on improving more sustainable technologies, and as the prices of fossil raw materials are set to rise, the coming decades will inevitably see an increase in renewable-based coatings, combining them with unique properties.

6. Future Perspectives for Liquid and Solid Bio-Based Coatings

The present review evidenced the needs of formulating new bio-based coatings, which can be highly compatible with cellulosic and bioplastic substrates, in which thermoplastic starch films are one of the main examples [213]. The use of proper food or agricultural waste for their formulation agrees with the circular economy principles, can keep the cost of new materials down and can result in evident environmental advantages.

It is easy to predict, on the basis of the present literature survey, that chitin/chitosan coatings could be interesting both in liquid and solid forms. Cellulosic substrates [26], but also bioplastic [214] and textile substrates [113], can be easily treated with liquid coatings. The penetration of the liquid in the cellulosic or textile tissue is an important aspect to be controlled. Whereas chitosan, dissolved in acidic water, can penetrate inside the tissues, the chitin nanofibrils, generally suspended in water, remain on the substrate surface. In both cases the antimicrobial action can be modulated by controlling the concentration of these biopolymers in the liquid product. Solid coatings in powder or in film can be highly innovative. CNs or chitosan could be properly dispersed in thermoplastic matrices, having a low melting temperature for an easy and not expensive coating in terms of energy application.

Proteins can actually be used more on plastic and cellulosic substrates for developing high oxygen barrier coating for plastic and cellulose packaging [81,93], but they could also be potentially employed in solid coating formulations, despite their difficult processability and temperature sensitivity [215].

A cutin lacquer was developed for metallic substrates [103], but it is potentially applicable by liquid coatings on cellulosic and bioplastic substrates to obtain coloured (i.e., not transparent) coatings. The high hydrorepellency of cutin could probably allow these properties to be modulated on many substrates. The application of cutin in solid coating would be very new and interesting for the same reason.

These last considerations are summarized in Table 4.

Table 4. Predictable perspectives for chitin/chitosan, protein and cutin on different substrates.

Biomolecule	Liquid	Solid
Chitin/chitosan	Antimicrobial coatings for cellulose, bioplastic and textile substrates.	Potentially antimicrobial and water barrier coatings for cellulose, plastic and textile substrates.
Protein	High oxygen barrier coatings for plastic and cellulose.	In blend with polyesters, oxygen barrier coatings for cellulose and plastic.
Cutin	Hydrorepellent coatings and potentially for cellulose, bioplastic, and textile substrates.	Potentially hydrorepellent coatings for cellulose, bioplastic, and textile substrates.

In general, the preparation of liquid coatings based on chitin/chitosan, protein or cutin is at a higher technological readiness level, with respect to solid coatings.

The latter are extremely promising but more challenging than liquid coatings, as the modulation of morphological features based on coating concentration is a complex issue, as well as for the possible thermal degradation that could occur during processing and further application.

The considered biopolymers are thus extremely promising for developing innovative and environmentally friendly coatings for several substrates with some pros and cons, shown in Figure 2. These coatings can be extremely useful for improving the properties of renewable products, thus boosting their use in several applications.

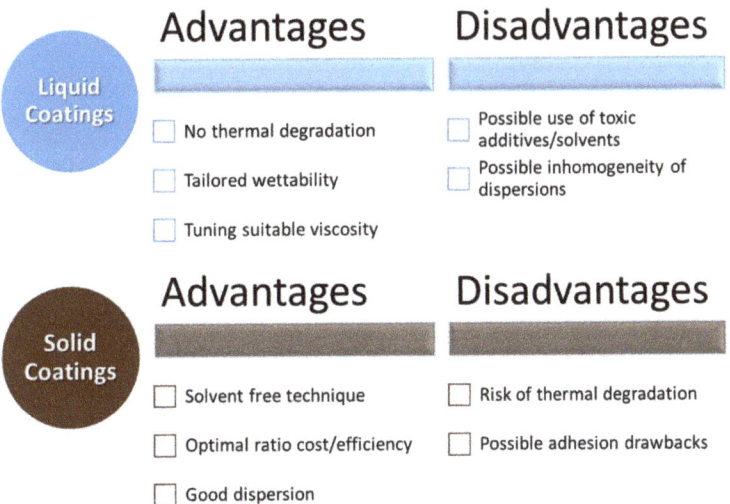

Figure 2. Advantages and disadvantages of liquid and solid coatings.

7. Conclusions

The objective of this review has been to summarize the main techniques for the application of bio-based coatings, differentiating between liquid and solid methods. Moreover, an in-depth literature search was necessary to evaluate some properties, which can be obtained starting from the dispersion of biomolecules within the coating itself. Chitosan/chitin, proteins and cutin were the main focus of this review paper, because of their complementary functional properties, antimicrobial, oxygen and water barrier, respectively. These properties are highly requested in novel functional bio-based coatings. Liquid and solid bio-based coatings showed advantages and disadvantages, but they can provide high flexibility to industry as well as drive specific innovations in the market, thus satisfying the exigencies of more sustainable yet performant products, than fossil-based counterparts.

In conclusion, this paper evidenced that the world of bio-based coatings is constantly evolving and expanding; several sectors are looking for a bio-based solution to improve the properties of their substrates and a considerable technological step forward has been made in this field.

Author Contributions: Conceptualization, M.-B.C. and P.C.; investigation, L.P. and V.G.; writing—original draft preparation, V.G., L.P., M.-B.C. and P.C.; writing—review and editing, V.G., L.P., M.-B.C., P.C., A.S., S.C., I.C., A.H., S.D. and K.A.O.; supervision, A.L.; project administration, P.C.; funding acquisition, P.C. All authors have read and agreed to the published version of the manuscript.

Funding: This research was funded by ECOFUNCO project, funded by Biobased Industries Joint Undertaking under the European Union Horizon 2020 research program (BBI-H2020): 837863.

Institutional Review Board Statement: Not applicable.

Informed Consent Statement: Not applicable.

Acknowledgments: Stefano Fiori of Condensia Quimica (Spain) is acknowledged for kind discussion.

Conflicts of Interest: The authors declare no conflict of interest.

References

1. Hatti-Kaul, R.; Nilsson, L.J.; Zhang, B.; Rehnberg, N.; Lundmark, S. Designing Biobased Recyclable Polymers for Plastics. *Trends Biotechnol.* **2020**, *38*, 50–67. [CrossRef] [PubMed]
2. Ren, K.; Fei, T.; Metzger, K.; Wang, T. Coating performance and rheological characteristics of novel soybean oil-based wax emulsions. *Ind. Crops Prod.* **2019**, *140*, 111654. [CrossRef]
3. Iwata, T. Biodegradable and Bio-Based Polymers: Future Prospects of Eco-Friendly Plastics. *Angew. Chem. Int. Ed.* **2015**, *54*, 3210–3215. [CrossRef]
4. Jubete, E.; Liauw, C.M.; Allen, N.S. Water uptake and tensile properties of carboxylated styrene butadiene rubber based water born paints: Models for water uptake prediction. *Prog. Org. Coat.* **2007**, *59*, 126–133. [CrossRef]
5. Tokiwa, Y.; Calabia, B.P. Biodegradable Polymers. In *BT—Encyclopedia of Polymeric Nanomaterials*; Kobayashi, S., Müllen, K., Eds.; Springer: Berlin/Heidelberg, Germany, 2015; pp. 145–155; ISBN 978-3-642-29648-2.
6. Vroman, I.; Tighzert, L. Biodegradable Polymers. *Materials* **2009**, *2*, 307–344. [CrossRef]
7. Van Tuil, R.; Fowler, P.; Lawther, M.; Weber, C.J. Properties of biobased packaging materials. In *Biobased Packaging Material for the Food Industry-Status and Perspectives*; Royal Veterinary and Agricultural University; Woodhead Publishing: Copenhagen, Denmark, 2000; pp. 13–44.
8. Narodoslawsky, M.; Shazad, K.; Kollmann, R.; Schnitzer, H. LCA of PHA production–Identifying the ecological potential of bio-plastic. *Chem. Biochem. Eng. Q.* **2015**, *29*, 299–305. [CrossRef]
9. Rastogi, V.K.; Samyn, P. Bio-based coatings for paper applications. *Coatings* **2015**, *5*, 887–930. [CrossRef]
10. Tharanathan, R.N. Biodegradable films and composite coatings: Past, present and future. *Trends Food Sci. Technol.* **2003**, *14*, 71–78. [CrossRef]
11. Cinelli, P.; Seggiani, M.; Coltelli, M.B.; Danti, S.; Righetti, M.C.; Gigante, V.; Sandroni, M.; Signori, F.; Lazzeri, A. Overview of Agro-Food Waste and By-Products Valorization for Polymer Synthesis and Modification for Bio-Composite Production. *Proceedings* **2021**, *69*, 22. [CrossRef]
12. Abdul Khalil, H.P.; Bhat, A.H.; Ireana Yusra, A.F. Green composites from sustainable cellulose nanofibrils: A review. *Carbohydr. Polym.* **2012**, *87*, 963–979. [CrossRef]
13. Gicquel, E.; Martin, C.; Yanez, J.G.; Bras, J. Cellulose nanocrystals as new bio-based coating layer for improving fiber-based mechanical and barrier properties. *J. Mater. Sci.* **2017**, *52*, 3048–3061. [CrossRef]
14. Coltelli, M.B.; Gigante, V.; Cinelli, P.; Lazzeri, A. Flexible Food Packaging Using Polymers from Biomass. In *Bionanotechnology to Save the Environment*; Morganti, P., Ed.; MDPI: Basel, Switzerland, 2019; pp. 272–298; ISBN 978-3-03842-693-6.
15. Molinari, G.; Gigante, V.; Fiori, S.; Aliotta, L.; Lazzeri, A. Dispersion of Micro Fibrillated Cellulose (MFC) in Poly(lactic acid) (PLA) from Lab-Scale to Semi-Industrial Processing Using Biobased Plasticizers as Dispersing Aids. *Chemistry* **2021**, *3*, 896–915. [CrossRef]
16. Aliotta, L.; Vannozzi, A.; Panariello, L.; Gigante, V.; Coltelli, M.B.; Lazzeri, A. Sustainable Micro and Nano Additives for Controlling the Migration of a Biobased Plasticizer from PLA-Based Flexible Films. *Polymers* **2020**, *12*, 1366. [CrossRef]
17. Gigante, V.; Coltelli, M.B.; Vannozzi, A.; Panariello, L.; Fusco, A.; Trombi, L.; Donnarumma, G.; Danti, S.; Lazzeri, A. Flat Die Extruded Biocompatible Poly(Lactic Acid) (PLA)/Poly(Butylene Succinate) (PBS) Based Films. *Polymers* **2019**, *11*, 1857. [CrossRef]

18. Scaffaro, R.; Botta, L.; Lopresti, F.; Maio, A.; Sutera, F. Polysaccharide nanocrystals as fillers for PLA based nanocomposites. *Cellulose* **2017**, *24*, 447–478. [CrossRef]
19. Sharmin, E.; Zafar, F.; Akram, D.; Alam, M.; Ahmad, S. Recent advances in vegetable oils based environment friendly coatings: A review. *Ind. Crops Prod.* **2015**, *76*, 215–229. [CrossRef]
20. Ashter, S.A. *Introduction to Bioplastics Engineering*; Andrew, W., Ed.; William Andrew: London, UK, 2016; ISBN 0323394078.
21. Shao, J.; Ünal, E. What do consumers value more in green purchasing? Assessing the sustainability practices from demand side of business. *J. Clean. Prod.* **2019**, *209*, 1473–1483. [CrossRef]
22. Cunningham, M.F.; Campbell, J.D.; Fu, Z.; Bohling, J.; Leroux, J.G.; Mabee, W.; Robert, T. Future green chemistry and sustainability needs in polymeric coatings. *Green Chem.* **2019**, *21*, 4919–4926. [CrossRef]
23. Balasubramaniam, V.M.; Chinnan, M.S. Role of packaging in quality preservation of frozen foods. In *Quality in Frozen Food*; Springer: New York, NY, USA, 1997; pp. 296–309.
24. Pietrosanto, A.; Scarfato, P.; Di Maio, L.; Incarnato, L. Development of Eco-Sustainable PBAT-Based Blown Packaging Applications. *Materials* **2020**, *13*, 5395. [CrossRef] [PubMed]
25. Álvarez, K.; Alvarez, V.A.; Gutiérrez, T.J. Biopolymer Composite Materials with Antimicrobial Effects Applied to the Food Industry BT—Functional Biopolymers. In *Handbook of Sustainable Polymers: Structure and Chemistry*; Thakur, V.K., Thakur, M.K., Eds.; Springer International Publishing: Cham, Switzerland, 2016; pp. 57–96; ISBN 978-3-319-66417-0.
26. Panariello, L.; Coltelli, M.B.; Buchignani, M.; Lazzeri, A. Chitosan and nano-structured chitin for biobased anti-microbial treatments onto cellulose based materials. *Eur. Polym. J.* **2019**, *113*, 328–339. [CrossRef]
27. Apicella, A.; Scarfato, P.; Di Maio, L.; Incarnato, L. Sustainable Active PET Films by Functionalization with Antimicrobial Bio-Coatings. *Front. Mater.* **2019**, *6*, 1–10. [CrossRef]
28. Shah, M.Y.; Ahmad, S. Waterborne vegetable oil epoxy coatings: Preparation and characterization. *Prog. Org. Coat.* **2012**, *75*, 248–252. [CrossRef]
29. Song, Z.; Xiao, H.; Li, Y. Effects of renewable materials coatings on oil resistant properties of paper. *Nord. Pulp Pap. Res. J.* **2015**, *30*, 344–349. [CrossRef]
30. Qin, Y.; Zhang, S.; Yu, J.; Yang, J.; Xiong, L.; Sun, Q. Effects of chitin nano-whiskers on the antibacterial and physicochemical properties of maize starch films. *Carbohydr. Polym.* **2016**, *147*, 372–378. [CrossRef] [PubMed]
31. Panariello, L.; Vannozzi, A.; Morganti, P.; Coltelli, M.B.; Lazzeri, A. Biobased and Eco-Compatible Beauty Films Coated with Chitin. *Cosmetics* **2021**, *8*, 27. [CrossRef]
32. Torlak, E.; Nizamlioğlu, M. Antimicrobial effectiveness of chitosan-essential oil coated plastic films against foodborne pathogens. *J. Plast. Film Sheeting* **2011**, *27*, 235–248. [CrossRef]
33. Ye, M.; Neetoo, H.; Chen, H. Control of *Listeria monocytogenes* on ham steaks by antimicrobials incorporated into chitosan-coated plastic films. *Food Microbiol.* **2008**, *25*, 260–268. [CrossRef]
34. Hult, E.L.; Ropponen, J.; Poppius-Levlin, K.; Ohra-Aho, T.; Tamminen, T. Enhancing the barrier properties of paper board by a novel lignin coating. *Ind. Crops Prod.* **2013**, *50*, 694–700. [CrossRef]
35. Hult, E.L.; Koivu, K.; Asikkala, J.; Ropponen, J.; Wrigstedt, P.; Sipilä, J.; Poppius-Levlin, K. Esterified lignin coating as water vapor and oxygen barrier for fiber-based packaging. *Holzforschung* **2013**, *67*, 899–905. [CrossRef]
36. Duan, H.; Shao, Z.; Zhao, M.; Zhou, Z. Preparation and properties of environmental-friendly coatings based on carboxymethyl cellulose nitrate ester & modified alkyd. *Carbohydr. Polym.* **2016**, *137*, 92–99. [CrossRef] [PubMed]
37. Obara, S.; Maruyama, N.; Nishiyama, Y.; Kokubo, H. Dry coating: An innovative enteric coating method using a cellulose derivative. *Eur. J. Pharm. Biopharm.* **1999**, *47*, 51–59. [CrossRef]
38. Lin, S.; Krochta, J.M. Plasticizer effect on grease barrier and color properties of whey-protein coatings on paperboard. *J. Food Sci.* **2003**, *68*, 229–233. [CrossRef]
39. Gällstedt, M.; Brottman, A.; Hedenqvist, M.S. Packaging-related properties of protein-and chitosan-coated paper. *Packag. Technol. Sci. Int. J.* **2005**, *18*, 161–170. [CrossRef]
40. Dabral, M.; Francis, L.F.; Scriven, L.E. Drying process paths of ternary polymer solution coating. *AIChE J.* **2002**, *48*, 25–37. [CrossRef]
41. Xu, L.C.; Vadillo-Rodriguez, V.; Logan, B.E. Residence time, loading force, pH, and ionic strength affect adhesion forces between colloids and biopolymer-coated surfaces. *Langmuir* **2005**, *21*, 7491–7500. [CrossRef] [PubMed]
42. Van der Wel, G.K.; Adan, O.C. Moisture in organic coatings—A review. *Prog. Org. Coat.* **1999**, *37*, 1–14. [CrossRef]
43. Mazela, B.; Broda, M.; Perdoch, W.; Ross Gobakken, L.; Ratajczak, I.; Cofta, G.; Grześkowiak, W.; Komasa, A.; Przybył, A. Bio-friendly preservative systems for enhanced wood durability–1st periodic report on DURAWOOD project. In *Proceedings of the 46th International Research Group on Wood Protection (IRG46), Viña del Mar, Chile, 10–14 May 2015*; IRG Secretariat: Stockholm, Sweden, 2015; pp. 10–14.
44. Bugnicourt, E.; Kehoe, T.; Latorre, M.; Serrano, C.; Philippe, S.; Schmid, M. Recent prospects in the inline monitoring of nanocomposites and nanocoatings by optical technologies. *Nanomaterials* **2016**, *6*, 150. [CrossRef]
45. Merklein, T.M. High resolution measurement of multilayer structures. *Appl. Opt.* **1990**, *29*, 505–511. [CrossRef]
46. Balouiri, M.; Sadiki, M.; Ibnsouda, S.K. Methods for in vitro evaluating antimicrobial activity: A review. *J. Pharm. Anal.* **2016**, *6*, 71–79. [CrossRef]

47. Reller, L.B.; Weinstein, M.; Jorgensen, J.H.; Ferraro, M.J. Antimicrobial susceptibility testing: A review of general principles and contemporary practices. *Clin. Infect. Dis.* **2009**, *49*, 1749–1755. [CrossRef]
48. Luangtongkum, T.; Morishita, T.Y.; El-Tayeb, A.B.; Ison, A.J.; Zhang, Q. Comparison of antimicrobial susceptibility testing of Campylobacter spp. by the agar dilution and the agar disk diffusion methods. *J. Clin. Microbiol.* **2007**, *45*, 590. [CrossRef]
49. Wang, J.; Gardner, D.J.; Stark, N.M.; Bousfield, D.W.; Tajvidi, M.; Cai, Z. Moisture and oxygen barrier properties of cellulose nanomaterial-based films. *ACS Sustain. Chem. Eng.* **2018**, *6*, 49–70. [CrossRef]
50. Wang, L.; Shogren, R.L.; Carriere, C. Preparation and properties of thermoplastic starch-polyester laminate sheets by coextrusion. *Polym. Eng. Sci.* **2000**, *40*, 499–506. [CrossRef]
51. Mittal, K.L. *Advances in Contact Angle, Wettability and Adhesion*; John Wiley & Sons: Hoboken, NJ, USA, 2015; ISBN 1119116996.
52. Fernandes, S.C.; Freire, C.S.; Silvestre, A.J.; Desbrières, J.; Gandini, A.; Neto, C.P. Production of coated papers with improved properties by using a water-soluble chitosan derivative. *Ind. Eng. Chem. Res.* **2010**, *49*, 6432–6438. [CrossRef]
53. Hershko, V.; Nussinovitch, A. The Behavior of Hydrocolloid Coatings on Vegetative Materials. *Biotechnol. Prog.* **1998**, *14*, 756–765. [CrossRef]
54. Dhillon, G.S.; Kaur, S.; Brar, S.K.; Verma, M. Green synthesis approach: Extraction of chitosan from fungus mycelia. *Crit. Rev. Biotechnol.* **2013**, *33*, 379–403. [CrossRef]
55. Coltelli, M.; Aliotta, L.; Vannozzi, A.; Morganti, P.; Fusco, A.; Donnarumma, G.; Lazzeri, A. Properties and Skin Compatibility of Films Based on Poly (Lactic Acid) (PLA) Bionanocomposites Incorporating Chitin Nanofibrils (CN). *J. Funct. Biomater.* **2020**, *11*, 21. [CrossRef]
56. Pottathara, Y.B.; Tiyyagura, H.R.; Ahmad, Z.; Thomas, S. Chapter 3—Chitin and chitosan composites for wearable electronics and energy storage devices. In *Handbook of Chitin and Chitosan*; Gopi, S., Thomas, S., Pius, A., Eds.; Elsevier: Amsterdam, The Netherlands, 2020; pp. 71–88; ISBN 978-0-12-817966-6.
57. Eswara, P. *Chitosan Market by Source, Application, Global Opportunity Analysis and Industry Forecast, 2020–2027*; Allied Analytics LLP: Pune, India, 2017.
58. Elieh-Ali-Komi, D.; Hamblin, M.R. Chitin and chitosan: Production and application of versatile biomedical nanomaterials. *Int. J. Adv. Res.* **2016**, *4*, 411.
59. Verlee, A.; Mincke, S.; Stevens, C.V. Recent developments in antibacterial and antifungal chitosan and its derivatives. *Carbohydr. Polym.* **2017**, *164*, 268–283. [CrossRef] [PubMed]
60. Szabo, K.; Teleky, B.E.; Mitrea, L.; Călinoiu, L.F.; Martău, G.A.; Simon, E.; Varvara, R.A.; Vodnar, D.C. Active Packaging—Poly(Vinyl Alcohol) Films Enriched with Tomato By-Products Extract. *Coatings* **2020**, *10*, 141. [CrossRef]
61. Zheng, L.Y.; Zhu, J.F. Study on antimicrobial activity of chitosan with different molecular weights. *Carbohydr. Polym.* **2003**, *54*, 527–530. [CrossRef]
62. Tampucci, S.; Castagna, A.; Monti, D.; Manera, C.; Saccomanni, G.; Chetoni, P.; Zucchetti, E.; Barbagallo, M.; Fazio, L.; Santin, M.; et al. Tyrosol-Enriched Tomatoes by Diffusion across the Fruit Peel from a Chitosan Coating: A Proposal of Functional Food. *Foods* **2021**, *10*, 335. [CrossRef]
63. Toan, N. Van Production of chitin and chitosan from partially autolyzed shrimp shell materials. *Open Biomater. J.* **2009**, *1*, 21–24. [CrossRef]
64. Duan, B.; Huang, Y.; Lu, A.; Zhang, L. Recent advances in chitin based materials constructed via physical methods. *Prog. Polym. Sci.* **2018**, *82*, 1–33. [CrossRef]
65. Ifuku, S.; Nogi, M.; Abe, K.; Yoshioka, M.; Morimoto, M.; Saimoto, H.; Yano, H. Preparation of Chitin Nanofibers with a Uniform Width as α-Chitin from Crab Shells. *Biomacromolecules* **2009**, *10*, 1584–1588. [CrossRef]
66. Tran, T.H.; Nguyen, H.L.; Hao, L.T.; Kong, H.; Park, J.M.; Jung, S.H.; Cha, H.G.; Lee, J.Y.; Kim, H.; Hwang, S.Y.; et al. A ball milling-based one-step transformation of chitin biomass to organo-dispersible strong nanofibers passing highly time and energy consuming processes. *Int. J. Biol. Macromol.* **2019**, *125*, 660–667. [PubMed]
67. Lu, Y.; Sun, Q.; She, X.; Xia, Y.; Liu, J.; Yang, D. Fabrication and characterisation of α-chitin nanofibers and highly transparent chitin films by pulsed ultrasonication. *Carbohydr. Polym.* **2013**, *98*, 1497–1504. [CrossRef] [PubMed]
68. Ifuku, S.; Nogi, M.; Yoshioka, M.; Morimoto, M.; Yano, H.; Saimoto, H. Fibrillation of dried chitin into 10–20 nm nanofibers by a simple grinding method under acidic conditions. *Carbohydr. Polym.* **2010**, *81*, 134–139. [CrossRef]
69. Fan, Y.; Saito, T.; Isogai, A. Individual chitin nano-whiskers prepared from partially deacetylated α-chitin by fibril surface cationization. *Carbohydr. Polym.* **2010**, *79*, 1046–1051. [CrossRef]
70. Aklog, Y.F.; Nagae, T.; Izawa, H.; Morimoto, M.; Saimoto, H.; Ifuku, S. Preparation of chitin nanofibers by surface esterification of chitin with maleic anhydride and mechanical treatment. *Carbohydr. Polym.* **2016**, *153*, 55–59. [CrossRef] [PubMed]
71. Ifuku, S. Chitin and Chitosan Nanofibers: Preparation and Chemical Modifications. *Molecules* **2014**, *19*, 18367–18380. [CrossRef]
72. Häckl, K.; Kunz, W. Some aspects of green solvents. *C. R. Chim.* **2018**, *21*, 572–580. [CrossRef]
73. Coltelli, M.B.; Gigante, V.; Panariello, L.; Morganti, P.; Cinelli, P.; Danti, S.; Lazzeri, A. Chitin nanofibrils in renewable materials for packaging and personal care applications. *Adv. Mater. Lett.* **2018**, *10*, 425–430. [CrossRef]
74. Coltelli, M.B.; Cinelli, P.; Gigante, V.; Aliotta, L.; Morganti, P.; Panariello, L.; Lazzeri, A. Chitin Nanofibrils in Poly(Lactic Acid) (PLA) Nanocomposites: Dispersion and Thermo-Mechanical Properties. *Int. J. Mol. Sci.* **2019**, *20*, 504. [CrossRef]
75. Danti, S.; Trombi, L.; Fusco, A.; Azimi, B.; Lazzeri, A.; Morganti, P.; Coltelli, M.B.; Donnarumma, G. Chitin Nanofibrils and Nanolignin as Functional Agents in Skin Regeneration. *Int. J. Mol. Sci.* **2019**, *20*, 2669. [CrossRef]

76. Morganti, P.; Danti, S.; Coltelli, M.B. Chitin and lignin to produce biocompatible tissues. *Res. Clin. Dermatol.* **2018**, *1*, 5–11. [CrossRef]
77. Morganti, P.; Coltelli, M.B. A new carrier for advanced cosmeceuticals. *Cosmetics* **2019**, *6*, 10. [CrossRef]
78. Danti, S.; Anand, S.; Azimi, B.; Milazzo, M.; Fusco, A.; Ricci, C.; Zavagna, L.; Linari, S.; Donnarumma, G.; Lazzeri, A.; et al. Chitin Nanofibril Application in Tympanic Membrane Scaffolds to Modulate Inflammatory and Immune Response. *Pharmaceutics* **2021**, *13*, 1440. [CrossRef]
79. Bugnicourt, E.; Schmid, M.; Nerney, O.M.; Wildner, J.; Smykala, L.; Lazzeri, A.; Cinelli, P. Processing and validation of whey-protein-coated films and laminates at semi-industrial scale as novel recyclable food packaging materials with excellent barrier properties. *Adv. Mater. Sci. Eng.* **2013**, *2013*, 496207. [CrossRef]
80. Schmid, M.; Dallmann, K.; Bugnicourt, E.; Cordoni, D.; Wild, F.; Lazzeri, A.; Noller, K. Properties of whey-protein-coated films and laminates as novel recyclable food packaging materials with excellent barrier properties. *Int. J. Polym. Sci.* **2012**, *2012*, 562381. [CrossRef]
81. Coltelli, M.B.; Aliotta, L.; Gigante, V.; Bellusci, M.; Cinelli, P.; Bugnicourt, E.; Schmid, M.; Staebler, A.; Lazzeri, A. Preparation and Compatibilization of PBS/Whey Protein Isolate Based Blends. *Molecules* **2020**, *25*, 3313. [CrossRef]
82. Kaewprachu, P.; Osako, K.; Benjakul, S.; Tongdeesoontorn, W.; Rawdkuen, S. Biodegradable protein-based films and their properties: A comparative study. *Packag. Technol. Sci.* **2016**, *29*, 77–90. [CrossRef]
83. Jost, V.; Kobsik, K.; Schmid, M.; Noller, K. Influence of plasticiser on the barrier, mechanical and grease resistance properties of alginate cast films. *Carbohydr. Polym.* **2014**, *110*, 309–319. [CrossRef] [PubMed]
84. Pommet, M.; Redl, A.; Morel, M.H.; Guilbert, S. Study of wheat gluten plasticization with fatty acids. *Polymer* **2003**, *44*, 115–122. [CrossRef]
85. Mo, X.; Sun, X. Plasticization of soy protein polymer by polyol-based plasticizers. *J. Am. Oil Chem. Soc.* **2002**, *79*, 197–202. [CrossRef]
86. Schmid, M.; Reichert, K.; Hammann, F.; Stäbler, A. Storage time-dependent alteration of molecular interaction–property relationships of whey protein isolate-based films and coatings. *J. Mater. Sci.* **2015**, *50*, 4396–4404. [CrossRef]
87. Schmid, M.; Prinz, T.K.; Stäbler, A.; Sängerlaub, S. Effect of sodium sulfite, sodium dodecyl sulfate, and urea on the molecular interactions and properties of whey protein isolate-based films. *Front. Chem.* **2017**, *4*, 49. [CrossRef]
88. Schmid, M.; Sängerlaub, S.; Wege, L.; Stäbler, A. Properties of transglutaminase crosslinked whey protein isolate coatings and cast films. *Packag. Technol. Sci.* **2014**, *27*, 799–817. [CrossRef]
89. Schmid, M.; Hinz, L.V.; Wild, F.; Noller, K. Effects of hydrolysed whey proteins on the techno-functional characteristics of whey protein-based films. *Materials* **2013**, *6*, 927–940. [CrossRef]
90. Schmid, M. Properties of cast films made from different ratios of whey protein isolate, hydrolysed whey protein isolate and glycerol. *Materials* **2013**, *6*, 3254–3269. [CrossRef] [PubMed]
91. Kristo, E.; Biliaderis, C.G. Water sorption and thermo-mechanical properties of water/sorbitol-plasticized composite biopolymer films: Caseinate–pullulan bilayers and blends. *Food Hydrocoll.* **2006**, *20*, 1057–1071. [CrossRef]
92. Mitrea, L.; Călinoiu, L.F.; Martău, G.A.; Szabo, K.; Teleky, B.E.; Mureșan, V.; Rusu, A.V.; Socol, C.T.; Vodnar, D.C. Poly(vinyl alcohol)-Based Biofilms Plasticized with Polyols and Colored with Pigments Extracted from Tomato By-Products. *Polymers* **2020**, *12*, 532. [CrossRef]
93. Coltelli, M.B.; Wild, F.; Bugnicourt, E.; Cinelli, P.; Lindner, M.; Schmid, M.; Weckel, V.; Müller, K.; Rodriguez, P.; Staebler, A. State of the art in the development and properties of protein-based films and coatings and their applicability to cellulose based products: An extensive review. *Coatings* **2016**, *6*, 1. [CrossRef]
94. Eibl, I.; von der Haar, D.; Jesdinszki, M.; Stäbler, A.; Schmid, M.; Langowski, H. Adhesive based on micellar lupin protein isolate exhibiting oxygen barrier properties. *J. Appl. Polym. Sci.* **2018**, *135*, 46383. [CrossRef]
95. Schmid, M.; Müller, K. Whey protein-based packaging films and coatings. In *Whey Proteins*; Elsevier: Amsterdam, The Netherlands, 2019; pp. 407–437.
96. Letendre, M.; D'aprano, G.; Lacroix, M.; Salmieri, S.; St-Gelais, D. Physicochemical properties and bacterial resistance of biodegradable milk protein films containing agar and pectin. *J. Agric. Food Chem.* **2002**, *50*, 6017–6022. [CrossRef]
97. Heredia, A. Biophysical and biochemical characteristics of cutin, a plant barrier biopolymer. *Biochim. Biophys. Acta BBA Gen. Subj.* **2003**, *1620*, 1–7. [CrossRef]
98. Riederer, M.; Schreiber, L. Protecting against water loss: Analysis of the barrier properties of plant cuticles. *J. Exp. Bot.* **2001**, *52*, 2023–2032. [CrossRef]
99. Pio, T.F.; Macedo, G.A. Optimizing the production of cutinase by Fusarium oxysporum using response surface methodology. *Enzyme Microb. Technol.* **2007**, *41*, 613–619. [CrossRef]
100. Martin, L.B.; Rose, J.K. There's more than one way to skin a fruit: Formation and functions of fruit cuticles. *J. Exp. Bot.* **2014**, *65*, 4639–4651. [CrossRef] [PubMed]
101. Kolattukudy, P.E. Polyesters in higher plants. In *Biopolyesters*; Springer: New York, NY, USA, 2001; pp. 1–49.
102. Cigognini, I.; Montanari, A.; De la Torre Carreras, R.; Cardoso, G. Extraction Method of a Polyester Polymer or Cutin from the Wasted Tomato Peels and Polyester Polimer so Extracted. WO Patent WO2015028299, 3 May 2015.

103. Montanari, A.; Bolzoni, L.; Cigognini, I.M.; Ciruelos, A.; Cardoso, M.G.; De La Torre, R. Tomato bio-based lacquer for sustainable metal packaging. In *Acta Horticulturae*; International Society for Horticultural Science (ISHS): Leuven, Belgium, 2017; Volume 1159, pp. 159–165.
104. Benítez, J.J.; Heredia-Guerrero, J.A.; Guzmán-Puyol, S.; Barthel, M.J.; Domínguez, E.; Heredia, A. Polyhydroxyester Films Obtained by Non-Catalyzed Melt-Polycondensation of Natural Occurring Fatty Polyhydroxyacids. *Front. Mater.* **2015**, *2*, 59. [CrossRef]
105. Tedeschi, G.; Benitez, J.J.; Ceseracciu, L.; Dastmalchi, K.; Itin, B.; Stark, R.E.; Heredia, A.; Athanassiou, A.; Heredia-Guerrero, J.A. Sustainable Fabrication of Plant Cuticle-Like Packaging Films from Tomato Pomace Agro-Waste, Beeswax, and Alginate. *ACS Sustain. Chem. Eng.* **2018**, *6*, 14955–14966. [CrossRef]
106. Manrich, A.; Moreira, F.K.; Otoni, C.G.; Lorevice, M.V.; Martins, M.A.; Mattoso, L.H. Hydrophobic edible films made up of tomato cutin and pectin. *Carbohydr. Polym.* **2017**, *164*, 83–91. [CrossRef]
107. Benítez, J.J.; Heredia-Guerrero, J.A.; de Vargas-Parody, M.I.; Cruz-Carrillo, M.A.; Morales-Flórez, V.; de la Rosa-Fox, N.; Heredia, A. Biodegradable polyester films from renewable aleuritic acid: Surface modifications induced by melt-polycondensation in air. *J. Phys. D Appl. Phys.* **2016**, *49*, 175601. [CrossRef]
108. Santos, D.; Maurício, A.C.; Sencadas, V.; Santos, J.D.; Fernandes, M.H.; Gomes, P.S. *Spray Drying: An Overview*; Pignatello, R., Ed.; IntechOpen: London, UK, 2018.
109. He, P.; Davis, S.S.; Illum, L. Chitosan microspheres prepared by spray drying. *Int. J. Pharm.* **1999**, *187*, 53–65. [CrossRef]
110. Muzzarelli, C.; Stanic, V.; Gobbi, L.; Tosi, G.; Muzzarelli, R.A. Spray-drying of solutions containing chitosan together with polyuronans and characterisation of the microspheres. *Carbohydr. Polym.* **2004**, *57*, 73–82. [CrossRef]
111. Ngan, L.T.; Wang, S.L.; Hiep, Ð.M.; Luong, P.M.; Vui, N.T.; Ðinh, T.M.; Dzung, N.A. Preparation of chitosan nanoparticles by spray drying, and their antibacterial activity. *Res. Chem. Intermed.* **2014**, *40*, 2165–2175. [CrossRef]
112. Jaworek, A.; Sobczyk, A.T.; Krupa, A. Electrospray application to powder production and surface coating. *J. Aerosol Sci.* **2018**, *125*, 57–92. [CrossRef]
113. Azimi, B.; Thomas, L.; Fusco, A.; Kalaoglu-Altan, O.I.; Basnett, P.; Cinelli, P.; De Clerck, K.; Roy, I.; Donnarumma, G.; Coltelli, M.B.; et al. Electrosprayed Chitin Nanofibril/Electrospun Polyhydroxyalkanoate Fiber Mesh as Functional Nonwoven for Skin Application. *J. Funct. Biomater.* **2020**, *11*, 62. [CrossRef] [PubMed]
114. Azimi, B.; Ricci, C.; Fusco, A.; Zavagna, L.; Linari, S.; Donnarumma, G.; Hadrich, A.; Cinelli, P.; Coltelli, M.B.; Danti, S. Electrosprayed Shrimp and Mushroom Nanochitins on Cellulose Tissue for Skin Contact Application. *Molecules* **2021**, *26*, 4374. [CrossRef]
115. Susanna, G.; Salamandra, L.; Brown, T.M.; Di Carlo, A.; Brunetti, F.; Reale, A. Airbrush spray-coating of polymer bulk-heterojunction solar cells. *Sol. Energy Mater. Sol. Cells* **2011**, *95*, 1775–1778. [CrossRef]
116. Zhong, C.; Kapetanovic, A.; Deng, Y.; Rolandi, M. A Chitin Nanofiber Ink for Airbrushing, Replica Molding, and Microcontact Printing of Self-assembled Macro-, Micro-, and Nanostructures. *Adv. Mater.* **2011**, *23*, 4776–4781. [CrossRef]
117. Benítez, J.J.; Osbild, S.; Guzman-Puyol, S.; Heredia, A.; Heredia-Guerrero, J.A. Bio-Based Coatings for Food Metal Packaging Inspired in Biopolyester Plant Cutin. *Polymers* **2020**, *12*, 942. [CrossRef] [PubMed]
118. Lawrence, C.J. The mechanics of spin coating of polymer films. *Phys. Fluids* **1988**, *31*, 2786–2795. [CrossRef]
119. Eichhorn, S.J. Cellulose nanowhiskers: Promising materials for advanced applications. *Soft Matter* **2011**, *7*, 303–315. [CrossRef]
120. Ren, Y.; Babaie, E.; Bhaduri, S.B. Nanostructured amorphous magnesium phosphate/poly (lactic acid) composite coating for enhanced corrosion resistance and bioactivity of biodegradable AZ31 magnesium alloy. *Prog. Org. Coat.* **2018**, *118*, 1–8. [CrossRef]
121. Rane, A.V.; Kanny, K. Manufacturing Process—Reinforced Rubber Sheet for Rubber Dam. In *Hydraulic Rubber Dam*; Plastics Design Library; Thomas, S., Rane, A.V., Abitha, V.K., Kanny, K., Dutta, A.B., Eds.; William Andrew Publishing: Norwich, NY, USA, 2019; pp. 37–46; ISBN 978-0-12-812210-5.
122. Aloui, H.; Ghazouani, Z.; Khwaldia, K. Bioactive Coatings Enriched with Cuticle Components from Tomato Wastes for Cherry Tomatoes Preservation. *Waste Biomass Valoriz.* **2021**, *12*, 6155–6163. [CrossRef]
123. Fooladi, S.; Kiahosseini, S.R. Creation and investigation of chitin/HA double-layer coatings on AZ91 magnesium alloy by dipping method. *J. Mater. Res.* **2017**, *32*, 2532–2541. [CrossRef]
124. Siemann, U. Solvent cast technology—A versatile tool for thin film production BT—Scattering Methods and the Properties of Polymer Materials. In *Scattering Methods and the Properties of Polymer Materials*; Stribeck, N., Smarsly, B., Eds.; Springer: Berlin/Heidelberg, Germany, 2005; pp. 1–14; ISBN 978-3-540-31510-0.
125. Garrido, T.; Leceta, I.; Cabezudo, S.; Guerrero, P.; de la Caba, K. Tailoring soy protein film properties by selecting casting or compression as processing methods. *Eur. Polym. J.* **2016**, *85*, 499–507. [CrossRef]
126. Taylor, J.; Taylor, J.R.; Dutton, M.F.; de Kock, S. Identification of kafirin film casting solvents. *Food Chem.* **2005**, *90*, 401–408. [CrossRef]
127. Fan, Y.; Fukuzumi, H.; Saito, T.; Isogai, A. Comparative characterization of aqueous dispersions and cast films of different chitin nanowhiskers/nanofibers. *Int. J. Biol. Macromol.* **2012**, *50*, 69–76. [CrossRef] [PubMed]
128. Tsuji, K.; Maeda, T.; Hotta, A. Polymer Surface Modifications by Coating. In *Printing on Polymers*; Izdebska, J., Thomas, S.B., Eds.; William Andrew Publishing: Norwich, NY, USA, 2016; pp. 143–160; ISBN 978-0-323-37468-2.

129. Bautista, L.; Molina, L.; Niembro, S.; García, J.M.; López, J.; Vílchez, A. Chapter Eight—Coatings and Inks for Food Packaging Including Nanomaterials. In *Emerging Nanotechnologies in Food Science*; Busquets, R., Ed.; Micro and Nano Technologies; Elsevier: Boston, MA, USA, 2017; pp. 149–173; ISBN 978-0-323-42980-1.
130. Webster, D.C.; Ryntz, R.A. Pigments, Paints, Polymer Coatings, Lacquers, and Printing Inks. In *Handbook of Industrial Chemistry and Biotechnology*; Kent, J.A., Bommaraju, T.V., Barnicki, S.D., Eds.; Springer International Publishing: Cham, Switzerland, 2017; pp. 805–822; ISBN 978-3-319-52287-6.
131. Kapur, N.; Hewson, R.; Sleigh, P.A.; Summers, J.L.; Thompson, H.M.; Abbott, S.J. A review of gravure coating systems. *Mater. Sci.* **2011**, *1*, 56–60.
132. Tambe, C.; Graiver, D.; Narayan, R. Moisture resistance coating of packaging paper from biobased silylated soybean oil. *Prog. Org. Coat.* **2016**, *101*, 270–278. [CrossRef]
133. Piergiovanni, L.; Li, F.; Farris, S. Coatings of Bio-based Materials on Flexible Food Packaging: Opportunities for Problem Solving and Innovations. In *Advances in Industrial Biotechnology*; Publishing House Pvt. Ltd, Ed.; IK International: Mumbai, India, 2014; pp. 233–251.
134. Spieser, H.; Denneulin, A.; Deganello, D.; Gethin, D.; Koppolu, R.; Bras, J. Cellulose nanofibrils and silver nanowires active coatings for the development of antibacterial packaging surfaces. *Carbohydr. Polym.* **2020**, *240*, 116305. [CrossRef]
135. Kallio, H.; Nieminen, R.; Tuomasjukka, S.; Hakala, M. Cutin composition of five Finnish berries. *J. Agric. Food Chem.* **2006**, *54*, 457–462. [CrossRef]
136. Picard, E.; Gauthier, H.; Gérard, J.F.; Espuche, E. Influence of the intercalated cations on the surface energy of montmorillonites: Consequences for the morphology and gas barrier properties of polyethylene/montmorillonites nanocomposites. *J. Colloid Interface Sci.* **2007**, *307*, 364–376. [CrossRef]
137. Burnett, D.J.; Thielmann, F.; Ryntz, R.A. Correlating thermodynamic and mechanical adhesion phenomena for thermoplastic polyolefins. *J. Coat. Technol. Res.* **2007**, *4*, 211–215. [CrossRef]
138. Dhakal, H.N.; Zhang, Z.Y.; Bennett, N. Influence of fibre treatment and glass fibre hybridisation on thermal degradation and surface energy characteristics of hemp/unsaturated polyester composites. *Compos. Part B Eng.* **2012**, *43*, 2757–2761. [CrossRef]
139. Belgacem, M.N.; Blayo, A.; Gandini, A. Surface Characterization of Polysaccharides, Lignins, Printing Ink Pigments, and Ink Fillers by Inverse Gas Chromatography. *J. Colloid Interface Sci.* **1996**, *182*, 431–436. [CrossRef]
140. Kendall, K. The adhesion and surface energy of elastic solids. *J. Phys. D Appl. Phys.* **1971**, *4*, 1186–1195. [CrossRef]
141. Packham, D.E. Surface energy, surface topography and adhesion. *Int. J. Adhes. Adhes.* **2003**, *23*, 437–448. [CrossRef]
142. Watts, J.F.; Critchlow, G.W.; Packham, D.E.; Kneafsey, B.; Sherriff, M.; Shanahan, M.E.; Cope, B.C.; Pascoe, M.W.; Sagar, A.J.; Allen, K.W.; et al. Handbook of Adhesion. In *Handbook of Adhesion*; John Wiley & Sons, Ltd: Hoboken, NJ, USA, 2005; pp. 1–58; ISBN 9780470014226.
143. Bogdanova, Y.G.; Dolzhikova, V.D.; Summ, B.D. Wetting of Solids by Aqueous Solutions of Surfactant Binary Mixtures: 2. Wetting of High-Energy Surface. *Colloid J.* **2003**, *65*, 290–294. [CrossRef]
144. Mourougou-Candoni, N.; Prunet-Foch, B.; Legay, F.; Vignes-Adler, M.; Wong, K. Retraction Phenomena of Surfactant Solution Drops upon Impact on a Solid Substrate of Low Surface Energy. *Langmuir* **1999**, *15*, 6563–6574. [CrossRef]
145. Kiani, S.; Rogers, S.E.; Sagisaka, M.; Alexander, S.; Barron, A.R. A New Class of Low Surface Energy Anionic Surfactant for Enhanced Oil Recovery. *Energy Fuels* **2019**, *33*, 3162–3175. [CrossRef]
146. Chen, D.; Jordan, E.H.; Gell, M. Effect of solution concentration on splat formation and coating microstructure using the solution precursor plasma spray process. *Surf. Coat. Technol.* **2008**, *202*, 2132–2138. [CrossRef]
147. Schubert, D.W.; Dunkel, T. Spin coating from a molecular point of view: Its concentration regimes, influence of molar mass and distribution. *Mater. Res. Innov.* **2003**, *7*, 314–321. [CrossRef]
148. Cisneros-Zevallos, L.; Krochta, J.M. Dependence of Coating Thickness on Viscosity of Coating Solution Applied to Fruits and Vegetables by Dipping Method. *J. Food Sci.* **2003**, *68*, 503–510. [CrossRef]
149. Kalin, M.; Polajnar, M. The correlation between the surface energy, the contact angle and the spreading parameter, and their relevance for the wetting behaviour of DLC with lubricating oils. *Tribol. Int.* **2013**, *66*, 225–233. [CrossRef]
150. Nakae, H.; Inui, R.; Hirata, Y.; Saito, H. Effects of surface roughness on wettability. *Acta Mater.* **1998**, *46*, 2313–2318. [CrossRef]
151. Kubiak, K.J.; Wilson, M.C.; Mathia, T.G.; Carval, P. Wettability versus roughness of engineering surfaces. *Wear* **2011**, *271*, 523–528. [CrossRef]
152. Encinas, N.; Pantoja, M.; Abenojar, J.; Martínez, M.A. Control of Wettability of Polymers by Surface Roughness Modification. *J. Adhes. Sci. Technol.* **2010**, *24*, 1869–1883. [CrossRef]
153. Al-Turaif, H.; Bousfield, D.W. The influence of substrate absorbency on coating surface energy. *Prog. Org. Coat.* **2004**, *49*, 62–68. [CrossRef]
154. Al-Turaif, H.; Bousfield, D.W.; LePoutre, P. The influence of substrate absorbency on coating surface chemistry. *Prog. Org. Coat.* **2002**, *44*, 307–315. [CrossRef]
155. Zheng, C.G.; Gall, B.L.; Gao, H.W.; Miller, A.E.; Bryant, R.S. Effects of Polymer Adsorption and Flow Behavior on Two-Phase Flow in Porous Media. *SPE Reserv. Eval. Eng.* **2000**, *3*, 216–223. [CrossRef]
156. Ogihara, H.; Xie, J.; Saji, T. Factors determining wettability of superhydrophobic paper prepared by spraying nanoparticle suspensions. *Colloids Surf. A Physicochem. Eng. Asp.* **2013**, *434*, 35–41. [CrossRef]

157. Holman, R.K.; Cima, M.J.; Uhland, S.A.; Sachs, E. Spreading and Infiltration of Inkjet-Printed Polymer Solution Droplets on a Porous Substrate. *J. Colloid Interface Sci.* **2002**, *249*, 432–440. [CrossRef]
158. Bose, S.; Keller, S.S.; Alstrøm, T.S.; Boisen, A.; Almdal, K. Process Optimization of Ultrasonic Spray Coating of Polymer Films. *Langmuir* **2013**, *29*, 6911–6919. [CrossRef]
159. Rowe, R.C. The measurement of the adhesion of film coatings to tablet surfaces: The effect of tablet porosity, surface roughness and film thickness. *J. Pharm. Pharmacol.* **1978**, *30*, 343–346. [CrossRef]
160. Collins, G.W.; Letts, S.A.; Fearon, E.M.; McEachern, R.L.; Bernat, T.P. Surface Roughness Scaling of Plasma Polymer Films. *Phys. Rev. Lett.* **1994**, *73*, 708–711. [CrossRef]
161. Mesic, B.; Cairns, M.; Järnstrom, L.; Joo Le Guen, M.; Parr, R. Film formation and barrier performance of latex based coating: Impact of drying temperature in a flexographic process. *Prog. Org. Coat.* **2019**, *129*, 43–51. [CrossRef]
162. Wang, J.; Law, C.L.; Nema, P.K.; Zhao, J.H.; Liu, Z.L.; Deng, L.Z.; Gao, Z.J.; Xiao, H.W. Pulsed vacuum drying enhances drying kinetics and quality of lemon slices. *J. Food Eng.* **2018**, *224*, 129–138. [CrossRef]
163. Alibas, I. Microwave, Vacuum, and Air Drying Characteristics of Collard Leaves. *Dry. Technol.* **2009**, *27*, 1266–1273. [CrossRef]
164. Signori, F.; Coltelli, M.B.; Bronco, S. Thermal degradation of poly(lactic acid) (PLA) and poly(butylene adipate-co-terephthalate) (PBAT) and their blends upon melt processing. *Polym. Degrad. Stab.* **2009**, *94*, 74–82. [CrossRef]
165. Ayu, R.S.; Khalina, A.; Harmaen, A.S.; Zaman, K.; Jawaid, M.; Lee, C.H. Effect of Modified Tapioca Starch on Mechanical, Thermal, and Morphological Properties of PBS Blends for Food Packaging. *Polymers* **2018**, *10*, 1187. [CrossRef]
166. Montano-Herrera, L.; Pratt, S.; Arcos-Hernandez, M.V.; Halley, P.J.; Lant, P.A.; Werker, A.; Laycock, B. In-line monitoring of thermal degradation of PHA during melt-processing by Near-Infrared spectroscopy. *New Biotechnol.* **2014**, *31*, 357–363. [CrossRef]
167. Brömme, H.J.; Peschke, E.; Israel, G. Photo-degradation of melatonin: Influence of argon, hydrogen peroxide, and ethanol. *J. Pineal Res.* **2008**, *44*, 366–372. [CrossRef] [PubMed]
168. Nonell, S.; Moncayo, L.; Trull, F.; Amat-Guerri, F.; Lissi, E.A.; Soltermann, A.T.; Criado, S.; García, N.A. Solvent influence on the kinetics of the photodynamic degradation of trolox, a water-soluble model compound for vitamin E. *J. Photochem. Photobiol. B Biol.* **1995**, *29*, 157–162. [CrossRef]
169. Auria, R.; Aycaguer, A.C.; Devinny, J.S. Influence of Water Content on Degradation Rates for Ethanol in Biofiltration. *J. Air Waste Manag. Assoc.* **1998**, *48*, 65–70. [CrossRef]
170. Zhao, F.; Yu, B.; Yue, Z.; Wang, T.; Wen, X.; Liu, Z.; Zhao, C. Preparation of porous chitosan gel beads for copper(II) ion adsorption. *J. Hazard. Mater.* **2007**, *147*, 67–73. [CrossRef]
171. Morganti, P.; Morganti, G. Chitin nanofibrils for advanced cosmeceuticals. *Clin. Dermatol.* **2008**, *26*, 334–340. [CrossRef] [PubMed]
172. Cabrera-Trujillo, M.A.; Filomena-Ambrosio, A.; Quintanilla-Carvajal, M.X.; Sotelo-Díaz, L.I. Stability of low-fat oil in water emulsions obtained by ultra turrax, rotor-stator and ultrasound homogenization methods. *Int. J. Gastron. Food Sci.* **2018**, *13*, 58–64. [CrossRef]
173. Ganta, S.; Paxton, J.W.; Baguley, B.C.; Garg, S. Formulation and pharmacokinetic evaluation of an asulacrine nanocrystalline suspension for intravenous delivery. *Int. J. Pharm.* **2009**, *367*, 179–186. [CrossRef]
174. Xu, J.; Zhou, Z.; Cai, J.; Tian, J. Conductive biomass-based composite wires with cross-linked anionic nanocellulose and cationic nanochitin as scaffolds. *Int. J. Biol. Macromol.* **2020**, *156*, 1183–1190. [CrossRef]
175. Lv, S.; Zhou, H.; Bai, L.; Rojas, O.J.; McClements, D.J. Development of food-grade Pickering emulsions stabilized by a mixture of cellulose nanofibrils and nanochitin. *Food Hydrocoll.* **2021**, *113*, 106451. [CrossRef]
176. Lin, D.; Li, R.; Lopez-Sanchez, P.; Li, Z. Physical properties of bacterial cellulose aqueous suspensions treated by high pressure homogenizer. *Food Hydrocoll.* **2015**, *44*, 435–442. [CrossRef]
177. Lee, S.Y.; Chun, S.J.; Kang, I.A.; Park, J.Y. Preparation of cellulose nanofibrils by high-pressure homogenizer and cellulose-based composite films. *J. Ind. Eng. Chem.* **2009**, *15*, 50–55. [CrossRef]
178. Youssef, N.H.; Duncan, K.E.; Nagle, D.P.; Savage, K.N.; Knapp, R.M.; McInerney, M.J. Comparison of methods to detect biosurfactant production by diverse microorganisms. *J. Microbiol. Methods* **2004**, *56*, 339–347. [CrossRef] [PubMed]
179. Md, F. Biosurfactant: Production and Application. *J. Pet. Environ. Biotechnol.* **2012**, *3*, 124. [CrossRef]
180. He, Q. Development of Waterborne Bio-Based Primers for Metal Application. Master's Thesis, KTH, School of Engineering Sciences in Chemistry, Biotechnology and Health (CBH), Stockholm, Sweden, 2020.
181. Atta, A.M.; Al-Hodan, H.A.; Hameed, R.S.; Ezzat, A.O. Preparation of green cardanol-based epoxy and hardener as primer coatings for petroleum and gas steel in marine environment. *Prog. Org. Coat.* **2017**, *111*, 283–293. [CrossRef]
182. Voirin, C.; Caillol, S.; Sadavarte, N.V.; Tawade, B.V.; Boutevin, B.; Wadgaonkar, P.P. Functionalization of cardanol: Towards biobased polymers and additives. *Polym. Chem.* **2014**, *5*, 3142–3162. [CrossRef]
183. Van Savage, G.; Rhodes, C.T. The sustained release coating of solid dosage forms: A historical review. *Drug Dev. Ind. Pharm.* **1995**, *21*, 93–118. [CrossRef]
184. Paul, C.W. Hot-melt adhesives. *Mrs Bull.* **2003**, *28*, 440–444. [CrossRef]
185. Park, Y.J.; Joo, H.S.; Kim, H.J.; Lee, Y.K. Adhesion and rheological properties of EVA-based hot-melt adhesives. *Int. J. Adhes. Adhes.* **2006**, *26*, 571–576. [CrossRef]
186. Moody, V.; Needles, H.L. Hot Melt Coating. In *Tufted Carpet: Textile Fibers, Dyes, Finishes and Processes*; William Andrew Publishing: Norwich, NY, USA, 2004; p. 202; ISBN 0815519400.

187. Li, W.; Bouzidi, L.; Narine, S.S. Current research and development status and prospect of hot-melt adhesives: A review. *Ind. Eng. Chem. Res.* **2008**, *47*, 7524–7532. [CrossRef]
188. Gilvary, G.C.; Ammar, A.; Li, S.; Senta-Loys, Z.; Tian, Y.; Kelleher, J.F.; Healy, A.M.; Jones, D.S.; Andrews, G.P. Hot-melt co-extrusion technology as a manufacturing platform for anti-hypertensive fixed-dose combinations. *Br. J. Pharm.* **2019**, *4*, S14–S15. [CrossRef]
189. Vynckier, A.K.; Dierickx, L.; Voorspoels, J.; Gonnissen, Y.; Remon, J.P.; Vervaet, C. Hot-melt co-extrusion: Requirements, challenges and opportunities for pharmaceutical applications. *J. Pharm. Pharmacol.* **2014**, *66*, 167–179. [CrossRef] [PubMed]
190. Farris, S. Main Manufacturing Processes for Food Packaging Materials. In *Reference Module in Food Science*; Elsevier: Amsterdam, The Netherlands, 2016; pp. 1–9; ISBN 978-0-08-100596-5.
191. Sarraf, A.G.; Tissot, H.; Tissot, P.; Alfonso, D.; Gurny, R.; Doelker, E. Influence of hot-melt extrusion and compression molding on polymer structure organization, investigated by differential scanning calorimetry. *J. Appl. Polym. Sci.* **2001**, *81*, 3124–3132. [CrossRef]
192. Barletta, M.; Bolelli, G.; Guarino, S.; Lusvarghi, L. Development of matte finishes in electrostatic (EFB) and conventional hot dipping (CHDFB) fluidized bed coating process. *Prog. Org. Coat.* **2007**, *59*, 53–67. [CrossRef]
193. Barletta, M.; Gisario, A.; Rubino, G. Scratch response of high-performance thermoset and thermoplastic powders deposited by the electrostatic spray and 'hot dipping'fluidised bed coating methods: The role of the contact condition. *Surf. Coat. Technol.* **2011**, *205*, 5186–5198. [CrossRef]
194. Kang, D.; Kim, J.; Kim, I.; Choi, K.H.; Lee, T.M. Experimental qualification of the process of electrostatic spray deposition. *Coatings* **2019**, *9*, 294. [CrossRef]
195. Kawakami, K.; Zhang, S.; Chauhan, R.S.; Ishizuka, N.; Yamamoto, M.; Masaoka, Y.; Kataoka, M.; Yamashita, S.; Sakuma, S. Preparation of fenofibrate solid dispersion using electrospray deposition and improvement in oral absorption by instantaneous post-heating of the formulation. *Int. J. Pharm.* **2013**, *450*, 123–128. [CrossRef] [PubMed]
196. Park, J.; Shin, K.; Lee, C. Roll-to-roll coating technology and its applications: A review. *Int. J. Precis. Eng. Manuf.* **2016**, *17*, 537–550. [CrossRef]
197. Moustafa, A.F. Release of a cohesively strong, general purpose hot-melt pressure sensitive adhesive from a silicone liner. *Int. J. Adhes. Adhes.* **2014**, *50*, 65–69. [CrossRef]
198. Achanta, A.S.; Adusumilli, P.S.; James, K.W.; Rhodes, C.T. Development of hot melt coating methods. *Drug Dev. Ind. Pharm.* **1997**, *23*, 441–449. [CrossRef]
199. Bose, S.; Bogner, R.H. Solventless pharmaceutical coating processes: A review. *Pharm. Dev. Technol.* **2007**, *12*, 115–131. [CrossRef]
200. Jones, D.M.; Percel, P.J. *Coating of Multiparticulates Using Molten Materials. Formulation and Process Considerations*; Ghebre-Sellasie, I., Ed.; Taylor & Francis Inc: Abingdon, UK, 1994; Volume 65.
201. Khobragade, D.S.; Wankar, J.; Patil, A.T.; Potbhare, M.S.; Lakhotiya, C.L.; Umathe, S.N. A novel practical approach for enhancement of bioavailability of a poorly water soluble drug by hot melt coating technique. *Int. J. Pharm. Sci. Rev. Res* **2014**, *26*, 258–263.
202. Jannin, V.; Berard, V.; N'diaye, A.; Andres, C.; Pourcelot, Y. Comparative study of the lubricant performance of Compritol 888 ATO either used by blending or by hot melt coating. *Int. J. Pharm.* **2003**, *262*, 39–45. [CrossRef]
203. Lopes, D.G.; Salar-Behzadi, S.; Zimmer, A. Designing optimal formulations for hot-melt coating. *Int. J. Pharm.* **2017**, *533*, 357–363. [CrossRef]
204. Maniruzzaman, M.; Boateng, J.S.; Snowden, M.J.; Douroumis, D. A review of hot-melt extrusion: Process technology to pharmaceutical products. *Int. Sch. Res. Not.* **2012**, *12*, 27. [CrossRef]
205. Maniruzzaman, M.; Rana, M.M.; Boateng, J.S.; Mitchell, J.C.; Douroumis, D. Dissolution enhancement of poorly water-soluble APIs processed by hot-melt extrusion using hydrophilic polymers. *Drug Dev. Ind. Pharm.* **2013**, *39*, 218–227. [CrossRef]
206. Correlo, V.M.; Boesel, L.F.; Bhattacharya, M.; Mano, J.F.; Neves, N.M.; Reis, R.L. Properties of melt processed chitosan and aliphatic polyester blends. *Mater. Sci. Eng. A* **2005**, *403*, 57–68. [CrossRef]
207. Takeshita, Y.; Sawada, T.; Handa, T.; Watanuki, Y.; Kudo, T. Influence of air-cooling time on physical properties of thermoplastic polyester powder coatings. *Prog. Org. Coat.* **2012**, *75*, 584–589. [CrossRef]
208. Wu, B.; Wang, Z. Powder Coating Compositions Containing Reactive Nanoparticles. U.S. Patent 10/450,399, 1 April 2004.
209. Kage, H.; Dohzaki, M.; Ogura, H.; Matsuno, Y. Powder coating efficiency of small particles and their agglomeration in circulating fluidized bed. *Korean J. Chem. Eng.* **1999**, *16*, 630–634. [CrossRef]
210. Takeshita, Y.; Miwa, T.; Ishii, A.; Sawada, T. Innovative thermoplastic powder coatings in telecommunication fields. *J. Curr. Issues Media Telecommun.* **2017**, *9*, 289–3131.
211. Crowley, M.M.; Zhang, F.; Repka, M.A.; Thumma, S.; Upadhye, S.B.; Battu, S.K.; McGinity, J.W.; Martin, C. Pharmaceutical applications of hot-melt extrusion: Part I. *Drug Dev. Ind. Pharm.* **2007**, *33*, 909–926. [CrossRef] [PubMed]
212. Van Haveren, J.; Oostveen, E.A.; Miccichè, F.; Noordover, B.A.; Koning, C.E.; Van Benthem, R.A.; Frissen, A.E.; Weijnen, J.G. Resins and additives for powder coatings and alkyd paints, based on renewable resources. *J. Coat. Technol. Res.* **2007**, *4*, 177–186. [CrossRef]
213. Dorigato, A.; Perin, D.; Pegoretti, A. Effect of the Temperature and of the Drawing Conditions on the Fracture Behaviour of Thermoplastic Starch Films for Packaging Applications. *J. Polym. Environ.* **2020**, *28*, 3244–3255. [CrossRef]

214. Miletić, A.; Ristić, I.; Coltelli, M.B.; Pilić, B. Modification of PLA-Based Films by Grafting or Coating. *J. Funct. Biomater.* **2020**, *11*, 30. [CrossRef] [PubMed]
215. Schmid, M.; Herbst, C.; Müller, K.; Stäbler, A.; Schlemmer, D.; Coltelli, M.B.; Lazzeri, A. Effect of Potato Pulp Filler on the Mechanical Properties and Water Vapor Transmission Rate of Thermoplastic WPI/PBS Blends. *Polym. Plast. Technol. Eng.* **2016**, *55*, 510–517. [CrossRef]

Article

Influence of Functional Bio-Based Coatings Including Chitin Nanofibrils or Polyphenols on Mechanical Properties of Paper Tissues

Luca Panariello [1,2], Maria-Beatrice Coltelli [1,2,*], Simone Giangrandi [3], María Carmen Garrigós [4], Ahdi Hadrich [5], Andrea Lazzeri [1,2,6] and Patrizia Cinelli [1,2,*]

1. National Interuniversity Consortium of Materials Science and Technology (INSTM), 50121 Firenze, Italy
2. Department of Civil and Industrial Engineering, University of Pisa, 56126 Pisa, Italy; luca.panariello@ing.unipi.it
3. LUCENSE SCaRL, 55100 Lucca, Italy; simone.giangrandi@lucense.it
4. Department of Analytical Chemistry, Nutrition and Food Sciences, University of Alicante, 03080 Alicante, Spain; mc.garrigos@ua.es
5. Biomass Valorization Platform-Materials, CELABOR s.c.r.l., 4650 Chaineux, Belgium; ahdi.hadrich@celabor.be
6. Planet Bioplastics s.r.l., 56127 Pisa, Italy; andrea.lazzeri@unipi.it
* Correspondence: maria.beatrice.coltelli@unipi.it (M.-B.C.); patrizia.cinelli@unipi.it (P.C.)

Abstract: The paper tissue industry is a constantly evolving sector that supplies markets that require products with different specific properties. In order to meet the demand of functional properties, ensuring a green approach at the same time, research on bio-coatings has been very active in recent decades. The attention dedicated to research on functional properties has not been given to the study of the morphological and mechanical properties of the final products. This paper studied the effect of two representative bio-based coatings on paper tissue. Coatings based on chitin nanofibrils or polyphenols were sprayed on paper tissues to provide them, respectively, with antibacterial and antioxidant activity. The chemical structure of the obtained samples was preliminarily compared by ATR-FTIR before and after their application. Coatings were applied on paper tissues and, after drying, their homogeneity was investigated by ATR-FTIR on different surface areas. Antimicrobial and antioxidant properties were found for chitin nanofibrils- and polyphenols-treated paper tissues, respectively. The mechanical properties of treated and untreated paper tissues were studied, considering as a reference the same tissue paper sample treated only with water. Different mechanical tests were performed on tissues, including penetration, tensile, and tearing tests in two perpendicular directions, to consider the anisotropy of the produced tissues for industrial applications. The morphology of uncoated and coated paper tissues was analysed by field emission scanning electron microscopy. Results from mechanical properties evidenced a correlation between morphological and mechanical changes. The addition of polyphenols resulted in a reduction in mechanical resistance, while the addition of chitin enhanced this property. This study evidenced the different effects produced by two novel coatings on paper tissues for personal care in terms of properties and structure.

Keywords: chitin nanofibrils; polyphenols; paper tissues; mechanical properties; bio-based coatings

Citation: Panariello, L.; Coltelli, M.-B.; Giangrandi, S.; Garrigós, M.C.; Hadrich, A.; Lazzeri, A.; Cinelli, P. Influence of Functional Bio-Based Coatings Including Chitin Nanofibrils or Polyphenols on Mechanical Properties of Paper Tissues. *Polymers* 2022, 14, 2274. https://doi.org/10.3390/polym14112274

Academic Editor: Fernão D. Magalhães

Received: 28 April 2022
Accepted: 29 May 2022
Published: 2 June 2022

Publisher's Note: MDPI stays neutral with regard to jurisdictional claims in published maps and institutional affiliations.

Copyright: © 2022 by the authors. Licensee MDPI, Basel, Switzerland. This article is an open access article distributed under the terms and conditions of the Creative Commons Attribution (CC BY) license (https://creativecommons.org/licenses/by/4.0/).

1. Introduction

The paper industry is increasingly becoming interested in developing efficient and innovative solutions to guarantee high-quality products [1]. For this purpose, research activities have always been focused on the development of additives capable of adding functional properties to cellulosic substrates. Typical properties required by the paper industry are related to the water and grease barrier, antimicrobial properties, or antioxidant activities [2–5].

In parallel with innovative pharmaceutical products [6,7], considering it a sustainable and circular economy perspective, the use of functional molecules from natural sources or industrial wastes is increasingly being used [8–10].

Extracts derived from agro-food wastes and forest residues represent a valid source of a wide range of functional molecules [11,12]. Polyphenols are considered interesting and widely available active molecules obtainable from wastes, and they are mainly used as natural antioxidants. For instance, different studies dealing with the extraction of polyphenols from tomato, lemon, orange, carrot peels or seeds [13,14], fennel stems, foils, and outer sheaths [15] but also from by-products such as cashew nuts, coconut shells, or groundnut hulls [16] have been reported.

Other products of great interest, extracted from natural wastes, are chitin nanofibrils and chitosan, which are used for their natural antimicrobial properties [17]. Chitin can be naturally obtained from marine sources, such as exoskeletons of crustaceans (crabs and shrimps) and molluscs (squid pens and mussel shells) [17–19], but also from terrestrial sources, such as insects [20] or mushrooms [21]. The extraction process generally involves demineralization (only for animal shells) and deproteinization steps [22–24]. The obtained chitin can be converted into chitosan by a deacetylation process or to chitin nanofibrils thanks to milder processes [25].

Moreover, the availability and yield of these active molecules from biomass waste have been recently improved by using innovative extraction techniques such as ultrasound-assisted extraction (UAE) [26], microwave-assisted extraction (MAE) [26–28], ultrafiltration (UF) and nanofiltration (NF) [29], and hydrodynamic cavitation [30]. The application of these molecules on cellulosic substrates has received great attention due to their intrinsic properties, which guarantees the biodegradability of the substrates and their compatibility with industrial-scale production [31,32]. Their use is very important, especially in skin care products [33,34], as they exhibit properties such as UV radiation protection [34,35] and anti-age action, acting as anticancer or moisturizer agents [36,37] and skin inflammatory reaction modulators [38,39].

One of the most used methods for the application of functional molecules involves using water as a solvent or suspension medium [40–44], thanks to its good environmental and economic advantages. Many techniques have been developed for the application of water dispersions or solutions of active molecules, such as flexography [45], roll-to-roll [46,47], wire-bar [48], blade [49], and spray [50,51]. In particular, in the paper industry, the application of a coating on the surface of paper substrates enables the increase in properties such as water vapor or gas barriers [52,53], and antimicrobial [54] or antioxidant activities [55]. Even if it is important to verify the effectiveness of functional additives to transfer the desired properties, it is also necessary to investigate their effect on the morphology and mechanical properties of the substrates. The thickness of the paper substrate and coating plays a key role in the analysis of the mechanical behaviour. Substrates constituted by paperboard for packaging are often affected by cracking of the coated layer during creasing and folding [56,57], but mechanical properties are minimally affected by the presence of a coating due to its negligible thickness compared to the substrate. Conversely, the mechanical properties of materials with limited thickness, such as paper tissues or towels, are affected by the presence of a coating and its application technique [58–60]. The main issues are represented by the change in coated tissues in terms of softness and dry/wet mechanical strength but also properties such as hydrophobicity, surface anisotropy, absorbency, and colour [44,61,62]. In solvent-mediated applications, it is pivotal consider that a greater interaction between the coating and substrate often means a greater modification of properties. It was reported that surface roughness and porosity distribution of the paper are the main factors that affect the interaction between the liquid and substrate and the absorption properties of the fluids [63,64]. Furthermore, it is also important to consider the molecular size of the active molecules and their affinity with the substrate. Small hydrosoluble molecules or micrometric particles, such as calcium carbonate, kaolin, talc, alumina, and titanium oxide [65–67], often used as pigments, can penetrate deeper than macromolecules such as polyphenols (tannic acid, catechin) [68,69], polysaccharides (starch, chitin) [70,71], or other polymers (natural rubbers, polyesters, polysiloxane) [58] used for surface modification [67].

In this paper, two coatings based on chitin nanofibrils and polyphenols were applied onto paper tissues by using spray techniques. FTIR spectra of treated tissues were recorded on different surface points and compared with the spectra of raw coatings to evaluate the homogeneity and penetration of the performed treatment on the tissues. Antioxidant and antibacterial properties of, respectively, polyphenols and chitin were measured to verify their activity as functional molecules. Their effect on the mechanical properties of tissues was investigated by different mechanical tests that included puncture resistance, and tensile and tearing tests. Mechanical properties were discussed correlating their trend with the microstructure, observed through field emission scanning electron microscopy (FESEM).

2. Materials and Methods

2.1. Materials

Cellulosic tissues were provided by Lucense SCaRL (Lucca, Italy) (thickness = 200 µm, 19.5 × 20.5 cm).

Polyphenols powder was extracted by the University of Alicante (Alicante, Spain) from tomato seeds obtained from agri-food wastes. Dried seeds were ground with a high-speed rotor mill at 12,000 rpm (Ultra Centrifugal Mill ZM 200, RETSCH, Haan, Germany), and particles passing through a 1 mm sieve were used. Then, microwave-assisted extraction (MAE) was applied by using a FLEXIWAVETM microwave oven (Milestone srl, Bergamo, Italy), as reported in the literature [72]. One gram of sample was introduced and mixed with 80 mL of 65% (v/v) ethanol at 400 rpm for 15 min at 80 °C. The obtained extract was cooled to room temperature and centrifuged at 5300 rpm for 10 min. The solid residue was washed twice with the extraction solvent and then discarded. Then, the supernatant was pooled with the washing solvent and stored overnight at −20 °C in order to remove possible interferences by precipitation. After that, the precipitate was removed by centrifugation at 5300 rpm and 4 °C for 10 min. The supernatant was collected and the ethanol was subsequently evaporated under reduced pressure. Afterward, the extract was frozen at −80 °C and freeze-dried until complete dryness. Finally, tomato seed extract was stored in vacuum-sealed packs at −20 °C in darkness.

A partially deacetylated chitin nanofibrils (ChNFs) suspension was supplied by Celabor (Chaineux, Belgium). The suspension was obtained through a chemical pre-treatment followed by a mechanical defibrillation process using an ultra-fine friction grinder Super masscolloider (Masuko® Sangyo Co. Ltd., Kawaguchi, Japan) equipped with two ceramic nonporous grinders adjustable at any clearance between the upper and lower grinder. Chemical pre-treatment was performed by a partial deacetylation of commercial chitin from shrimp shells (Sigma-Aldrich, St. Louis, MO, USA) using concentrated sodium hydroxide. The reaction was stopped when a degree of deacetylation (DDA) of 16% was reached. The product was then purified until the pH value reached 6.5~7. After the partial deacetylation, the resultant chitin suspensions were then prepared for mechanical defibrillation by dispersing them in acidified water at a concentration of 1.5 wt%. The solution was then manually poured into the grinder and the partially deacetylated chitin suspensions fed into the hopper were dispersed by centrifugal force into the clearance between the grinding stones, where they were ground into ultra-fine particles, after being subjected to massive compression, shearing, and rolling friction forces. ChNFs were thus obtained and stored at 4 °C until further use.

2.2. Raw Materials and Substrates Characterization

The tissue substrate was analysed by thermogravimetric analysis using a TA Q-500 (TA Instruments, Waters LLC, New Castle, DE, USA) to evaluate its qualitative composition and thermal degradation profile. This analysis was performed from room temperature to 900 °C under a nitrogen atmosphere at 10 °C/min. At 900 °C, an isothermal treatment of 30 min under air was performed to evaluate the residual weight under oxidative atmosphere.

Extracted active molecules were analysed by infrared spectroscopy using a Nicolet T380 Thermo Scientific instrument equipped with a Smart ITX ATR accessory with a diamond plate (Thermo Fisher Scientific, Waltham, MA, USA), collecting 256 scans at 4 cm^{-1} resolutions.

The polyphenols main composition was determined by high-performance liquid chromatography coupled to mass spectrometry (HPLC-DAD-MS) at 294 nm. An Agilent 1100 HPLC system coupled to a LC/MSD ion trap mass spectrometer with an electrospray ionization (ESI) source (Agilent Technologies, Palo Alto, CA, USA) was used. A HALO C18 column (100 mm × 4.6 mm × 2.7 µm) coupled to a HALO C18 guard column 90 Å (4.6 × 5 mm × 2.7 µm) operating at 25 °C was used. The mobile phase was composed of milli-Q water (solvent A) and acetonitrile (solvent B), both added with 0.1% acetic acid. A gradient elution program was used at a flow rate of 0.5 mL/min: starting A at 85% to 60% in 15 min to 30% in 3 min to 10% in 1 min and returning to the initial composition in 3 min. Mass spectra were recorded in the negative ionization mode (m/z 50–900). The electrospray chamber was set at 3.5 kV with a drying gas temperature of 350 °C. The N2 pressure and flow rate of the nebulizer were 50 psi and 10 L/min, respectively. Polyphenols were identified by comparing MS experimental data with those of standard compounds prepared in ethanol:water (60%, v/v). All solutions were filtered through a 0.22 µm nylon membrane prior to injection (6 µL).

The ChNF nature was investigated to verify the effect of the defibrillation process. A drop of diluted ChNF dispersion (1:1000 v/v in water) was poured on a microscopy slide and dried at room temperature. The slide was mounted on a SEM stub and observed under a field emission scanning electron microscopy (FESEM) using a FEG-Quanta 450 instrument (FEI, Hillsboro, OR, USA).

2.3. Coating Application

The chitin suspension was applied as received while the polyphenols powder was dissolved in water. Both coatings were applied by the spray technique with a hand sprayer on one side of the paper tissue. Different amounts of ChNFs (1 wt%, 7.5 wt%) and polyphenols (0.1 wt%, 1 wt%, 10 wt%), with respect to the tissue weight, were applied on the substrate. In addition to the tissues treated with active molecules, a reference sample treated with pure water was prepared. The same amount (30 mL) of water was used in all samples. In order to apply uniformly the coating, the application was performed at 10 cm from the surface of the tissue, equally dividing the spray between 9 application points as reported in Figure 1.

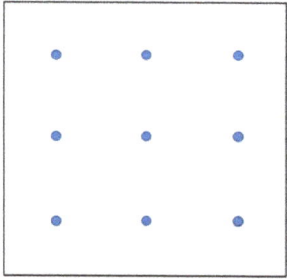

Figure 1. Application points selected for spray coating.

2.4. Coating Homogeneity Evaluation

The homogeneity of treated tissues was evaluated by performing five ATR-FTIR spectra on each side (selected points are reported in Figure 2) by using the same conditions of extracted active molecules described in Section 2.2.

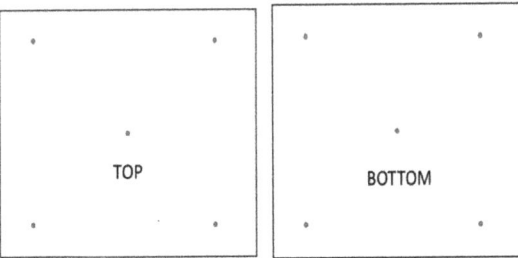

Figure 2. Scheme of selected points (blue dots) for ATR-FTIR analysis for evaluating coating homogeneity.

The two samples treated with the highest concentrations of each molecule (7.5 wt% ChNFs and 10 wt% polyphenols) were applied and dried on aluminium foil. Samples were analysed by ATR-FTIR to evaluate the effect of the application technique for eliminating the presence of the paper tissue signals.

2.5. Morphological Evaluation

The morphology of all samples was investigated by field emission scanning electron microscopy (FESEM) using a FEG-Quanta 450 instrument (FEI, Hillsboro, OR, USA). From all the samples constituted by the chitin and polyphenols sprayed on aluminium foils and the coated and uncoated paper tissues, a 1 × 1 cm square was cut with a hand precision scissor. Cut samples were stuck on an aluminium stub with a diameter of 12 mm through conductive carbon scotch. Stabs were sputtered with Pt using a LEICA EM ACE600 high-vacuum sputter coater (LEICA, Buccinasco, Italy) to make them conductive.

2.6. Antioxidant and Antibacterial Assays

Antibacterial properties of chitin suspensions were tested on treated tissues according to the BS ISO 8784-1:2014 Standard for determining the total number of colony-forming units of bacteria on disintegration on at least 5 plates.

The measurement of the radical scavenging activity (RSA) of polyphenols was performed according to the DPPH assay, as a simple and widely used method to determine the antioxidant properties of active molecules [73]. The analysis was conducted on an aqueous solution of polyphenols at concentrations of 0.07 wt%, 0.3 wt%, and 3 wt% to verify their antioxidant activity. The analysis was performed on three samples for each concentration. The absorbance of reference and sample solutions was measured with a UV-VIS spectrophotometer Perkin Elmer L60000B (Perkin Elmer, Waltham, MA, USA). The RSA was calculated as reported in Equation (1):

$$RSA(\%) = [(A_{reference} - A_{sample})/A_{reference}] \times 100 \tag{1}$$

2.7. Mechanical Properties

The mechanical properties of treated tissues were measured by different tests. All the specimens were previously conditioned at least 48 h in a 50% RH chamber.

The puncture resistance test was performed with a universal testing machine model 3365 (Instron, Norwood, MA, USA) equipped with a load cell of 100 N. A penetration probe with a hemispherical tip was used. As reported in the literature, this type of tip showed the highest reproducibility (lowest variability coefficient) on tests performed on paper laminates [74]. The test was performed at 1 mm/min until a complete break of the tissue. For each test, at least seven specimens were measured. As the puncture test was conducted orthogonally with respect to the surface of the tissues, the eventual orientation of the fibre did not influence the test.

Other mechanical tests were performed on specimens cut in two perpendicular ways, namely cross direction (CD) and machine direction (MD). The tissue was ripped by hand in the two directions. The direction where the rip had a straight progression was called the machine direction (MD), while the direction where the rip went through a deviation was called the cross direction (CD).

The Elmendorf tear test was performed with an Elmendorf tearing tester model 275A (Mesdan-lab, Raffa di Pugenago, Italy) equipped with a 1600 g pendulum, according to BS EN ISO 13937-1:2000 Standard. Specimens were cut with a manual scissor with the help of a jig according to the shape and dimensions reported in the Standard. A pre-notch was performed with the appropriate blade on the instrument. For each test, at least four specimens were measured.

A trouser test was conducted with a universal testing machine model 3365 (Instron, Norwood, MA, USA) equipped with a load cell of 100 N according to ASTM D1938 Standard with a grip-separation speed of 250 mm/min. Specimens were prepared with a manual scissor using a 45° blade from the top of the pre-crack to the bottom of trousers, achieving legs dimensions of 65 mm × 12.5 mm and an uncracked area of 25 mm × 25 mm. For each test, at least five specimens were measured.

The tensile test was conducted with a universal testing machine model 3365 (Instron, Norwood, MA, USA) equipped with a load cell of 100 N according to ASTM D882 Standard with a grip-separation speed of 10 mm/min. Tests were performed on ISO 527-2/5A dumbbell specimens obtained with a Manual Cutting Press EP 08 (Elastocon, Brahmult, Sweden). For each test, at least five specimens were measured.

The analysed samples are reported in Table 1.

Table 1. Samples studied by mechanical tests.

Name	Composition
TP	Pure tissue
TW	Tissue treated with sole water
TA	Tissue treated with 10 wt% antioxidant polyphenols
TC	Tissue treated with 7.5 wt% ChNFs

2.8. Statistical Analysis

The significance of differences between mean results in mechanical properties were analysed through a Tukey HSD post hoc test. Means that were identified as not significantly different were grouped under the same letter.

The test was performed on Minitab® software (Gmsl S.r.l., Nerviano, Italy) using a one-way analysis of variance assuming a confidence level of 95%.

3. Results

3.1. Raw Materials Characterization

Different biomolecules, such as ChNFs and polyphenols, from natural sources were applied onto the tissues to obtain functional tissues with enhanced properties. Before the application on the substrate, active molecules were characterized to identify their main composition. In Table 2, the main results obtained from the TGA thermogram are reported.

Table 2. Main components of cellulosic tissues elucidated by TGA.

Sample	Water Content (wt%)	Cellulose Degradation Onset Temperature (°C)	Cellulose Degradation (wt%)	Residue in N_2 (900 °C) (wt%)	Residue in Air (900 °C) (wt%)
Tissue	5.66	275	81.63	12.71	1.10

TGA analysis confirmed that the tissue was mainly composed of cellulose (81.63%), with a water content of 5.66%. The residue obtained under nitrogen atmosphere accounted for 12.71 wt%, which was mostly attributed to carbonaceous derivatives or organic additives that can be oxidized under air (final residue in air was 1.1 wt%).

Powders of active molecules were analysed by ATR-FTIR to qualitatively determine their composition.

The polyphenols spectrum showed typical bands associated with amide (1650 and 1540 cm^{-1}) and lipid (1720–1650 cm^{-1} and 3000–2800 cm^{-1}) groups. Other bands occurring at 1440–1400 cm^{-1} (C-H bending) and 1240–1400 cm^{-1} (C-C and C-C-H stretching) indicated the presence of methyl groups of proteins, and 1170–1115 cm^{-1} was attributed to C-O stretching. Broad absorption bands of OH group were shown in the 3500–3000 cm^{-1} range. The two peaks at 2920 and 2850 cm^{-1} were related, respectively, to asymmetric and symmetric C-H stretching. The region between 1040 and 990 cm^{-1} showed intense bands attributed to C-O-C vibrational modes of various carbohydrates and acids, which are abundant groups in tomatoes [75]. Finally, a small peak related to C=O stretching (1720 cm^{-1}) showed the presence of acetate groups.

In Figure 3, the FTIR spectrum of the polyphenol powder was compared with the one obtained from the sprayed polyphenols solution on an aluminium foil. This comparison evidenced a correspondence between main bands except for the band appearing at 990 cm^{-1}, which was attributed to the C-O-C vibrational modes of carbohydrates and organic acids derived from tomato [76]. This difference can be ascribed to the different surface composition detected by ATR-FTIR in the case of the powder or spray suspension. In the latter case, electrolytic substances are reasonably more concentrated on the surface than in the case of the powder.

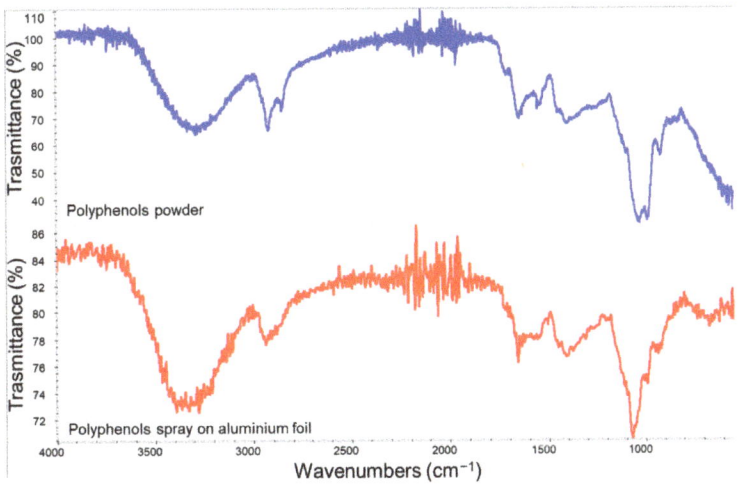

Figure 3. Spectra of polyphenols powder (top) and its sprayed film deposited on an aluminium foil.

Typical peaks related to acetyl and amide groups of chitin were observed in the spectrum of Figure 4. C-H stretching was identified at 1850 cm^{-1}, and the bands observed at 1548 cm^{-1}, 1615 cm^{-1}, and 1650 cm^{-1} were typical of amide groups. N-H stretching band relatives to deacetylated groups of chitin were observed at 3270 cm^{-1}. Other typical peaks related to the chitin carbohydrate backbone were the C-H stretching band shown at 2870–2880 cm^{-1}, the O-H wide band at 3450 cm^{-1}, and C-O-C stretching at 1025 and 1075 cm^{-1} [77]. The comparison between ATR-FTIR results for ChNF powder and its spray on aluminium in Figure 4 evidenced a good correspondence between both spectra.

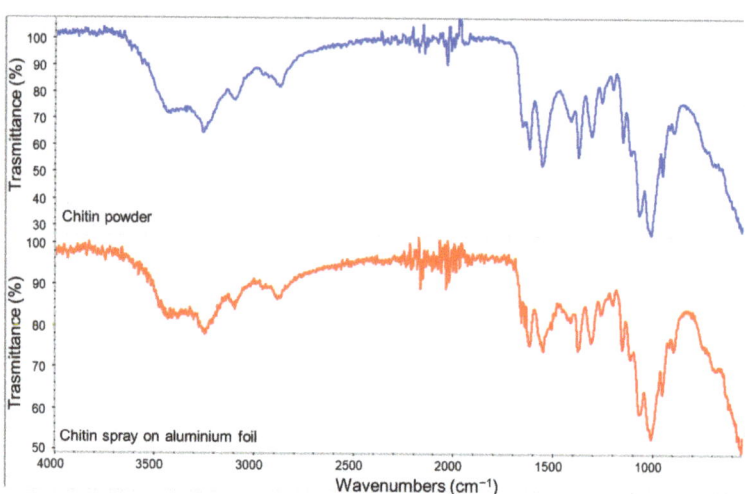

Figure 4. Spectra of chitin powder (top) and its sprayed film deposited on an aluminium foil.

The ChNF structure was studied by FESEM. The micrograph reported in Figure 5 showed a nanometric structure of chitin composed of fibrils with a diameter <50 nm.

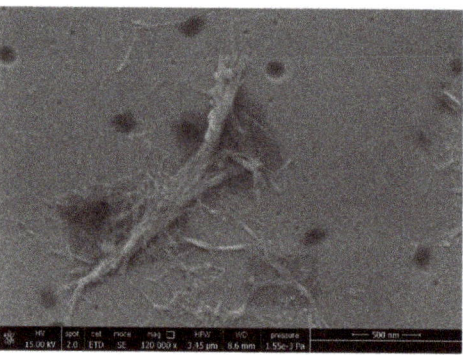

Figure 5. ESEM micrograph obtained for ChNFs.

Regarding the composition of polyphenols determined by HPLC-DAD-MS, chlorogenic acid, rutin, and naringenin were mainly identified in the tomato seeds powder. The quantification of these polyphenols was performed, in triplicate, based on integrated peak areas of samples and standards using external calibration. As a result, 2.99 ± 0.11 mg/100 g, 1.38 ± 0.02 mg/100 g, and 1.11 ± 0.35 mg/100 g were obtained for naringenin, rutin, and chlorogenic acid, respectively. These results are in agreement with other authors who reported similar polyphenols contents in tomato seeds [78].

3.2. Homogeneity Evaluation

The homogeneity of the performed treatment was investigated by ATR-FTIR analysis in five points on each side of tissue samples. This analysis was carried out only on tissues treated with the highest concentration of the active molecules to have enough intense IR signals to be detected.

Figures 6 and 7 show the spectra acquired for the tissues treated, respectively, with chitin and polyphenols. The spectrum of the untreated paper tissue was also considered as a reference, in order to select the bands that could reveal the presence of the active compounds in the coating. The intense signals appearing at 1650 cm^{-1} and 1615 cm^{-1} (related to amide groups) were, respectively, used to identify the presence of antioxidant polyphenols and chitin on the surface of the tissues. Similar intensities were obtained for this band in all five points analysed of each side, denoting a good homogeneity of the treatment. The comparison between the top and bottom (opposite side) parts of each tissue evidenced the prevalence of the active molecules on the top (application side), but they were detectable also on the bottom (opposite side). This difference was more evident in the chitin-treated tissue than in the polyphenols one, indicating that this latter had a higher penetration in the tissue. This phenomenon can be attributed to the similar molecular structure of chitin and cellulose (they both are polysaccharides) compared to polyphenols, to the lowest molecular weight of polyphenols with respect to the chitin, and to the fact that ChNFs were dispersed as a solid suspension in water, while polyphenols were at least partially dissolved. Hence, ChNFs were applied mainly on the application side where nanofibrils tend to be deposited forming a superficial tissue.

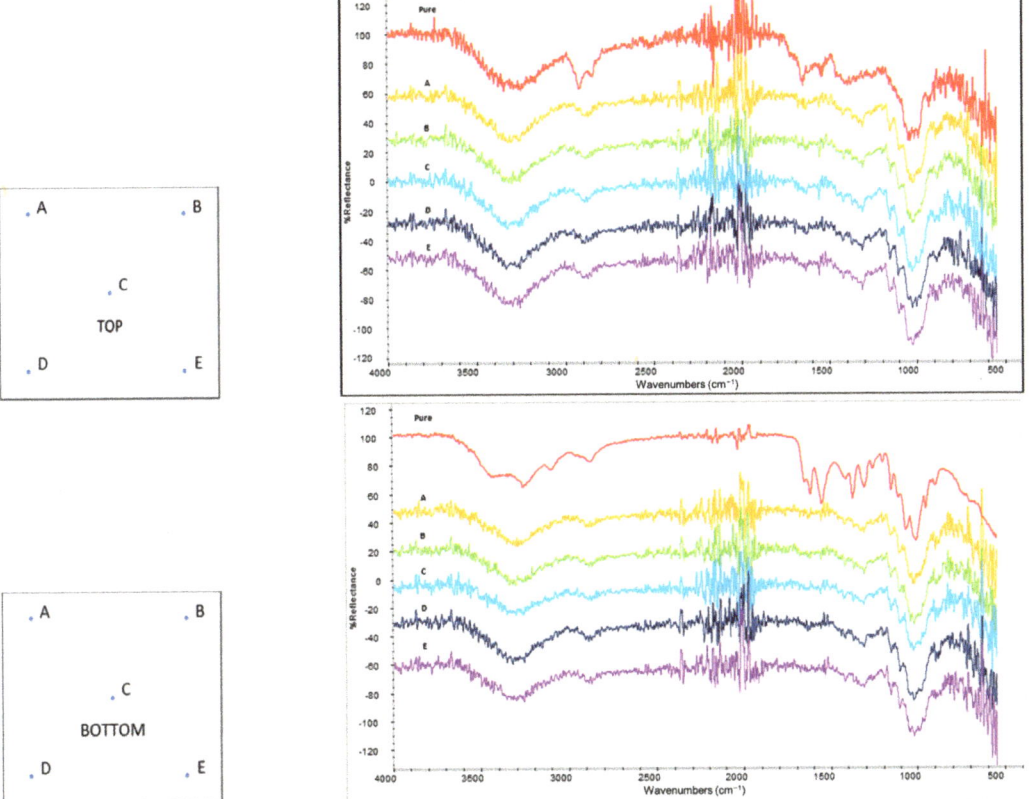

Figure 6. Spectra of the top and bottom part of chitin-treated tissues compared with IR spectrum of pure chitin (red). Capital letter on each spectrum indicate the point where it was acquired according to the relative picture on its left.

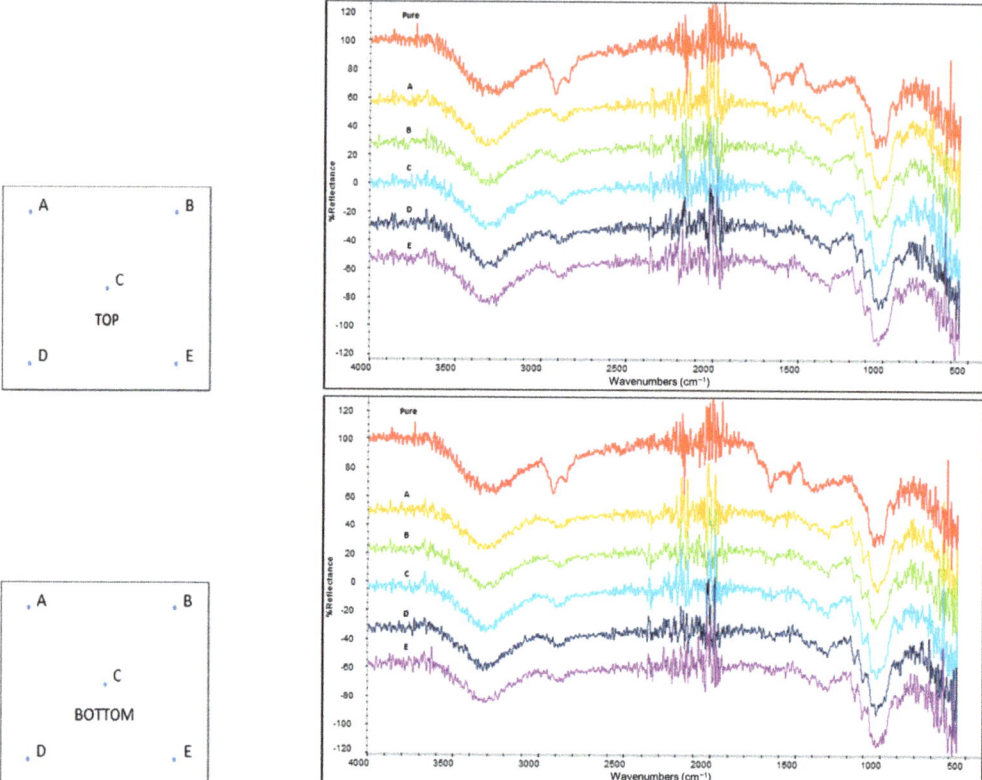

Figure 7. Spectra of the top and bottom part of antioxidant-treated tissues compared with IR spectrum of pure antioxidant extract (red). Capital letter on each spectrum indicate the point where it was acquired according to the relative picture on its left.

3.3. Antibacterial Test

Antibacterial properties of tissues sprayed with chitin were analysed through the enumeration of bacteria on the surface according to ISO/CD 8784-2 Standard. The obtained results (Table 3) showed a clear effect of chitin on bacteria proliferation that was almost halved at 1 wt% and was a fifth at 7.5 wt%. Although bacterial proliferation was not inhibited even at the highest concentration of chitin, its significant decrease can be considered an improvement in the capacity of the prepared tissue to limit bacteria growth.

Table 3. Results for bacteria enumeration on tissues treated with ChNFs (percentage of chitin was referred to the weight of raw tissue).

	Untreated	Tissue + 1 wt% Chitin	Tissue + 7.5 wt% Chitin
Enumeration of bacteria on surface (CFU/g)	1390 ± 120	867 ± 33	280 ± 17

3.4. Antioxidant Test

Antioxidant properties of tissues sprayed with polyphenols were evaluated by using the DPPH method. The antioxidant activity was expressed in radical scavenging activity RSA (%) values that were calculated as reported in Equation (1) and represent the capacity of the active molecules to reduce the free-radical activity. From the results shown in Figure 8, a clear antioxidant activity of the active tissues was observed, which significantly grew with the concentration of polyphenols. In addition, it was shown that the lowest concentration of polyphenols needed more than 30 minutes to reveal antioxidant properties, while the other concentrations used showed a significant antioxidant effect in less than 30 minutes.

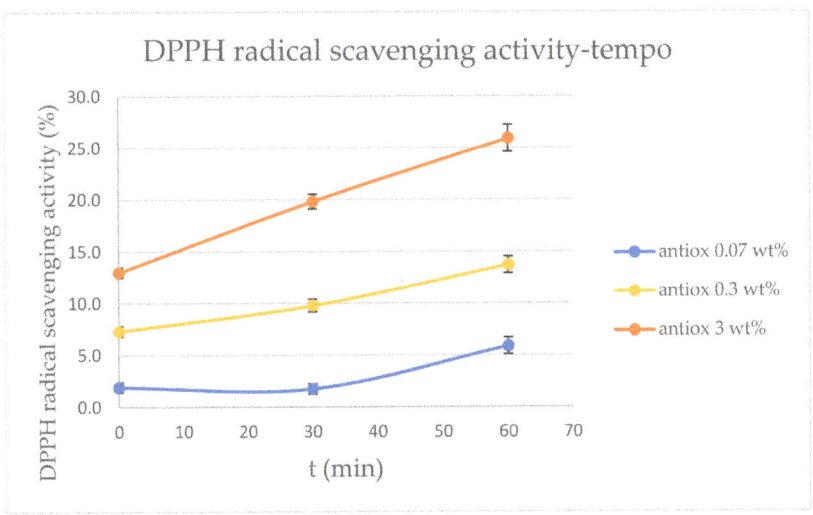

Figure 8. RSA (%) values of DPPH for polyphenol solutions at different concentrations after 0, 30, and 60 minutes.

3.5. Mechanical Properties

In addition to functional and compositional properties of the treated tissues, mechanical tests were also performed to describe exhaustively the effect of the coatings on the substrates. In addition to the untreated tissues and tissues treated with the maximum concentration of active molecules, the tissue treated with sole water was also studied.

3.5.1. Puncture Resistance Test

The puncture resistance of coated and uncoated tissues was determined by measuring the force needed to penetrate the paper with a blunt probe and its maximum deflection before breakage. This test can be used as a simple and quick preliminary study to predict the mechanical properties of nonwoven fabrics under quasi-static deformation [79–81].

In Figure 9a, differences between the force needed to penetrate treated and untreated tissues are shown. It was clear that the application technique strongly influenced the mechanical properties of the tissue. In fact, the sample treated with water (TW) showed a maximum force that was almost doubled with respect to the untreated tissue (TP). As a result, treated tissues should then be compared with this tissue and not with the untreated one. Considering treated tissues, TA appeared to be more easily penetrable compared to TW, while TC showed a similar behaviour.

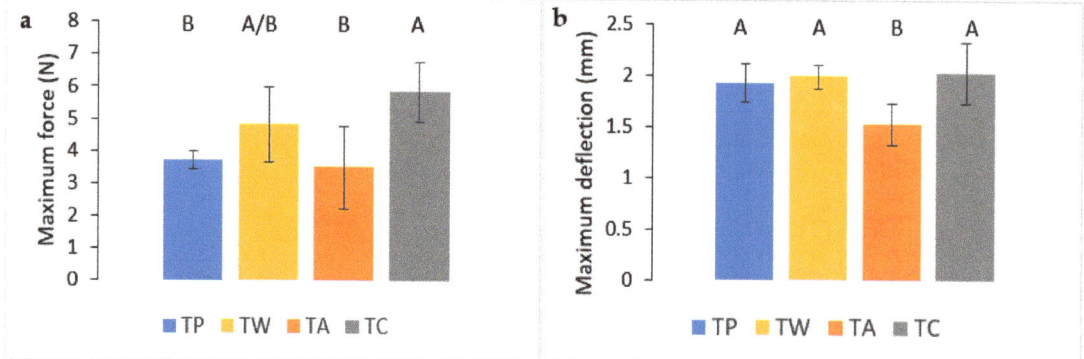

Figure 9. Maximum force (**a**) and maximum deflection (**b**) that tissues can sustain before breakage. Significance of standard deviation was investigated with a Tukey HSD post hoc test performed on 7 different specimens. On the top of each column, a letter was reported. Means that were identified as not significantly different were grouped under the same letter.

Maximum deflection values of tissues reported in Figure 9b represent the maximum displacement of the probe before tissues breakage. Nonsignificant differences between TP, TW, and TC values were observed.

In general, all measured values were similar except for TC that showed a maximum deflection slightly higher compared to the other samples. This behaviour could be compatible with the formation of a more compact structure thanks to the chitin nanofibrils presence that can sustain a greater deflection and force. The analysis of variance of the mean maximum force (Figure 9a) evidenced a significant difference between TC with respect to TA or the pure tissue, confirming its effect on reducing the penetrability of the coated tissue. Conversely, the analysis of mean maximum deflection (Figure 9b) showed a significant negative effect of TA with respect to the other samples, reducing the maximum deflection before breakage of the tissue.

3.5.2. Tensile Test

The tensile test performed on the two directions of the film (MD and CD) is a simple method to analyse the anisotropy in the mechanical properties of the tissues and the effect of the coating on these properties [61]. Moreover, the comparison between treated and untreated samples gave information on the effect of the coating on the resistance of the fibre texture, regarding the complex mechanism that consists of different steps, such as the alignment of the fibre, the reduction in the intermesh voids, and sliding of the fibres [82].

In Figure 10a, the stress at break of treated and untreated tissues in both directions are shown and represent the maximum values of stress that the mesh can sustain before the first fibre (or bundle of fibres) started to break. A clear difference between TC and the other samples can be noticed. Similar values of stress at break were observed in TP, TW, and TA samples, with a higher resistance at break in MD with respect to CD. The high stress in MD can be explained considering the alignment of the fibres in that direction. The comparison between TP, TW, and TA samples showed that water treatment seems to slightly reinforce the fibres, while polyphenols reduce their ability to resist the tensile force. Regarding TC, stress values significantly higher with respect to the other samples were noticed and similar values were observed for both directions. Therefore, chitin was able to override the anisotropy of the material and increase its properties. This behaviour could be compatible with the filling of the intermesh space with nanofibrils and the formation of higher-density materials that can sustain a greater force, in agreement with puncture tests results.

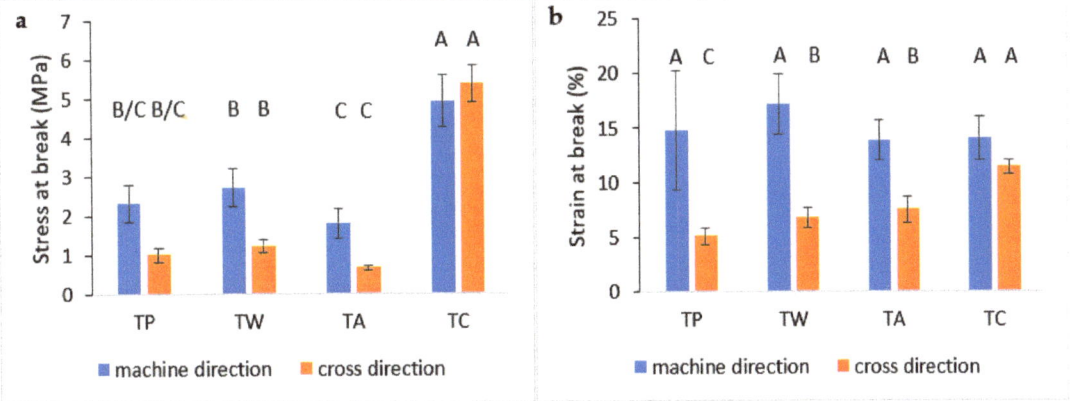

Figure 10. Stress (**a**) and strain (**b**) at break of treated and untreated tissues for tensile test. Significance of standard deviation was investigated with a Tukey HSD post hoc test performed on 5 different specimens. On the top of each column, a letter was reported. Means that were identified as not significantly different were grouped under the same letter.

In Figure 10b, strain values related to maximum stress values are reported. Compared to stress at break values (Figure 10a), all samples showed similar results, although in TC, the difference between MD and CD was less evident than in the other samples. Strain at break is caused by a complex mechanism that comprehends the alignment and movement of the fibres in the tissue. Hence, the presence of chitin, filming on the treated surface and in between the different fibres, and thus enhancing inter-fibre linkages makes the materials more resistant but less deformable.

ANOVA of stress at break values (Figure 10a) showed the same grouping in both MD and CD, confirming a significant strengthening effect of chitin with respect to the pure and other coatings. Instead, the statistical analysis of strain at break (Figure 10b) showed some significant differences between MD and CD. The values measured in CD with respect to the same grouping observed in stress at break values showed a predominance of the chitin with respect to the other coating, while, in MD, all the means were not significantly different.

3.5.3. Tearing Tests

The tearing test is classified as a type III mode of fracture mechanics (out-of-plane fracturing test) and allows the fracture propagation on the plane to be studied by applying an out-of-plane shear [83]. Two types of tearing tests were performed on treated and untreated tissues: Elmendorf and trouser tearing tests. The differences between these tests were mainly in the test speed; the Elmendorf apparatus is composed of a swinging pendulum that was released, causing a very fast propagation of the crack, while the trouser test was a test with a controlled strain rate.

In Figure 11a, the trouser tear propagation force is shown, which can be defined as the force required to propagate the crack. TP and TW samples showed similar values, denoting that, in this case, the application technique did not affect the measured property. Instead, TA and TC showed, respectively, lower and higher values with respect to TP and TW. Regarding the analysis direction in all samples, it was observed that the crack required more force to propagate in CD compared to MD, conversely to the behaviour observed in the tensile test. This behaviour agreed with a partially oriented fibre structure where a tear can easily propagate parallel to the fibres with respect to the perpendicular direction.

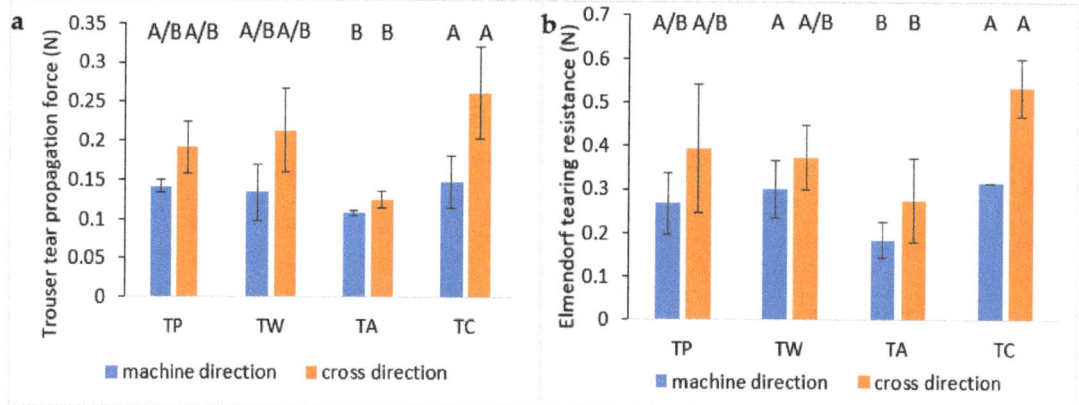

Figure 11. Tear propagation force (**a**) measured through trouser test and tear propagation resistance (**b**) measured with Elmendorf pendulum. Significance of standard deviation was investigated with a Tukey HSD post hoc test performed on 4 different specimens for Elmendorf test and on 5 different specimens for tear propagation test. On the top of each column, a letter was reported. Means that were identified as not significantly different were grouped under the same letter.

Elmendorf tearing test results reported in Figure 11b showed a trend similar to the trouser test but with some differences. The obtained results were all higher compared to the respective values observed in the trouser test due to the higher test speed. Differences between TP and TW were shown, with an increment in tear resistance caused by the application technique, but in these two samples, the difference between MD and CD was no longer visible. Conversely, in TA and TC, higher values in CD were observed compared to MD, maintaining the same trend observed in the trouser test with propagation tear resistance values of TA and TC, respectively, lower and higher with respect to TP.

Analysis of variance of both trouser and Elmendorf tearing tests showed the same grouping of means, evidencing a significant difference between TC and TA tissues. In particular, TC showed the highest tearing resistance values in all tests and directions, while TA showed the lowest ones.

3.6. FESEM Analysis

FESEM analysis was performed to observe the structure of ChNFs and polyphenols and their effect on the paper tissue structure. The brittle dusty sample deposited on aluminium foil by spraying the polyphenol-based treatment and the homogeneous and compact film formed by depositing the treatment based on ChNFs were both analysed.

In Figure 12, the different microstructures of polyphenols (Figure 12a) and ChNFs (Figure 12b) were clearly observed. Polyphenols showed the aspect of a paste consisting of round submicrometric partially united particles deposited on the surface of aluminium foil forming a nonhomogeneous film with the presence of cracks in correspondence of the particles joining lines. Instead, chitin showed a nanostructured morphology consisting of nanofibrils that form a compact structure fully covering the aluminium foil. In good agreement, a good film-forming capacity was evidenced for ChNFs [54,84].

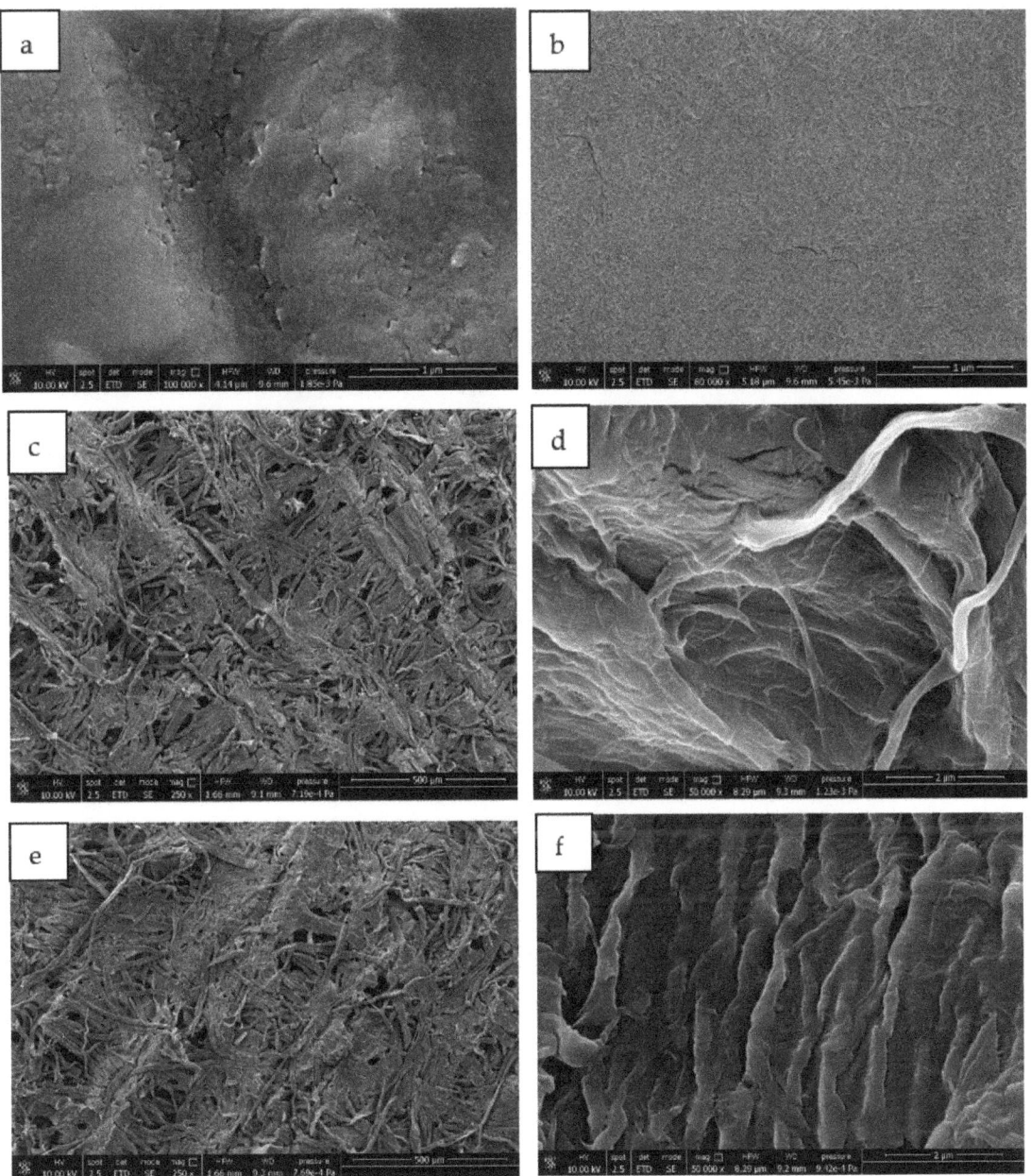

Figure 12. *Cont.*

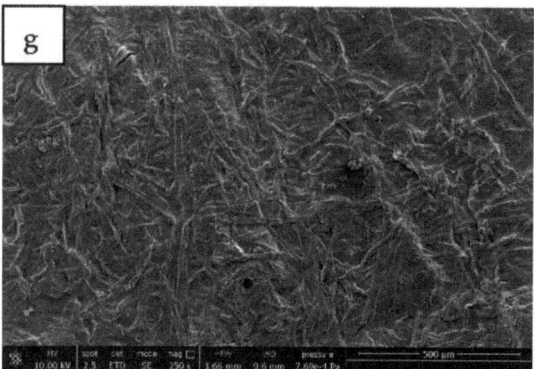
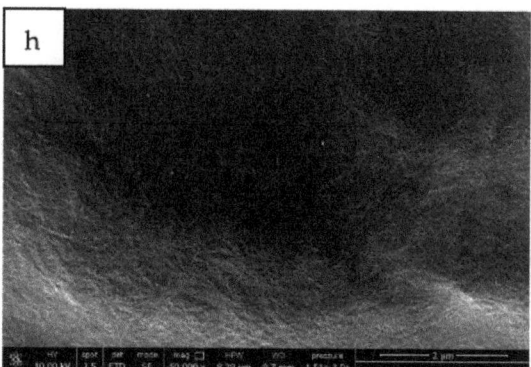

Figure 12. FESEM micrographs of: (**a**) polyphenols and (**b**) chitin sprayed on aluminium foil, (**c**,**d**) paper tissue, (**e**,**f**) paper tissue treated with polyphenols, and (**g**,**h**) paper tissue treated with ChNFs.

Pure tissue, as observed in the micrograph of Figure 12c,d, showed a fibrillar structure grouped in bundles that were oriented in different directions. However, the structure was not compact and it showed many empty spaces having dimensions of tenth of micrometres.

In Figure 12e,f, micrographs of paper tissues treated with polyphenols are shown. At low magnification (250×), it was not possible to identify the presence of antioxidant molecules on the fibres and the microstructure appeared unmodified by the treatment. The comparison between the high-magnification micrographs of pure tissue (Figure 12c,d) and the one treated with polyphenols (Figure 12e,f) allowed the deposition of polyphenols onto the fibrillar structure of the paper to be observed.

Micrographs of tissues coated with chitin (Figure 12g,h) showed a completely different structure with respect to the one treated with polyphenols. In fact, it was also possible to observe the coating at low magnification (250×) where the chitin completely filled the empty space between the paper fibrils, forming a continue surface. At high magnification, it was not possible to distinguish the fibrillar structure of the paper but only the nanofibrillated structure of chitin covering the treated surface.

4. Conclusions

In this paper, two coatings based on chitin and polyphenols were successfully applied on paper tissues. As a result, antibacterial properties due to ChNFs and antioxidant properties due to polyphenols were positively detected. In particular, the correlation between the concentration of functional coatings and the expected properties was confirmed, resulting in an increase in antibacterial and antioxidant properties at higher contents of active molecules. ATR-IR analysis confirmed the presence of a homogeneous coating on the surface of the paper tissues and showed a higher concentration of chitin on the top surface with respect to the bottom part, revealing its affinity for the cellulosic substrate. Mechanical properties of treated and untreated tissues were also studied considering the effect of the application technique with water. The results showed that, in all mechanical tests, the water treatment had a toughening generalized effect. In the literature, it was reported that the change in swelling and moisture can induce anisotropic shrinkage or relaxation of the microcompression created during the manufacturing [85]. The latter described phenomenon is in agreement with the generalized toughening effect. Regarding the tissues treated with ChNFs or polyphenols, it was also possible to observe different effects. In fact, the addition of polyphenols resulted in a reduction in mechanical resistance, while the addition of ChNFs resulted in its enhancement, which was confirmed by the statistical analysis of results. The observation of mechanical properties in the different directions for tensile and tearing tests confirmed the general orientation of paper fibres in the machine direction, in agreement with industrial methodologies for paper production. The effect of coatings on mechanical properties can be better explained by observing FESEM micrographs. ChNFs increased the density of the tissue, filling the empty spaces between the paper fibres and obtaining a more compact structure that can justify the increase in mechanical resistance. Conversely, polyphenols did not significantly affect the structure of the tissue, as they were impregnated but had no film-forming capacity. Therefore, the decrease in mechanical properties of tissues treated with polyphenols cannot be attributed to a change in the structure but to a detrimental effect due to the presence, along all the tissue thickness, of the polyphenols extract showing a limited compatibility with the cellulosic fibres of the paper tissue. As polyphenols derive from a natural extract that contains several molecules, it might be reasonable to assume a chemical degradation of the fibres. On the whole, the results of the present paper can be useful for developing, in the near future, functional coatings for cellulosic products in the skin care or packaging sectors considering the effects of interesting biomass waste derivatives.

Author Contributions: Conceptualization, L.P. and M.-B.C.; methodology, L.P.; investigation, L.P.; data curation, L.P.; writing—original draft preparation, L.P. and M.-B.C.; writing—review and editing, S.G., M.C.G. and A.H.; supervision, A.L. and P.C.; project administration, P.C.; funding acquisition P.C. All authors have read and agreed to the published version of the manuscript.

Funding: This research was funded by the Bio-Based Industries Joint Undertaking under the European Union Horizon 2020 research program (BBI-H2020), ECOFUNCO project, grant number G.A 837863.

Institutional Review Board Statement: Not applicable.

Informed Consent Statement: Not applicable.

Conflicts of Interest: The authors declare no conflict of interest.

References

1. Xiong, S.J.; Pang, B.; Zhou, S.J.; Li, M.K.; Yang, S.; Wang, Y.Y.; Shi, Q.; Wang, S.F.; Yuan, T.Q.; Sun, R.C. Economically Competitive Biodegradable PBAT/Lignin Composites: Effect of Lignin Methylation and Compatibilizer. *ACS Sustain. Chem. Eng.* **2020**, *8*, 5338–5346. [CrossRef]
2. Kjellgren, H.; Gällstedt, M.; Engström, G.; Järnström, L. Barrier and surface properties of chitosan-coated greaseproof paper. *Carbohydr. Polym.* **2006**, *65*, 453–460. [CrossRef]
3. Zhang, W.; Xiao, H.; Qian, L. Enhanced water vapour barrier and grease resistance of paper bilayer-coated with chitosan and beeswax. *Carbohydr. Polym.* **2014**, *101*, 401–406. [CrossRef] [PubMed]
4. Zhang, D.; Xiao, H. Dual-Functional Beeswaxes on Enhancing Antimicrobial Activity and Water Vapor Barrier Property of Paper. *ACS Appl. Mater. Interfaces* **2013**, *5*, 3464–3468. [CrossRef] [PubMed]
5. Cabañas-Romero, L.V.; Valls, C.; Valenzuela, S.V.; Roncero, M.B.; Pastor, F.I.J.; Diaz, P.; Martínez, J. Bacterial Cellulose–Chitosan Paper with Antimicrobial and Antioxidant Activities. *Biomacromolecules* **2020**, *21*, 1568–1577. [CrossRef]
6. Wang, J.; Liu, X.; Milcovich, G.; Chen, T.-Y.; Durack, E.; Mallen, S.; Ruan, Y.; Weng, X.; Hudson, S.P. Co-reductive fabrication of carbon nanodots with high quantum yield for bioimaging of bacteria. *Beilstein J. Nanotechnol.* **2018**, *9*, 137–145. [CrossRef]
7. Bartelmess, J.; Milcovich, G.; Maffeis, V.; d'Amora, M.; Bertozzi, S.M.; Giordani, S. Modulation of Efficient Diiodo-BODIPY in vitro Phototoxicity to Cancer Cells by Carbon Nano-Onions. *Front. Chem.* **2020**, *8*, 868. [CrossRef]
8. Reichert, C.L.; Bugnicourt, E.; Coltelli, M.-B.; Cinelli, P.; Lazzeri, A.; Canesi, I.; Braca, F.; Martínez, B.M.; Alonso, R.; Agostinis, L.; et al. Bio-Based Packaging: Materials, Modifications, Industrial Applications and Sustainability. *Polymers* **2020**, *12*, 1558. [CrossRef]
9. De Barros, C.H.N.; Cruz, G.C.F.; Mayrink, W.; Tasic, L. Bio-based synthesis of silver nanoparticles from orange waste: Effects of distinct biomolecule coatings on size, morphology, and antimicrobial activity. *Nanotechnol. Sci. Appl.* **2018**, *11*, 1–14. [CrossRef]
10. Kocaman, S.; Karaman, M.; Gursoy, M.; Ahmetli, G. Chemical and plasma surface modification of lignocellulose coconut waste for the preparation of advanced biobased composite materials. *Carbohydr. Polym.* **2017**, *159*, 48–57. [CrossRef]
11. Kuppusamy, S.; Thavamani, P.; Megharaj, M.; Naidu, R. Bioremediation potential of natural polyphenol rich green wastes: A review of current research and recommendations for future directions. *Environ. Technol. Innov.* **2015**, *4*, 17–28. [CrossRef]
12. Castro-Muñoz, R.; Orozco-Álvarez, C.; Yáñez-Fernández, J. Recovery of bioactive compounds from food processing wastewaters by Ultra and Nanofiltration: A review. *Adv. Bio Res.* **2015**, *6*, 152–158.

13. Mohamed Khalith, S.B.; Ramalingam, R.; Karuppannan, S.K.; Dowlath, M.J.H.; Kumar, R.; Vijayalakshmi, S.; Uma Maheshwari, R.; Arunachalam, K.D. Synthesis and characterization of polyphenols functionalized graphitic hematite nanocomposite adsorbent from an agro waste and its application for removal of Cs from aqueous solution. *Chemosphere* **2022**, *286*, 131493. [CrossRef] [PubMed]
14. Di Donato, P.; Taurisano, V.; Tommonaro, G.; Pasquale, V.; Jiménez, J.M.S.; de Pascual-Teresa, S.; Poli, A.; Nicolaus, B. Biological Properties of Polyphenols Extracts from Agro Industry's Wastes. *Waste Biomass Valorization* **2018**, *9*, 1567–1578. [CrossRef]
15. Castaldo, L.; Izzo, L.; De Pascale, S.; Narváez, A.; Rodriguez-Carrasco, Y.; Ritieni, A. Chemical Composition, In Vitro Bioaccessibility and Antioxidant Activity of Polyphenolic Compounds from Nutraceutical Fennel Waste Extract. *Molecules* **2021**, *26*, 1968. [CrossRef]
16. Prakash, A.; Vadivel, V.; Banu, S.F.; Nithyanand, P.; Lalitha, C.; Brindha, P. Evaluation of antioxidant and antimicrobial properties of solvent extracts of agro-food by-products (cashew nut shell, coconut shell and groundnut hull). *Agric. Nat. Resour.* **2018**, *52*, 451–459. [CrossRef]
17. Pokhrel, S.; Yadav, P.N.; Adhikari, R. Applications of Chitin and Chitosan in Industry and Medical Science: A Review. *Nepal J. Sci. Technol.* **2016**, *16*, 99–104. [CrossRef]
18. Abdou, E.S.; Nagy, K.S.A.; Elsabee, M.Z. Extraction and characterization of chitin and chitosan from local sources. *Bioresour. Technol.* **2008**, *99*, 1359–1367. [CrossRef]
19. Abdelmalek, B.E.; Sila, A.; Haddar, A.; Bougatef, A.; Ayadi, M.A. β-Chitin and chitosan from squid gladius: Biological activities of chitosan and its application as clarifying agent for apple juice. *Int. J. Biol. Macromol.* **2017**, *104*, 953–962. [CrossRef]
20. Triunfo, M.; Tafi, E.; Guarnieri, A.; Salvia, R.; Scieuzo, C.; Hahn, T.; Zibek, S.; Gagliardini, A.; Panariello, L.; Coltelli, M.B.; et al. Characterization of chitin and chitosan derived from Hermetia illucens, a further step in a circular economy process. *Sci. Rep.* **2022**, *12*, 6213. [CrossRef]
21. Hahn, T.; Tafi, E.; Paul, A.; Salvia, R.; Falabella, P.; Zibek, S. Current state of chitin purification and chitosan production from insects. *J. Chem. Technol. Biotechnol.* **2020**, *95*, 2775–2795. [CrossRef]
22. Percot, A.; Viton, C.; Domard, A. Optimization of Chitin Extraction from Shrimp Shells. *Biomacromolecules* **2003**, *4*, 12–18. [CrossRef] [PubMed]
23. Fazli Wan Nawawi, W.M.; Lee, K.-Y.; Kontturi, E.; Murphy, R.J.; Bismarck, A. Chitin Nanopaper from Mushroom Extract: Natural Composite of Nanofibers and Glucan from a Single Biobased Source. *ACS Sustain. Chem. Eng.* **2019**, *7*, 6492–6496. [CrossRef]
24. Hassainia, A.; Satha, H.; Boufi, S. Chitin from Agaricus bisporus: Extraction and characterization. *Int. J. Biol. Macromol.* **2018**, *117*, 1334–1342. [CrossRef] [PubMed]
25. Muzzarelli, C.; Morganti, P. Preparation of Chitin and Derivatives Thereof for Cosmetic and Therapeutic Use. US Patent WO2006048829A2, 21 September 2006.
26. Alexandru, L.; Binello, A.; Mantegna, S.; Boffa, L.; Chemat, F.; Cravotto, G. Efficient green extraction of polyphenols from post-harvested agro-industry vegetal sources in Piedmont. *C. R. Chim.* **2014**, *17*, 212–217. [CrossRef]
27. Mellinas, A.C.; Jiménez, A.; Garrigós, M.C. Optimization of microwave-assisted extraction of cocoa bean shell waste and evaluation of its antioxidant, physicochemical and functional properties. *LWT* **2020**, *127*, 109361. [CrossRef]
28. Zhao, D.; Huang, W.-C.; Guo, N.; Zhang, S.; Xue, C.; Mao, X. Two-Step Separation of Chitin from Shrimp Shells Using Citric Acid and Deep Eutectic Solvents with the Assistance of Microwave. *Polymers* **2019**, *11*, 409. [CrossRef]
29. Cassano, A.; Conidi, C.; Ruby-Figueroa, R.; Castro-Muñoz, R. Nanofiltration and Tight Ultrafiltration Membranes for the Recovery of Polyphenols from Agro-Food By-Products. *Int. J. Mol. Sci.* **2018**, *19*, 351. [CrossRef]
30. Cravotto, G.; Mariatti, F.; Gunjevic, V.; Secondo, M.; Villa, M.; Parolin, J.; Cavaglià, G. Pilot Scale Cavitational Reactors and Other Enabling Technologies to Design the Industrial Recovery of Polyphenols from Agro-Food By-Products, a Technical and Economical Overview. *Foods* **2018**, *7*, 130. [CrossRef]
31. Hu, Z.; Berry, R.M.; Pelton, R.; Cranston, E.D. One-Pot Water-Based Hydrophobic Surface Modification of Cellulose Nanocrystals Using Plant Polyphenols. *ACS Sustain. Chem. Eng.* **2017**, *5*, 5018–5026. [CrossRef]
32. Hai, L.; Choi, E.S.; Zhai, L.; Panicker, P.S.; Kim, J. Green nanocomposite made with chitin and bamboo nanofibers and its mechanical, thermal and biodegradable properties for food packaging. *Int. J. Biol. Macromol.* **2020**, *144*, 491–499. [CrossRef] [PubMed]
33. Danti, S.; Trombi, L.; Fusco, A.; Azimi, B.; Lazzeri, A.; Morganti, P.; Coltelli, M.-B.M.-B.; Donnarumma, G. Chitin Nanofibrils and Nanolignin as Functional Agents in Skin Regeneration. *Int. J. Mol. Sci.* **2019**, *20*, 2669. [CrossRef] [PubMed]
34. De Andrade Arruda Fernandes, I.; Maciel, G.M.; Ribeiro, V.R.; Rossetto, R.; Pedro, A.C.; Haminiuk, C.W.I. The role of bacterial cellulose loaded with plant phenolics in prevention of UV-induced skin damage. *Carbohydr. Polym. Technol. Appl.* **2021**, *2*, 100122. [CrossRef]
35. Rahman Liman, M.L.; Islam, M.T.; Repon, M.R.; Hossain, M.M.; Sarker, P. Comparative dyeing behavior and UV protective characteristics of cotton fabric treated with polyphenols enriched banana and watermelon biowaste. *Sustain. Chem. Pharm.* **2021**, *21*, 100417. [CrossRef]
36. Sajadimajd, S.; Bahramsoltani, R.; Iranpanah, A.; Kumar Patra, J.; Das, G.; Gouda, S.; Rahimi, R.; Rezaeiamiri, E.; Cao, H.; Giampieri, F.; et al. Advances on Natural Polyphenols as Anticancer Agents for Skin Cancer. *Pharmacol. Res.* **2020**, *151*, 104584. [CrossRef] [PubMed]

37. Casadidio, C.; Peregrina, D.V.; Gigliobianco, M.R.; Deng, S.; Censi, R.; Di Martino, P. Chitin and Chitosans: Characteristics, Eco-Friendly Processes, and Applications in Cosmetic Science. *Mar. Drugs* **2019**, *17*, 369. [CrossRef]
38. Azimi, B.; Ricci, C.; Fusco, A.; Zavagna, L.; Linari, S.; Donnarumma, G.; Hadrich, A.; Cinelli, P.; Coltelli, M.-B.; Danti, S. Electrosprayed Shrimp and Mushroom Nanochitins on Cellulose Tissue for Skin Contact Application. *Molecules* **2021**, *26*, 4374. [CrossRef]
39. Khan, M.K.; Paniwnyk, L.; Hassan, S. Polyphenols as Natural Antioxidants: Sources, Extraction and Applications in Food, Cosmetics and Drugs. In *Plant Based "Green Chemistry 2.0": Moving from Evolutionary to Revolutionary*; Li, Y., Chemat, F., Eds.; Springer Singapore: Singapore, 2019; pp. 197–235, ISBN 978-981-13-3810-6.
40. JiangLian, D.; ShaoYing, Z. Application of chitosan based coating in fruit and vegetable preservation: A review. *J. Food Process. Technol.* **2013**, *4*, 227. [CrossRef]
41. Hu, X.; Sun, Z.; Zhu, X.; Sun, Z. Montmorillonite-Synergized Water-Based Intumescent Flame Retardant Coating for Plywood. *Coatings* **2020**, *10*, 109. [CrossRef]
42. Han, Y.; Chen, S.; Yang, M.; Zou, H.; Zhang, Y. Inorganic matter modified water-based copolymer prepared by chitosan-starch-CMC-Na-PVAL as an environment-friendly coating material. *Carbohydr. Polym.* **2020**, *234*, 115925. [CrossRef]
43. LeCorre, D.; Dufresne, A.; Rueff, M.; Khelifi, B.; Bras, J. All starch nanocomposite coating for barrier material. *J. Appl. Polym. Sci.* **2014**, *131*, 39826. [CrossRef]
44. Mates, J.E.; Schutzius, T.M.; Bayer, I.S.; Qin, J.; Waldroup, D.E.; Megaridis, C.M. Water-Based Superhydrophobic Coatings for Nonwoven and Cellulosic Substrates. *Ind. Eng. Chem. Res.* **2014**, *53*, 222–227. [CrossRef]
45. Rentzhog, M.; Fogden, A. Print quality and resistance for water-based flexography on polymer-coated boards: Dependence on ink formulation and substrate pretreatment. *Prog. Org. Coatings* **2006**, *57*, 183–194. [CrossRef]
46. Murthy, S.; Matschuk, M.; Huang, Q.; Mandsberg, N.K.; Feidenhans'l, N.A.; Johansen, P.; Christensen, L.; Pranov, H.; Kofod, G.; Pedersen, H.C.; et al. Fabrication of Nanostructures by Roll-to-Roll Extrusion Coating. *Adv. Eng. Mater.* **2016**, *18*, 484–489. [CrossRef]
47. Gregory, B.H. *Extrusion Coating: A Process Manual*; Trafford Publishing: Bloomington, IN, USA, 2005; ISBN 9781412040723.
48. Arulkumar, S.; Parthiban, S.; Goswami, A.; Varma, R.S.; Naushad, M.; Gawande, M.B. Low temperature processed titanium oxide thin-film using scalable wire-bar coating. *Mater. Res. Express* **2019**, *6*, 126427. [CrossRef]
49. Kanwal, M.; Wang, X.; Shahzad, H.; Chen, Y.; Chai, H. Blade coating analysis of viscous nanofluid having Cu–water nanoparticles using flexible blade coater. *J. Plast. Film Sheeting* **2020**, *36*, 348–367. [CrossRef]
50. Panariello, L.; Vannozzi, A.; Morganti, P.; Coltelli, M.-B.; Lazzeri, A. Biobased and Eco-Compatible Beauty Films Coated with Chitin Nanofibrils, Nanolignin and Vitamin E. *Cosmetics* **2021**, *8*, 27. [CrossRef]
51. Kimpimäki, T.; Savolainen, A.V. Barrier dispersion coating of paper and board. In *Surface Application of Paper Chemicals*; Brander, J., Thorn, I., Eds.; Springer: Dordrecht, The Netherlands, 1997; pp. 208–228, ISBN 978-94-009-1457-5.
52. Sun, Q.; Schork, F.J.; Deng, Y. Water-based polymer/clay nanocomposite suspension for improving water and moisture barrier in coating. *Compos. Sci. Technol.* **2007**, *67*, 1823–1829. [CrossRef]
53. Tyagi, P.; Lucia, L.A.; Hubbe, M.A.; Pal, L. Nanocellulose-based multilayer barrier coatings for gas, oil, and grease resistance. *Carbohydr. Polym.* **2019**, *206*, 281–288. [CrossRef]
54. Panariello, L.; Coltelli, M.-B.M.B.; Buchignani, M.; Lazzeri, A. Chitosan and nano-structured chitin for biobased anti-microbial treatments onto cellulose based materials. *Eur. Polym. J.* **2019**, *113*, 328–339. [CrossRef]
55. Battisti, R.; Fronza, N.; Vargas Júnior, Á.; da Silveira, S.M.; Damas, M.S.P.; Quadri, M.G.N. Gelatin-coated paper with antimicrobial and antioxidant effect for beef packaging. *Food Packag. Shelf Life* **2017**, *11*, 115–124. [CrossRef]
56. Kim, C.-K.; Lim, W.-S.; Lee, Y.K. Studies on the fold-ability of coated paperboard (I): Influence of latex on fold-ability during creasing/folding coated paperboard. *J. Ind. Eng. Chem.* **2010**, *16*, 842–847. [CrossRef]
57. Nagasawa, S.; Fukuzawa, Y.; Yamaguchi, T.; Tsukatani, S.; Katayama, I. Effect of crease depth and crease deviation on folding deformation characteristics of coated paperboard. *J. Mater. Process. Technol.* **2003**, *140*, 157–162. [CrossRef]
58. Sun, J.; Bi, H.; Jia, H.; Su, S.; Dong, H.; Xie, X.; Sun, L. A low cost paper tissue-based PDMS/SiO2 composite for both high efficient oil absorption and water-in-oil emulsion separation. *J. Clean. Prod.* **2020**, *244*, 118814. [CrossRef]
59. Nhuapeng, W.; Thamjaree, W. Fabrication and Mechanical Properties of Hybrid Composites between Pineapple fiber/Styrofoam Particle/Paper Tissue. *Mater. Today Proc.* **2019**, *17*, 1444–1450. [CrossRef]
60. Ferreira, A.C.S.; Aguado, R.; Bértolo, R.; Carta, A.M.M.S.; Murtinho, D.; Valente, A.J.M. Enhanced water absorption of tissue paper by cross-linking cellulose with poly(vinyl alcohol). *Chem. Pap.* **2022**. [CrossRef]
61. Spina, R.; Cavalcante, B. Characterizing materials and processes used on paper tissue converting lines. *Mater. Today Commun.* **2018**, *17*, 427–437. [CrossRef]
62. De Assis, T.; Reisinger, L.W.; Pal, L.; Pawlak, J.; Jameel, H.; Gonzalez, R.W. Understanding the Effect of Machine Technology and Cellulosic Fibers on Tissue Properties—A Review. *BioResources* **2018**, *13*, 4593–4629. [CrossRef]
63. Preston, J.; Elton, N.J.; Legrix, A.; Nutbeem, C.; Husband, J.C. The role of pore density in the setting of offset printing ink on coated paper. *Tappi J.* **2002**, *1*, 3–5.
64. Schoelkopf, J.; Matthews, G.P. Influence of inertia on liquid absorption into paper coating structures. *Nord. Pulp Pap. Res. J.* **2000**, *15*, 422–430. [CrossRef]

65. Lee, H.K.; Joyce, M.K.; Fleming, P.D. Influence of Pigment Particle Size and Pigment Ratio on Printability of Glossy Ink Jet Paper Coatings. *J. Imaging Sci. Technol.* **2005**, *49*, 54–60.
66. Li, Y.; He, B. Characterization of Ink Pigment Penetration and Distribution Related to Surface Topography of Paper Using Confocal Laser Scanning Microscopy. *BioResources* **2011**, *6*, 2690–2702.
67. Ginebreda, A.; Guillén, D.; Barceló, D.; Darbra, R.M. Additives in the Paper Industry. In *Global Risk-Based Management of Chemical Additives I: Production, Usage and Environmental Occurrence*; Bilitewski, B., Darbra, R.M., Barceló, D., Eds.; Springer: Berlin/Heidelberg, Germany, 2012; pp. 11–34, ISBN 978-3-642-24876-4.
68. Yong, H.; Liu, J. Active packaging films and edible coatings based on polyphenol-rich propolis extract: A review. *Compr. Rev. Food Sci. Food Saf.* **2021**, *20*, 2106–2145. [CrossRef] [PubMed]
69. Kandirmaz, E.A. Fabrication of rosemary essential oil microcapsules and using in active packaging. *Nord. Pulp Pap. Res. J.* **2021**, *36*, 323–330. [CrossRef]
70. Nechita, P.; Roman, M. Review on Polysaccharides Used in Coatings for Food Packaging Papers. *Coatings* **2020**, *10*, 566. [CrossRef]
71. Coltelli, M.-B.; Wild, F.; Bugnicourt, E.; Cinelli, P.; Lindner, M.; Schmid, M.; Weckel, V.; Müller, K.; Rodriguez, P.; Staebler, A.; et al. State of the Art in the Development and Properties of Protein-Based Films and Coatings and Their Applicability to Cellulose Based Products: An Extensive Review. *Coatings* **2016**, *6*, 1. [CrossRef]
72. Quiles-Carrillo, L.; Mellinas, C.; Garrigos, M.C.; Balart, R.; Torres-Giner, S. Optimization of Microwave-Assisted Extraction of Phenolic Compounds with Antioxidant Activity from Carob Pods. *Food Anal. Methods* **2019**, *12*, 2480–2490. [CrossRef]
73. Sirivibulkovit, K.; Nouanthavong, S.; Sameenoi, Y. Paper-based DPPH Assay for Antioxidant Activity Analysis. *Anal. Sci.* **2018**, *34*, 795–800. [CrossRef]
74. Lange, J.; Mokdad, H.; Wysery, Y. Understanding Puncture Resistance and Perforation Behavior of Packaging Laminates. *J. Plast. Film Sheeting* **2002**, *18*, 231–244. [CrossRef]
75. Irudayaraj, J.; Xu, F.; Tewari, J. Rapid determination of invert cane sugar adulteration in honey using FTIR spectroscopy and multivariate analysis. *J. Food Sci.* **2003**, *68*, 2040–2045. [CrossRef]
76. Barakat, H.H.; Hussein, S.A.M.; Marzouk, M.S.; Merfort, I.; Linscheid, M.; Nawwar, M.A.M. Polyphenolic metabolites of Epilobium hirsutum. *Phytochemistry* **1997**, *46*, 935–941. [CrossRef]
77. Kasaai, M.R. A review of several reported procedures to determine the degree of N-acetylation for chitin and chitosan using infrared spectroscopy. *Carbohydr. Polym.* **2008**, *71*, 497–508. [CrossRef]
78. Kalogeropoulos, N.; Chiou, A.; Pyriochou, V.; Peristeraki, A.; Karathanos, V.T. Bioactive phytochemicals in industrial tomatoes and their processing byproducts. *LWT—Food Sci. Technol.* **2012**, *49*, 213–216. [CrossRef]
79. Li, T.-T.; Wang, R.; Lou, C.W.; Lin, J.-H. Evaluation of high-modulus, puncture-resistance composite nonwoven fabrics by response surface methodology. *J. Ind. Text.* **2013**, *43*, 247–263. [CrossRef]
80. Hassim, N.; Ahmad, M.R.; Ahmad, W.Y.W.; Samsuri, A.; Yahya, M.H.M. Puncture resistance of natural rubber latex unidirectional coated fabrics. *J. Ind. Text.* **2012**, *42*, 118–131. [CrossRef]
81. Wang, Q.-S.; Sun, R.-J.; Tian, X.; Yao, M.; Feng, Y. Quasi-static puncture resistance behaviors of high-strength polyester fabric for soft body armor. *Results Phys.* **2016**, *6*, 554–560. [CrossRef]
82. Kumar, P.; Vasita, R. Understanding the relation between structural and mechanical properties of electrospun fiber mesh through uniaxial tensile testing. *J. Appl. Polym. Sci.* **2017**, *134*, 45012. [CrossRef]
83. Yamauchi, T.; Tanaka, A. Tearing test for paper using a tensile tester. *J. Wood Sci.* **2002**, *48*, 532–535. [CrossRef]
84. Zhong, T.; Wolcott, M.P.; Liu, H.; Wang, J. Developing chitin nanocrystals for flexible packaging coatings. *Carbohydr. Polym.* **2019**, *226*, 115276. [CrossRef]
85. Haslach, H.W. The Moisture and Rate-Dependent Mechanical Properties of Paper: A Review. *Mech. Time-Dependent Mater.* **2000**, *4*, 169–210. [CrossRef]

Article

Analysis, Development, and Scaling-Up of Poly(lactic acid) (PLA) Biocomposites with Hazelnuts Shell Powder (HSP)

Laura Aliotta [1,2,*], Alessandro Vannozzi [1,2], Daniele Bonacchi [3], Maria-Beatrice Coltelli [1,2,*] and Andrea Lazzeri [1,2]

[1] Department of Civil and Industrial Engineering, Pisa University, 56122 Pisa, Italy; alessandrovannozzi91@hotmail.it (A.V.); andrea.lazzeri@unipi.it (A.L.)
[2] National Interuniversity Consortium of Material Science and Technology (INSTM), 50121 Florence, Italy
[3] Arianna Fibers s.r.l, 51100 Pistoia, Italy; d.bonacchi@ariannafibers.com
* Correspondence: laura.aliotta@dici.unipi.it (L.A.); maria.beatrice.coltelli@unipi.it (M.-B.C.)

Abstract: In this work, two different typologies of hazelnuts shell powders (HSPs) having different granulometric distributions were melt-compounded into poly(lactic acid) (PLA) matrix. Different HSPs concentration (from 20 up to 40 wt.%) were investigated with the aim to obtain final biocomposites with a high filler quantity, acceptable mechanical properties, and good melt fluidity in order to be processable. For the best composition, the scale-up in a semi-industrial extruder was then explored. Good results were achieved for the scaled-up composites; in fact, thanks to the extruder venting system, the residual moisture is efficiently removed, guaranteeing to the final composites improved mechanical and melt fluidity properties, when compared to the lab-scaled composites. Analytical models were also adopted to predict the trend of mechanical properties (in particular, tensile strength), also considering the effect of HSPs sizes and the role of the interfacial adhesion between the fillers and the matrix.

Keywords: biocomposites; natural fibers; poly(lactic acid) (PLA); extrusion compounding

1. Introduction

Due to their complex end-of life management, petroleum-based plastics have caused a serious environmental problem, mainly related to their disposal. It was observed that from 1950 to 2015, less than 10% of the total plastic produced amount was recycled [1]. A possible solution to the waste management problem caused by non-degradable plastics can be obtained by replacing these materials with biodegradable polymers obtained from renewable resources compounded with agro-food waste. In this context, biobased and biodegradable polymers are an interesting solution to preserve petroleum resources and to decrease CO_2 emissions [2].

Agro-industry generates large biomass amounts that are not sufficiently and adequately exploited. For example, in the European Union alone, about 700 million tons of agriculture waste is annually produced [3]. The use of plant waste materials as raw materials in the production of biocomposites materials represents an exceptional opportunity for sustainable technological development. In fact, fruit shells and other agricultural waste are potentially important sources for the production of sustainable and competitive biocomposites. These plant by-products are produced in high quantities and crop wastes are rich in different nutritional components that can be valorized. Recently, the utilization of by-products has been increased by food and pharmaceutical manufacturers to produce valuable compounds from such inexpensive resources. In particular, nuts are one of the most important agricultural products due to their different uses within the food industry [4].

Walnut and hazelnut shells have great potential due to their large scale production; considering that about 67% of the total product weight consists of the shell, 646,818 tons of walnut shells, and 353,807 tons of hazelnut shells are produced each year [5]. After the separation of the kernel from the external parts of the fruit, large quantities of peel and shell are generated. These materials are the main part (over 60%) of the nut fruit and are discarded or burned as fuel without any useful application. Unfortunately, this waste material is typically burned directly in situ for heating purposes, while it could potentially be used for the production of both high added-value chemicals and biocomposites. Hazelnut shells are cost-effective byproducts [6] and their exploitation represents a stimulating challenge [7]. To better exploit their potentialities, it is necessary to find other better uses for hazelnut shells [8,9]. Hazelnut shells' composition is very similar to that of other wood-based biomass because cellulose, hemicellulose, and lignin are the main components. Shell grinding allows to produce hazelnut shell powder (HSP) of different sizes and morphologies. HSPs consist of lignin (40–50% by weight), cellulose (25–28%), and hemicellulose (22–30%), but they also contain a fraction of polyphenols (flavonoids and tannins), which can be recovered by hydroalcoholic extraction [10–12]. The shell extracts can be used as natural antioxidants in polymeric matrices as they can act as thermal and photo oxidative stabilizers for different types of polymers, including biopolymers like poly(lactic) acid (PLA) [13,14]. Moreover, the HSP addition enables light biocomposites to be obtained that, in some cases, possess improved mechanical and thermal properties and have enhanced biodegradability, when compared to the pure matrices [15–19]. Furthermore, the incorporation of HSPs into a biopolymeric matrix contributes to reducing the overall biocomposite cost [20]. However, some drawbacks must be mentioned in using agricultural waste for the production of lignocellulosic composites: unstable fiber availability over the year, absence of industrialized processing, and the need for big storage facilities and different necessary pre-treatments [21–24]. For this purpose, in order to take a step forward, the extrusion and injection molding processes considering the biocomposites scaling-up ability were investigated in this paper.

The polymeric matrix chosen for this study was poly(lactic acid) (PLA). In fact, among the biopolymeric matrices commercially available in the market, poly(lactic acid) (PLA) is one of the most attractive and its use in the production of green composites is gaining great importance [25]. PLA can be considered the front runner of the bioplastic market with an annual consumption of about 140,000 tons [26]. What has pushed up the increasing PLA demand are its excellent starting mechanical properties ($\approx$3 GPa of Young's modulus, $\approx$60 MPa of tensile strength, $\approx$3% of elongation at break and an impact strength close to 2.5 kJ/m^2) that are comparable to those of polystyrene (PS) [27].

Song et al. investigated the addition of walnut shell powder into PLA; they noticed during the biocomposites processing that an increase in the melt fluidity was correlated to the fiber powder addition [28]. This melt fluidity increment can lead to problems during the processing, making impossible or very difficult, for example, the extrusion compounding, the injection molding, the casting extrusion, etc. The evaluation of the fiber/matrix adhesion plays an important role and must be considered. From the processing point of view, fiber-matrix adhesion improvement can be done by chemical fiber pre-treatments or in-situ reactive blending. The last option is very interesting for the scaling-up point of view and involves the use, during the extrusion compounding, of coupling agents that are able to modify the polarity and surface tension of the fibers, enhancing the fiber-matrix adhesion [29,30]. The main coupling agents added to improve the fiber-matrix adhesion are maleic anhydride (MA), silane, isocyanate, and peroxide [29,31,32]. Commercial chain extender represents another way to improve the fiber-matrix adhesion, thanks to their easy processability during the extrusion compounding; however, they are not bio based and not biodegradable and even if they are introduced in very few amounts, they compromise the totally full bio-based origin of the final biocomposites.

The addition of HSPs into a PLA matrix must be deeply investigated and little work has been done regarding the scaling-up of these biocomposites into semi-industrial extrusion compounding process. For this reason, in this work, firstly the effect of the addition of different amounts (from 20 up to 40 wt.%) of two HSPs with different values of granulometry was investigated. The effect on melt fluidity, and thermal and mechanical properties was investigated on a lab-scale. Analytical models were also adopted to evaluate the powder size effect and adhesion between HSPs and PLA matrix. Then, the best selected compositions were extruded into a semi-industrial twin screw extruder, evaluating scale-up feasibility, focusing on the change of melt fluidity and mechanical properties of the scaled-up composites.

2. Materials and Methods

2.1. Materials

The materials used in this work are:

- PLA3251D from Natureworks is a PLA designed for injection-molding applications. This polymer grade is very stable in the molten state and can be processed on conventional injection molding equipment [density: 1.24 g/cm^3; MFR (210 °C, 2.16 kg): 80 g/10 min].
- Two different KERN hazelnut shell powders (HSPs) with different granulometry were provided by Arianna Fibers. Empty hazelnut shells were grounded by an impact mill. HSP with coarser grain size are named H0210, while those with finer grain size are named HM200 [ρ = 0.954 to 1.08 g/cm^3 with HR 5 to 30%].

2.2. Hazelnut Shell Powders (HSPs) Characterization

In order to quantify the humidity present in the HSPs, about 0.5 g of HSP for each sample were put in a Petri dish (previously weighed) and they were weighed before and after the drying process in a ventilated oven at 60 °C for 16 h. For each fiber typology, at least 3 measurements were carried out.

To investigate the possible degradation of the fillers during the extrusion compounding and to evaluate differences in chemical compositions between H0210 and HM200 HSPs, thermogravimetric (TGA) and FT-IR analysis were carried out.

TGA was performed on a TA Q-500 instrument (TA Instruments, Waters LLC, New Castle, DE, USA). Few milligrams were heated at 10 °C/min from room temperature up to 700 °C at 10 °C/min in nitrogen atmosphere.

FT-IR analysis was carried out on a Nicolet T380 FT-IR (Thermo Scientific, Madison, WI, USA) spectrometer equipped with an ATR Smart iTX accessory. Infrared spectrum of HSP was recorded in the 550–4000 cm^{-1} range, collecting 256 scans at 4 cm^{-1} resolutions.

The powders morphology was investigated by scanning electron microscopy (SEM) analysis using a FEI Quanta 450 FEG (Thermo Fisher Scientific, Waltham, MA, USA). The samples were prior sputtered with platinum to enhance their conductivity and generate the images, thanks to the secondary electrons. For each fiber typology, different images were acquired in order to obtain the filler distributions. The HSPs distributions were obtained, according to literature [33,34], measuring the dimensions of at least 200 filler particles by using Image-J software.

2.3. Lab-Scale and Semi-Industrial Scale-Up Extrusion Compounding and Injection Molding

PLA based composites containing different HSP amounts (from 20 up to 40 wt.%) were extruded at laboratory scale with a Haake Minilab II (HAAKE, Vreden, Germany) twin-screw mini-compounder. Before the extrusion, all materials were dried in a Piovan DP 604–615 dryers (Piova S.p.A., Verona, Italy) at 60 °C for 16 h. The extrusion temperature was set at 190 °C with a mixing residence time inside the extrusion chamber of 40 s and a screw speed of 60 rpm. The strand coming out from the mini extruder was then cooled and pelletized to obtain granules. The composites name and their compositions are reported in Table 1.

Table 1. Blends name and compositions.

Blend Name	PLA wt.%	HSP wt.%
PLA	100	0
PLA_20_H0210	80	20
PLA_30_H0210	70	30
PLA_40_H0210	60	40
PLA_20_HM200	80	20
PLA_30_HM200	70	30
PLA_40_HM200	60	40
PLA *	100	0
PLA_30_H0210 *	70	30
PLA_30_HM200 *	70	30

* Blends extruded with a semi-industrial COMAC twin-screw extruder (up-scaled).

To the best composition of both HSP typologies, the extrusion compounding was scale-upped on a semi-industrial Comac EBC 25HT (L/D = 44) (Comac, Cerro Maggiore, Italy), twin screw extruder. Also, in this case, the materials were dried following the same procedure adopted for the mini-compounding. PLA pellets were introduced into the main feeder while HSPs were fed with a specific lateral feeder that allows, once that the weight concentration was set, a constant feeding rate during the extrusion. A schematization of the extrusion feeder configurations, as well as the temperature profile adopted in the 11 extruder zones, is reported in Figure 1.

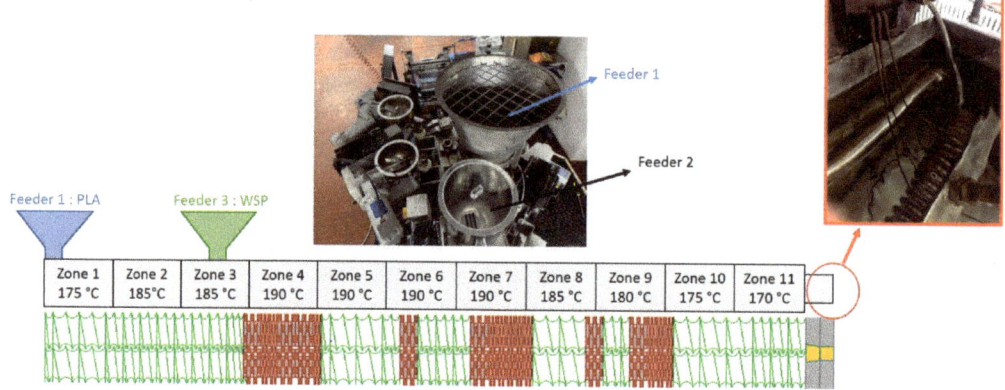

Figure 1. Schematization of the semi-industrial Comac twin screw extruder. In the figure are highlighted the feeder position, the screw configuration and the profile temperature along the 11 extruder zones.

The strands coming out from the extruder were cooled in a water bath and then pelletized by an automatic cutter. After the extrusion (both in lab-scale and in scale-up process), the pellets (dried in the before mentioned Piovan dryer at 60 °C for 16 h) were injection molded with a Megatech H10/18 injection molding machine (TECNICA DUEBI s.r.l., Fabriano, Italy) to obtain ISO 527-1A dog-bone specimens (width 10 mm, thickness 4 mm, useful length 80 mm) and ISO 179 Charpy impact specimens (width 10 mm, thickness 4 mm, length 80 mm). The injection molding was carried out in order to minimize any change in the processing parameters (reported in Table 2) for a better understanding of melt viscosity variation induced by the addition of different quantities and different HSP typology (H0210 and HM200). Consequently, the temperature profile, the mold temperature, the injection time, and the cooling time were fixed and only the injection pressure was modified when necessary.

Table 2. Injection-molding parameters.

Blend Name	Temperature Profile (°C)	Mold Temperature (°C)	Injection Time and Cooling Time (sec)	Injection Pressure (bar)
PLA	185–190–190	60	5	120
PLA_20_H0210				90
PLA_30_H0210				90
PLA_40_H0210				95
PLA_20_HM200				70
PLA_30_HM200				70
PLA_40_HM200				70
PLA *				120
PLA_30_H0210 *				95
PLA_30_HM200 *				95

* Blends extruded with a semi-industrial COMAC twin-screw extruder (up-scaled).

2.4. Melt Flow Rate (MFR)

In order to evaluate the melt fluidity variation caused by the addition of HSP, the melt flow rate (MFR) were measured on the biocomposites pellets by a CEAST Melt Flow Tester M20 (Instron, Canton, MA, USA) equipped with an encoder. The standard ISO1133D method was used: the sample was preheated without any weight for 30 s at 190 °C and then a weight of 2.16 kg was applied. The molten material quantity that flows for 30 s was then weighted and the MFR calculated. At least three measurements for each composition were carried out and the mean MFR value reported. Before the test, the materials were kept in a ventilated oven at 60 °C to avoid the pellets water uptake.

2.5. Mechanical and Thermal Characterization

Tensile tests were carried out on the ISO 527-1A extrusion molded specimen using an MTS Criterion model 43 (MTS Systems Corporation, Eden Prairie, MN, USA) universal testing machine. The MTS was equipped with a 10 kN load cell and the crosshead speed was set at 10 mm/min. Tensile tests were performed, at room temperature, after 3 days after the sample injection molding and during this time, the sample were stored in a dry keep at 25 °C and 50% of relative humidity. At least six specimens for each composition were tested.

Charpy impact tests were carried on the injection molded specimen pre-notched with a V-notch of 2 mm. A CEAST 9050 machine (INSTRON, Canton, MA, USA) was used and at least six specimens, at room temperature, were tested. The impact tests, also in this case, were carried out after 3 days of the injection molding keeping the samples in a controlled atmosphere.

The main biocomposites; thermal properties were calculated by differential scanning calorimetry (DSC) using a Q200-TA DSC (TA Instruments, New Castle, DE, USA) equipped with an RSC 90 cooling system. Nitrogen was used as purge gas set at 50 mL/min. Few milligrams (about 12 mg) were cut from the injection molded samples and the heating program was set in order to consider the thermal history of the samples and thus considering the injection molding history. In this way it was possible to calculate the crystallinity reached by the samples after the injection molding process. The thermal program was: heating at 10 °C/min from room temperature up to 200 °C, the final temperature was kept for 1 min. The melting and crystallization temperatures were calculated in correspondence of the maximum and minimum of the melting peak and cold crystallization peak, respectively. As far as the melting and cold crystallization enthalpies were concerned, they were calculated integrating the peak areas of the melting and crystallization peaks, respectively. The

PLA crystallinity percentage of PLA was calculated according to the following equation (Equation (1)) [27]:

$$X_{cc} = \frac{\Delta H_m - \Delta H_{cc}}{\Delta H°_m \cdot wt.\%_{PLA}} \quad (1)$$

where, ΔH_m and ΔH_{cc} are the melting and cold crystallization PLA enthalpies of PLA, $\Delta H°_m$ is the theoretical melting heat of 100% crystalline PLA (taken equal to 93 J/g [35]).

2.6. Composite Morphology Investigation

The composites morphology was investigated on the fractured cross-sections of the Charpy samples prior the sputtering with platinum. A FEI Quanta 450 FEG scanning electron microscope (SEM) equipped with a Large Field Detector for low kV imaging simultaneous secondary electron (SE) was used.

3. Theoretical Analysis

During the lab scale investigation, different analytical models were applied on the HSP/PLA based composites to estimate the fiber/matrix adhesion and to predict the tensile strength trend as a function of the HSPs volumetric content. The addition of rigid particles into a polymeric matrix can affect the strength in two ways. The tensile strength prediction of particulate filled composites is not easy because it is affected by different parameters, such as interface adhesion, stress concentration, and defect size/spatial fillers distribution [36].

For particulate fillers and for fibers with low aspect ratio, the prediction of the tensile strength can be expressed quantitatively by the following equation, proposed by Pukánszky [37]:

$$\sigma_c = \sigma_m \left[\frac{1 - V_f}{1 + 2.5 V_f} \right] \exp(BV_f) \quad (2)$$

where, σ_c and σ_m are the stress at break of the composite and matrix, respectively, while V_f is the volume fiber fraction. The term in square bracket is correlated to a decrement of the tensile strength of the composite caused by the fillers addition that reduce the load-bearing cross-section of the composite. The parameter B is an interaction parameter that takes into account the efficiency of the stress transmission between the matrix and the filler and can be indirectly correlated to the filler/matrix adhesion [38]. Simplifying Equation (2), a linear correlation can be obtained (Equation (3)) in which the B parameter is found as the slope of the natural logarithm of reduced strength (σ_{red}) against the volume filler fraction.

$$ln\sigma_{red} = ln\frac{\sigma_c(1 - V_f)}{\sigma_m(1 + 2.5 V_f)} = BV_f \quad (3)$$

For particulate fillers, in the case that the stress cannot be transferred from the matrix to the filler and the final composite tensile strength is determined from the effective sectional area of the load-bearing matrix, the tensile strength of the composites lies between an upper and lower bound [36]. Based on the hypothesis that poor adhesion exists between the filler and the polymer and the load is sustained completely by the polymer matrix, the following equation (Equation (4)) formulated by Nicolais and Nicodemo [39] gives the lower-bound strength of the composite.

$$\sigma_c = \sigma_m \left(1 - 1.21 V_f^{\frac{2}{3}}\right) \quad (4)$$

The upper bound is immediately obtained as follows (Equation (5)):

$$\sigma_c = \sigma_m \left(1 - V_f\right) \quad (5)$$

Equation (5) generally has been considered as an ideal unattainable upper bound since, in addition to a matrix area reduction, critical effects are also induced by the filler particles in the system, with a further decrease of the composite strength.

4. Results

4.1. HSPs Characterization Results

The results of the HSP drying tests showed that H0210 had a humidity loss of about 9.05%, while for HM200 it was 7.64%. HSPs having lower particle size release less moisture after the drying.

From the TGA results reported in Figure 2, it can be observed that, for both HSP typologies, the thermograms are characterized by a first mass drop (completed below 100 °C) that is correlated to the humidity loss of the HSPs. The moisture loss is greater for the H0210 sample, indicating that the HSP having higher particle size dimension releases easily the up taken water. The second mass drop for H0210 corresponds to the thermal degradation of the material and presents a similar magnitude for the two HSP typologies. The final residue is also similar in magnitude for both samples. The residue includes both inorganic compounds and carbon, normally generated when thermal degradation occurs in a nitrogen atmosphere. The superimposed derivatives of the curve show the inflection point (where the mass loss occurs) as a maximum. The main maximum peak is about the same for the two samples; however, HM200 shows an additive peak at around 198 °C probably indicating a major quantity of water highly linked on the surface or substances with lower thermal resistance.

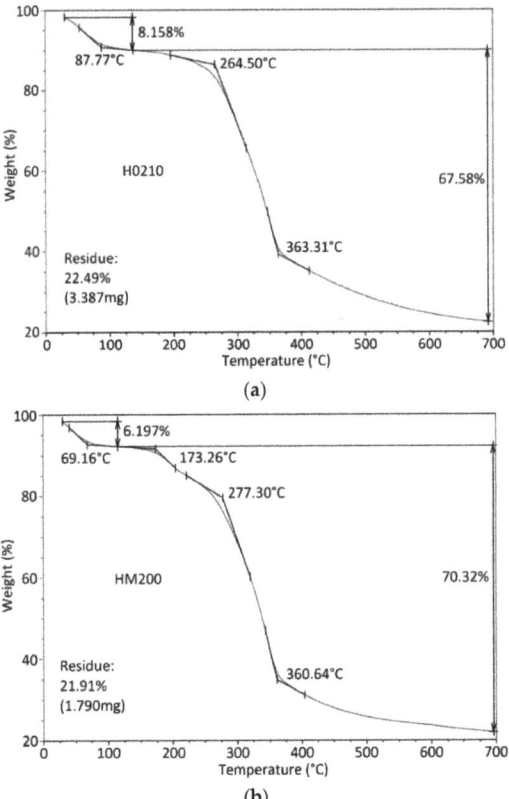

Figure 2. Cont.

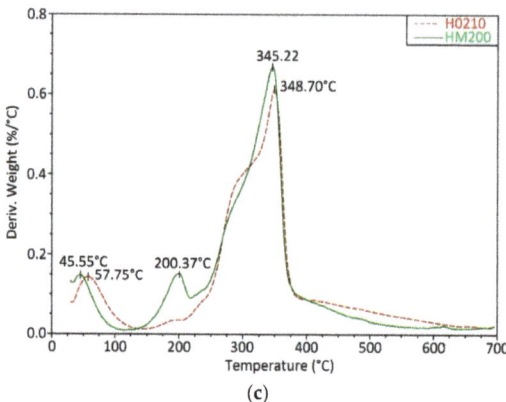

Figure 2. TGA curves of (**a**) H0210 and (**b**) HM200 hazelnut shell powder. Figure (**c**) illustrates the overlay of the TGA curves derivative.

Hazelnut shells are composed of cellulose, hemicellulose, and lignin. However, there is a significant amount of low molecular weight compounds. In literature [11], it was observed that hazelnut shell contains about 10.6% of low molecular weight extractable substances, about 30.1% of lignin and about 49.7% of polysaccharides (cellulose and hemicellulose). From the ATR spectra reported in Figure 3, for H0210 a wide band at around 3327 cm^{-1} was observed, attributed to the surface hydroxyl groups (-OH) mainly related to the presence of water as well as alcoholic, phenolic groups but also amino acids and carboxylic derivatives.

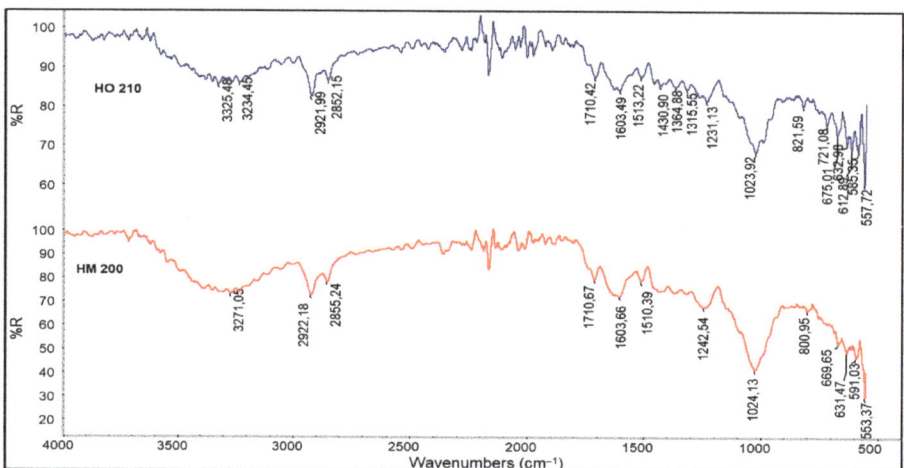

Figure 3. ATR spectra of H0210 and HM200 hazelnut shell powders.

The peak at 2920 cm^{-1} is assigned to the asymmetric stretching band C-H; also, that at 2850 cm^{-1} is related to the symmetrical stretching of the same bonds. These groups are also present in the structure of lignin [40]. The peak associated with the stretching of C=O (carbonyl compounds) is located at 1708 cm^{-1}, but a shoulder is noted at 1743 cm^{-1}. While the main peak is attributable to carboxylic acids, the second is attributable to the presence of ester groups. The presence of unsaturations and C=C bonds that occurred in the widened bands between 1606–1640 cm^{-1} is attributable to alkenes, aromatic groups, but also amide groups (C=O stretching); while the peaks at 1400 and 1240 cm^{-1} may be due to C−O, C−H or C−C elongation vibrations. The peak observed at 1024 cm^{-1} is due to C–O, present in the ethereal, alcoholic, and carboxylic groups. The band of the C–O group is more intense than that of the C=O group, and this shows that the polysaccharide component is certainly dominated in the sample. The peak at 588 cm^{-1} is due to the folding vibration in the aromatic compounds typical of lignin, highlighting their presence.

The spectrum of HM200 was acquired in a similar way to that of H0210, but the signals are more intense. This is attributable to the lower particle size of the powder, which allows better adhesion of the sample to the crystal. The observed bands are completely similar to those of the H0210 sample, suggesting that the only difference between H0210 and HM200 is in the particle size.

The SEM micrographs (Figure 4) show, especially for H0210, the presence of irregularly shaped particles having a rough surface attributed to the external part of the hazelnut shell, which has a different morphology, depending on the filler layers. For smaller-sized samples, the amount of rough surface particles is reduced.

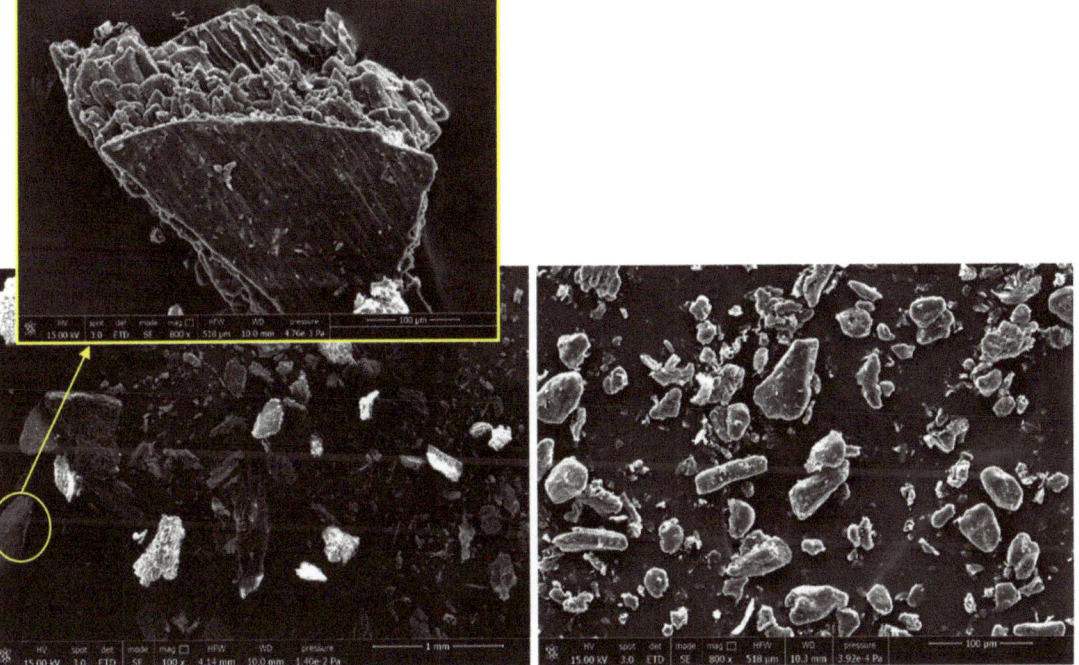

Figure 4. SEM images of H0210 (**left side**) and HM200 (**right side**).

Also cavities and reliefs are visible that correspond to the cell walls and lumens. In any case, for both HSPs, a greater variability in the filler shape and size can be observed. The morphological results are consistent to what can be found in literature and despite the different surface roughness, both HSPs can be considered as a typical particle-shaped fillers [41,42]. The "elliptical approach" was adopted to determine the diameter distribution; according to this model, the major axis of the ellipse corresponds to the length of the filler while the minor axis corresponds to the width. With this method, the length and aspect ratio are overestimated by about 10% for the fiber-shaped filler while this overestimation is practically negligible for the particulate filler [34]. Since from the SEM images the greater quantity of HSPs tends to be particle, with little presence of elongated fibers, it was preferred to adopt this elliptical model. In Figure 5 the diameter distribution curves are shown that confirm the great differences in diameters dimension between H0210 and

HM200. In particular, an average diameter of 206.7 μm and 25.8 μm was obtained for H0210 and HM200, respectively.

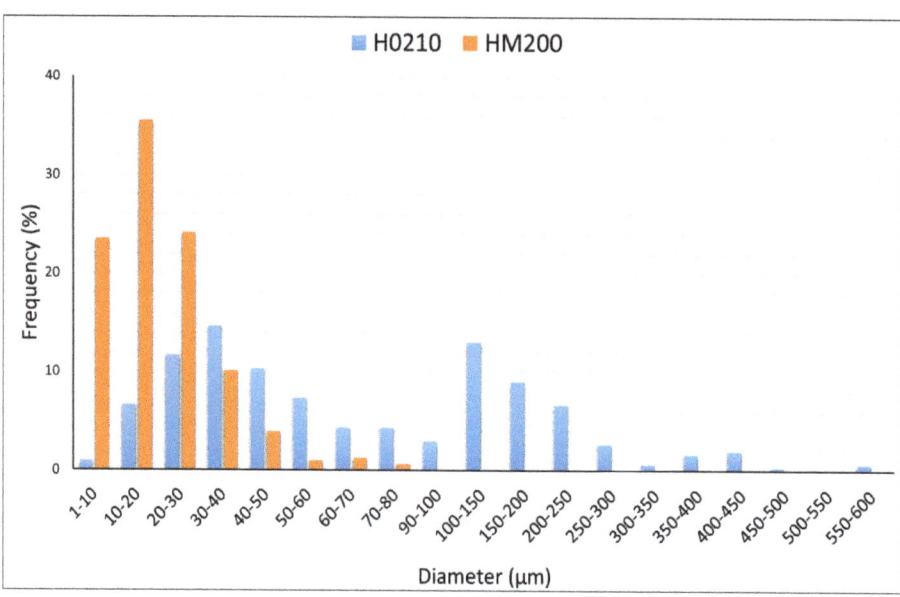

Figure 5. Diameter distributions for H0210 and HM200 HSP.

4.2. Lab-Scaled Composites Results

The results of mechanical tests and MFR are summarized in Table 3. From the point of view of tensile tests, the powders' addition makes the material more brittle with a decrement of both stress and elongation at break. The HSPs addition, on the other hand, significantly increases the elastic modulus compared to pure PLA. This is a common trend [43] and it is due to the introduction of fillers having higher elastic modulus than pure matrix. In general, a decrement of the mechanical properties increasing the HSPs amount can be observed; however, for H0210, the tensile decrement is less marked than HM200 and the impact resistance is not worsened with respect to pure PLA.

Table 3. Mechanical and MFR results of lab-scaled composites with different amounts of H0210 and HM200 HSP.

Blend Name	Elastic Modulus (GPa)	Stress at Break (MPa)	Elongation at Break (%)	Charpy Impact Resistance (C.I.S.) (kJ/m^2)	MFR (g/10 min)
PLA	3.56 ± 0.21	58.94 ± 1.16	2.30 ± 0.33	2.53 ± 0.29	3.80 ± 0.51
PLA_20_H0210	4.03 ± 0.15	40.85 ± 0.76	1.35 ± 0.15	2.60 ± 0.32	22.69 ± 1.76
PLA_30_H0210	4.16 ± 0.03	33.77 ± 3.16	0.95 ± 0.24	2.73 ± 1.22	13.78 ± 1.98
PLA_40_H0210	4.26 ± 0.12	27.38 ± 0.97	0.89 ± 0.10	2.94 ± 0.61	8.49 ± 1.82
PLA_20_HM200	3.88 ± 0.12	30.34 ± 1.85	1.11 ± 0.24	2.45 ± 0.20	32.91 ± 3.17
PLA_30_HM200	4.13 ± 0.30	26.85 ± 2.19	1.10 ± 0.11	1.74 ± 0.24	34.08 ± 3.26
PLA_40_HM200	4.44 ± 0.20	16.80 ± 4.60	0.45 ± 0.17	1.73 ± 0.26	33.41 ± 2.93

The better mechanical response achieved with H0210 can be attributed to several factors. First of all, H0210 possesses a greater diameter distribution that represents a more efficient obstacle towards the crack that advances during the Charpy test, compared to

HM200 with a finer diameter distribution [36]. Furthermore, residual moisture content must also be considered because it also affects the final mechanical response. H0210, under the same drying conditions, lost a greater amount of moisture that could potentially degrade the PLA (it must be considered that in this first lab-scale step, no venting for the humidity stripping is present in the mini-extruder). Finally, the filler/matrix adhesion must also be considered. It is well known from the literature that natural fibers have poor adhesion with PLA [44]. However, comparing HM200 and H0210, the fillers with higher grain sizes will have greater adhesion (the stress able to cause the fiber detachment is in fact a function of various parameters, including the aspect ratio) [45]. The different adhesion is also confirmed by the B parameter obtained as the slope of the Pukanszky's plot (Figure 6). A decrement of the B value (from 1.41 for H0210 to 0.56 for HM200) can be observed, indicating a worsening of the matrix/filler adhesion.

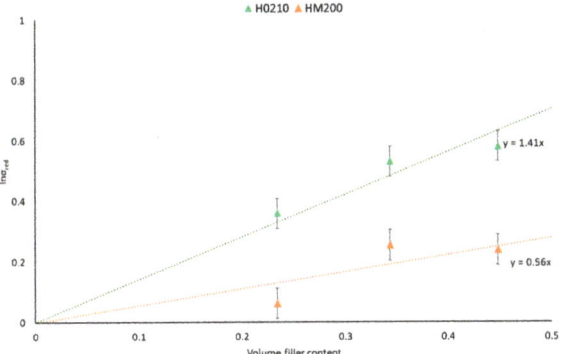

Figure 6. Pukanszky's plot for PLA-HSPs composites.

Observing in addition the experimental values of composites tensile strength (Figure 7), a different interaction between H0210 and HM200 with the PLA matrix can be observed. The experimental data in fact, lies between the upper and lower bound; however, HM200 are much closer to the Nicolais and Nicodemo lower-bound equation indicating a weaker adhesion respect to H0210 that are closer to the upper bound. The particles with smaller size have a great tendency to agglomerate, causing greater weakening of the matrix.

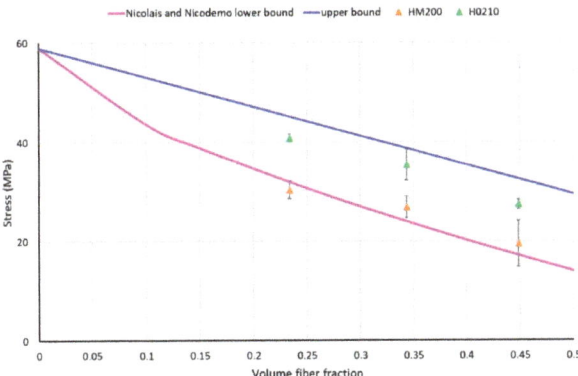

Figure 7. Comparison between the experimental composite strength and the values predicted according to the upper and lower bound equations.

The SEM images reported in Figure 8 confirm the prediction of the analytical models and of the mechanical results obtained. A better adhesion is registered for H0210, respect to HM200. In particular, for H0210, it can be observed that at 20 wt.% (Figure 8a), the fillers are fairly well distributed, and few agglomerations can be observed with 30 wt.% of HSP (Figure 8b). At 40 wt.% however, a greater agglomeration tendency, due to the greater HSPs amount introduced, is registered. The agglomerates are also less adherent to the PLA matrix and in Figure 8c, holes due to the detachment of these agglomerates are clearly visible; the presence of agglomerates is also responsible for the marked drop down of the mechanical properties recorded for the PLA_40_H0210 composite. HM200 show worse adhesion and already at 20 wt.%, voids can be observed due to HSPs' detachment (Figure 8d). However, for all the compositions (Figure 8d–f), a HSPs detachment can be recorded, which is very marked compared to H0210, confirming the results of mechanical tests and analytical models adopted.

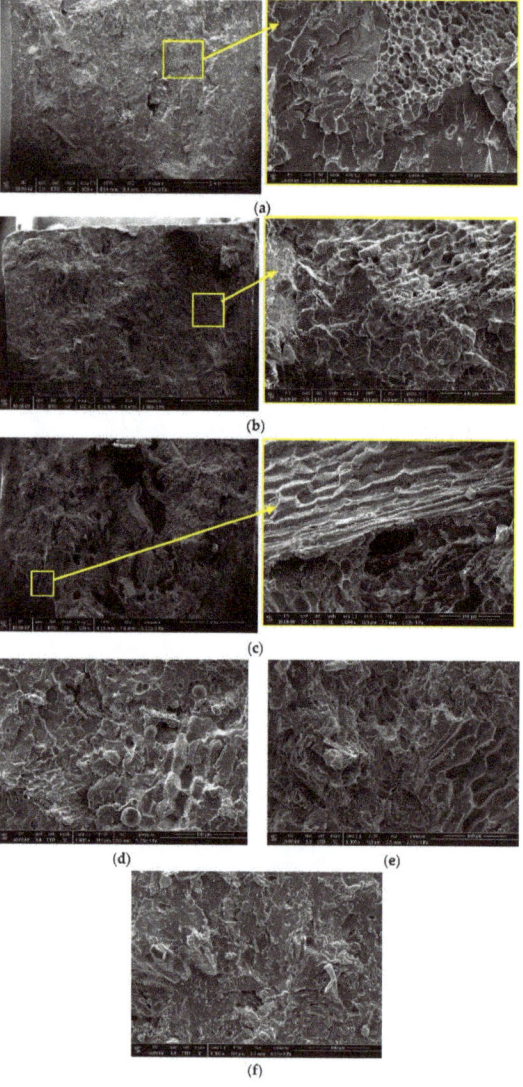

Figure 8. SEM micrographs of the fractured Charpy surface of: (**a**) PLA_20_H0210, (**b**) PLA_30_H0210, (**c**) PLA_40_H0210, (**d**) PLA_20_HM200, (**e**) PLA_30_HM200, (**f**) PLA_40_HM200.

Regarding the MFR values, it can be observed that the viscosity increased (so the fluidity decreased) on adding the HSPs (Figure 9). However, this occurred only for H0210. For HM200 the MFR values are higher if compared to those obtained with H0210, but no trend with the HSPs amount was detected. These results are in agreement with those reported by Song et al. [28] and with the injection pressure, reported in Table 2, where the injection pressure was increased with the H0210 content while it was decreased by increasing the HM200 content. These MFR trends can be attributable to the probable partial PLA hydrolysis caused by the filler moisture that is greater for HM200 due to its larger surface area and thus humidity content. However, as reported in literature, the general decrement of stress at break is mainly correlated to the poor interfacial adhesion of lignocellulosic fiber with the biopolymeric matrix [45,46].

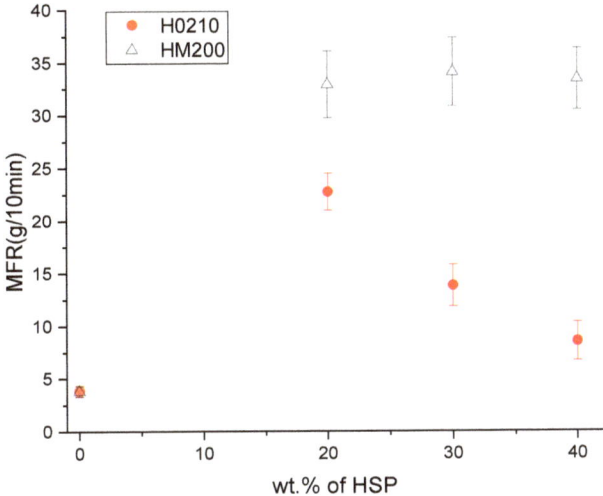

Figure 9. Trend of Melt Flow Rate (MFR) as a function of HSP content.

From a thermal point of view, a decrement of both melting temperature and glass transition temperature caused by the addition of HSPs can be observed (from Table 4); this decrement seems to be correlated to the HSP content. However, the HSP typologies also affect the melting temperature and glass transition temperature differently with a decrement that is more marked with HM200. This behavior can be correlated to the different granulometry between H0210 and HM200. HM200 have a higher surface area than H0210 and it adsorbs more moisture that can lead to decrease in the average molecular weight (resulting in a decrement of the glass transition and melting temperatures). The HSPs' addition increases the crystallinity of the PLA, causing a shift of the cold crystallization temperature towards lower temperatures. HSP seems to act as a nucleating agent, providing heterogeneous nucleation sites similar to other systems filled with natural fibers [27,47,48]. In particular, as the HM200 are finer and more homogeneous with a tighter diameter distribution curve, they are more effective in crystallizing the PLA, when compared to their H0210 counterparts.

4.3. Scaled-Up Composites Results

From the lab-scale data, 30 wt.% seems the most promising HSPs amount granting both a high fiber content and acceptable mechanical properties. The results of the scaled-up composites are summarized in Table 5.

Table 4. DSC first heating results for H0210 and HM200 PLA-based composites.

Blend Name	T_g (°C)	T_{cc} (°C)	T_m (°C)	ΔH_{cc} (J/g)	ΔH_m (J/g)	X_{cc} (%)
PLA	61.8	105.7	172.2	32.4	44.9	13.5
PLA_20_H0210	58.2	94.3	170.9	21.8	32.4	14.2
PLA_30_H0210	57.2	93.3	169.3	22.4	33.3	16.8
PLA_40_H0210	57.2	94.2	168.8	18.9	27.8	16.0
PLA_20_HM200	55.2	91.0	168.0	26.8	38.2	15.2
PLA_30_HM200	54.3	88.6	167.1	23.1	35.2	18.6
PLA_40_HM200	53.7	87.4	166.8	20.9	33.0	21.7

Table 5. Mechanical and MFR results of the scaled-up HSPs composites.

Blend Name	Elastic Modulus (GPa)	Stress at Break (MPa)	Elongation at Break (%)	Charpy Impact Resistance (C.I.S.) (kJ/m^2)	MFR (g/10 min)
PLA *	3.64 ± 0.19	64.60 ± 2.61	2.69 ± 0.14	2.51 ± 0.23	3.21 ± 0.55
PLA_30_H0210 *	4.30 ± 0.16	36.45 ± 1.00	1.09 ± 0.10	2.63 ± 0.35	6.23 ± 0.26
PLA_30_HM200 *	4.45 ± 0.11	38.42 ± 0.68	1.39 ± 0.18	2.29 ± 0.29	4.00 ± 0.59

* Blends extruded with a semi-industrial COMAC twin-screw extruder (up-scaled).

All scaled-up formulations show a lower MFR, respect to their corresponding lab-scale formulations. This MFR decrement is also reflected in the injection pressure increment during the injection molding process (Table 2). The marked viscosity decrement observed during the lab-scale step is limited, thanks to the coupling of the low extruder residence time and the presence of the venting system connected to a vacuum pump that guarantees the humidity stripping during the melt extrusion, avoiding or limiting any eventual PLA degradation [49,50]. The mechanical results are noteworthy. In fact, it can be observed that the scaled-up composites show an increment of elastic modulus and tensile strength. In particular, the reached tensile stress is very similar, confirming the efficiency of the venting system in removing the fillers humidity.

The thermal properties (Table 6) of the scaled-up composite remains almost unchanged, confirming the nucleation effect of HSPs.

Table 6. DSC first heating results of scaled-up HSPs composites.

Blend Name	T_g (°C)	T_{cc} (°C)	T_m (°C)	ΔH_{cc} (J/g)	ΔH_m (J/g)	X_{cc} (%)
PLA *	63.4	101.7	174.6	29.2	38.8	10.3
PLA_30_H0210 *	57.5	94.5	172.5	17.1	29.0	18.2
PLA_30_HM200 *	55.7	93.8	169.9	18.8	27.6	13.5

* Blends extruded with a semi-industrial COMAC twin-screw extruder (up-scaled).

5. Discussion

In attempt to better correlate the obtained results with the mechanical properties, it was noticed that in general the tensile strength of the prepared biocomposites decreased by increasing the melt fluidity, as shown in Figure 10, where the data related to composites containing 20, 30, and 40% of HSP are reported. Interestingly, for the 30% HSP biocomposites, the data obtained for the scaled-up samples follow a similar trend, but as yet observed, the tensile strength is higher and the MFR is lower. Moreover, finer HSP (HM200) results in the highest value of tensile strength and lowest value of MFR. Hence, by avoiding the chain scission of PLA thanks to the optimized processing conditions, the fluidity is greatly decreased.

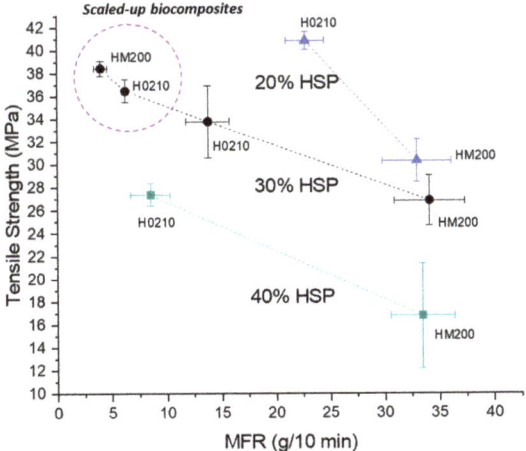

Figure 10. Tensile strength as a function of MFR for biocomposites containing 20, 30, and 40% of HSP.

The production of the biocomposites including PLA and HSP resulted in strong interactions or reactions (Figure 11, reaction 1) between the polymer matrix and the functional groups on the HSP surface. Hydroxyl groups, belonging to cellulose and hemicellulose, that represent the major component of HSP, were mainly considered.

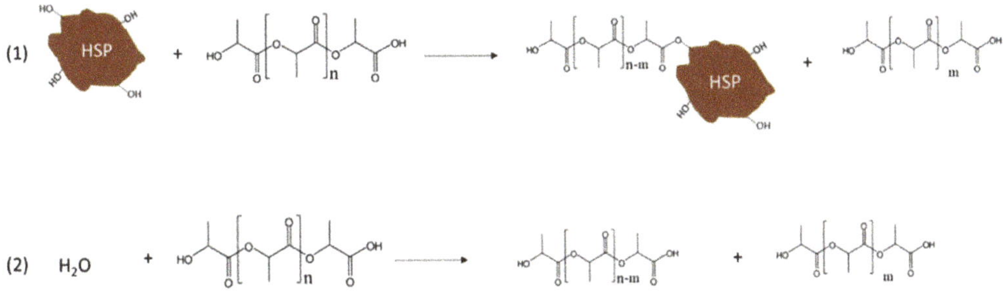

Figure 11. Main reactions occurring during the preparation of PLA/HSP biocomposites; reaction 1 involves an HSP solid particle.

Reaction 1's occurrence depends on the surficial area of HSP and can induce an increase in tensile strength, thanks to the improved matrix-filler adhesion. On the other hand, reaction 2 (Figure 11, reaction 2) is PLA hydrolysis due to humidity, occurring more in the composites containing the finer HSP. In a lab-scale extruder configuration, reaction 2 affects properties more than reaction 1 because of the higher residence time and absence of devolatilization. Thus, as demonstrated by the study of B parameter obtained as the slope of the Pukanszky's plot (Figure 6), the dispersion in the matrix of the HSP with the lowest dimension was less efficient. On the contrary, when the preparation is scaled-up, reaction 1 as well as the fibre-matrix interaction are more significant. In good agreement, the finer HSP, with the highest surface area, resulted in the highest tensile strength and highest melt viscosity.

6. Conclusions

In this study the possibility to process successfully, at the semi-industrial scale, PLA-based composites containing hazelnut shell powder (HSP) was investigated. A first lab-scale production was carried out in order to individuate the best HSPs amount for the subsequent scaling-up step. Two different HSPs typologies of different sizes were added from 20 up to 40 wt.%. The thermal, mechanical, and melt fluidity analysis showed poor stress transfer, which led to a decrement in tensile strength. The fillers seem to act as nucleating sites for PLA that increased its crystallinity; however, a marked decrement of the melt viscosity was recorded, especially for fillers small in size due to their major water uptake. The composition including 30 wt.% of HSP was selected for the successive scale-up in a semi-industrial extruder. Interesting results were obtained considering the scaled-up composites, as their melt fluidity was decreased thanks to the presence of the venting system in the extruder that efficiently removed the residual humidity. The scaled-up composites showed improved mechanical properties, respect to the lab-scaled composites, demonstrating that these composites are effectively processable and can be easily scaled-up The prepared biocomposites showed the possibility of achieving an optimized balance between improvement of mechanical properties and the valorization of a significantly high HSP content.

In future work, a further step towards more efficient exploitation of HSPs should concern their functionalization. The HSPs' superficial modification, coupled with the optimization of the extrusion process parameters, would allow to obtain biocomposites with further improved mechanical properties.

Author Contributions: Conceptualization, M.-B.C. and L.A.; experimental work, A.V. and L.A.; theoretical analysis, L.A. and A.L.; data curation and elaboration, A.V., L.A. and M.-B.C.; writing—original draft preparation, L.A. and A.V.; writing—review and editing M.-B.C.; supervision, D.B. and A.L. All authors have read and agreed to the published version of the manuscript.

Funding: Arianna Fiber s.r.l. received partial funding by Tuscany Region on POR FESR 2018–2020 thanks to the Finap Bio project and by Fondazione Cassa di Risparmio di Pistoia (CRPT), thanks to project GREEN SPEAR.

Institutional Review Board Statement: Not applicable.

Informed Consent Statement: Not applicable.

Data Availability Statement: The data presented in this study are available on request from the corresponding author.

Conflicts of Interest: The authors declare no conflict of interest.

References

1. Geyer, R.; Jambeck, J.R.; Law, K.L. Production, use, and fate of all plastics ever made. *Sci. Adv.* **2017**, *3*, e1700782. [CrossRef]
2. Folino, A.; Karageorgiou, A.; Calabrò, P.S.; Komilis, D. Biodegradation of wasted bioplastics in natural and industrial environments: A review. *Sustainability* **2020**, *12*, 6030. [CrossRef]
3. Pawelczyk, A. EU Policy and Legislation on recycling of organic wastes to agriculture. *ISAH* **2005**, *1*, 64–71.
4. Lourenço, S.C.; Moldão-Martins, M.; Alves, V.D. Antioxidants of natural plant origins: From sources to food industry applications. *Molecules* **2019**, *24*, 4132. [CrossRef] [PubMed]
5. Soleimani, M.; Kaghazchi, T. Agricultural waste conversion to activated carbon by chemical activation with phosphoric acid. *Chem. Eng. Technol. Ind. Chem. Equip.-Process Eng.* **2007**, *30*, 649–654. [CrossRef]
6. Balart, J.F.; Fombuena, V.; Fenollar, O.; Boronat, T.; Sánchez-Nacher, L. Processing and characterization of high environmental efficiency composites based on PLA and hazelnut shell flour (HSF) with biobased plasticizers derived from epoxidized linseed oil (ELO). *Compos. Part B Eng.* **2016**, *86*, 168–177. [CrossRef]
7. Müller, M.; Valášek, P.; Linda, M.; Petrásek, S. Exploitation of hazelnut (*Corylus avellana*) shell waste in the form of polymer-particle biocomposite. *Sci. Agric. Bohem.* **2018**, *49*, 53–59. [CrossRef]
8. Han, H.; Wang, S.; Rakita, M.; Wang, Y.; Han, Q.; Xu, Q. Effect of Ultrasound-Assisted Extraction of Phenolic Compounds on the Characteristics of Walnut Shells. *Food Nutr. Sci.* **2018**, *9*, 1034–1045. [CrossRef]
9. Pirayesh, H.; Khazaeian, A.; Tabarsa, T. The potential for using walnut (*Juglans regia* L.) shell as a raw material for wood-based particleboard manufacturing. *Compos. Part B Eng.* **2012**, *43*, 3276–3280. [CrossRef]
10. Demirbaş, A. Estimating of structural composition of wood and non-wood biomass samples. *Energy Sources* **2005**, *27*, 761–767. [CrossRef]
11. Queirós, C.S.G.P.; Cardoso, S.; Lourenço, A.; Ferreira, J.; Miranda, I.; Lourenço, M.J.V.; Pereira, H. Characterization of walnut, almond, and pine nut shells regarding chemical composition and extract composition. *Biomass Convers. Biorefinery* **2020**, *10*, 175–188. [CrossRef]

12. Herrera, R.; Hemming, J.; Smeds, A.; Gordobil, O.; Willför, S.; Labidi, J. Recovery of bioactive compounds from hazelnuts and walnuts shells: Quantitative–qualitative analysis and chromatographic purification. *Biomolecules* **2020**, *10*, 1363. [CrossRef] [PubMed]
13. Agustin-Salazar, S.; Gamez-Meza, N.; Medina-Juárez, L.A.; Malinconico, M.; Cerruti, P. Stabilization of polylactic acid and polyethylene with nutshell extract: Efficiency assessment and economic evaluation. *ACS Sustain. Chem. Eng.* **2017**, *5*, 4607–4618. [CrossRef]
14. Moccia, F.; Agustin-Salazar, S.; Verotta, L.; Caneva, E.; Giovando, S.; D'Errico, G.; Panzella, L.; d'Ischia, M.; Napolitano, A. Antioxidant properties of agri-food byproducts and specific boosting effects of hydrolytic treatments. *Antioxidants* **2020**, *9*, 438. [CrossRef] [PubMed]
15. Pradhan, P.; Nanda, B.P.; Satapathy, A. Polyester composites filled with walnut shell powder: Preparation and thermal characterization. *Polym. Compos.* **2020**, *41*, 3294–3308. [CrossRef]
16. Akıncıoğlu, G.; Akıncıoğlu, S.; Öktem, H.; Uygur, İ. Wear response of non-asbestos brake pad composites reinforced with walnut shell dust. *J. Aust. Ceram. Soc.* **2020**, *56*, 1061–1072. [CrossRef]
17. Yu, Y.; Guo, Y.; Jiang, T.; Li, J.; Jiang, K.; Zhang, H. Study on the ingredient proportions and after-treatment of laser sintering walnut shell composites. *Materials* **2017**, *10*, 1381. [CrossRef]
18. Kuciel, S.; Mazur, K.; Jakubowska, P. Novel biorenewable composites based on poly (3-hydroxybutyrate-co-3-hydroxyvalerate) with natural fillers. *J. Polym. Environ.* **2019**, *27*, 803–815. [CrossRef]
19. Kuciel, S.; Mazur, K.; Jakubowska, P.; Pradhan, P.; Nanda, B.P.; Satapathy, A.; Akıncıoğlu, G.; Akıncıoğlu, S.; Öktem, H.; Uygur, İ.; et al. Effects of walnut shell powders on the morphology and the thermal and mechanical properties of poly (lactic acid). *Polym. Compos.* **2020**, *33*, 803–815.
20. Orue, A.; Eceiza, A.; Arbelaiz, A. The use of alkali treated walnut shells as filler in plasticized poly(lactic acid) matrix composites. *Ind. Crops Prod.* **2020**, *145*. [CrossRef]
21. Bowyer, J.L.; Stockmann, V.E. Agricultural residues. *For. Prod. J.* **2001**, *51*, 10–21.
22. Suhaily, S.S.; Khalil, H.P.S.A.; Asniza, M.; Fazita, M.R.N.; Mohamed, A.R.; Dungani, R.; Zulqarnain, W.; Syakir, M.I. Design of green laminated composites from agricultural biomass. In *Lignocellulosic Fibre and Biomass-Based Composite Materials*; Elsevier: Amsterdam, The Netherlands, 2017; pp. 291–311.
23. Papadopoulou, E.; Chrissafis, K. Particleboards from agricultural lignocellulosics and biodegradable polymers prepared with raw materials from natural resources. In *Natural Fiber-Reinforced Biodegradable and Bioresorbable Polymer Composites*; Elsevier: Amsterdam, The Netherlands, 2017; pp. 19–30.
24. Barbu, M.C.; Sepperer, T.; Tudor, E.M.; Petutschnigg, A. Walnut and hazelnut shells: Untapped industrial resources and their suitability in lignocellulosic composites. *Appl. Sci.* **2020**, *10*, 6340. [CrossRef]
25. Cheng, S.; Lau, K.; Liu, T.; Zhao, Y.; Lam, P.-M.; Yin, Y. Mechanical and thermal properties of chicken feather fiber/PLA green composites. *Compos. Part B Eng.* **2009**, *40*, 650–654. [CrossRef]
26. Madhavan Nampoothiri, K.; Nair, N.R.; John, R.P. An overview of the recent developments in polylactide (PLA) research. *Bioresour. Technol.* **2010**, *101*, 8493–8501. [CrossRef] [PubMed]
27. Aliotta, L.; Gigante, V.; Coltelli, M.; Cinelli, P.; Lazzeri, A.; Seggiani, M. Thermo-Mechanical Properties of PLA / Short Flax Fiber Biocomposites. *Appl. Sci.* **2019**, *9*, 3797. [CrossRef]
28. Song, X.; He, W.; Yang, S.; Huang, G.; Yang, T. Fused Deposition Modeling of Poly (Lactic Acid)/Walnut Shell Biocomposite Filaments—Surface Treatment and Properties. *Appl. Sci.* **2019**, *9*, 4892. [CrossRef]
29. Kabir, M.M.; Wang, H.; Lau, K.T.; Cardona, F. Chemical treatments on plant-based natural fibre reinforced polymer composites: An overview. *Compos. Part B Eng.* **2012**, *43*, 2883–2892. [CrossRef]
30. Phuong, V.T.; Gigante, V.; Aliotta, L.; Coltelli, M.-B.; Cinelli, P.; Lazzeri, A. Reactively extruded ecocomposites based on poly(lactic acid)/bisphenol A polycarbonate blends reinforced with regenerated cellulose microfibers. *Compos. Sci. Technol.* **2017**, *139*, 127–137. [CrossRef]
31. Nanthananon, P.; Seadan, M.; Pivsa-Art, S.; Hiroyuki, H.; Suttiruengwong, S. Biodegradable polyesters reinforced with eucalyptus fiber: Effect of reactive agents. *AIP Conf. Proc.* **2017**, *1914*, 1–6. [CrossRef]
32. Coltelli, M.; Bertolini, A.; Aliotta, L.; Gigante, V.; Vannozzi, A.; Lazzeri, A. Chain Extension of Poly (Lactic Acid) (PLA)—Based Blends and Composites Containing Bran with Biobased Compounds for Controlling Their Processability and Recyclability. *Polymers* **2021**, *13*, 3050. [CrossRef]
33. Gigante, V.; Aliotta, L.; Phuong, V.T.; Coltelli, M.B.; Cinelli, P.; Lazzeri, A. Effects of waviness on fiber-length distribution and interfacial shear strength of natural fibers reinforced composites. *Compos. Sci. Technol.* **2017**, *152*, 129–138. [CrossRef]
34. Le Moigne, N.; Van Den Oever, M.; Budtova, T. A statistical analysis of fibre size and shape distribution after compounding in composites reinforced by natural fibres. *Compos. Part A Appl. Sci. Manuf.* **2011**, *42*, 1542–1550. [CrossRef]
35. Fischer, E.W.; Sterzel, H.J.; Wegner, G. Investigation of the structure of solution grown crystals of lactide copolymers by means of chemical reactions. *Kolloid-Z. Z. Polym.* **1973**, *251*, 980–990. [CrossRef]
36. Fu, S.Y.; Feng, X.Q.; Lauke, B.; Mai, Y.W. Effects of particle size, particle/matrix interface adhesion and particle loading on mechanical properties of particulate-polymer composites. *Compos. Part B Eng.* **2008**, *39*, 933–961. [CrossRef]
37. Pukánszky, B. Influence of interface interaction on the ultimate tensile properties of polymer composites. *Composites* **1990**, *21*, 255–262. [CrossRef]

38. Aliotta, L.; Gigante, V.; Cinelli, P.; Coltelli, M.-B.; Lazzeri, A. Effect of a Bio-Based Dispersing Aid (Einar ®101) on PLA-Arbocel ®Biocomposites: Evaluation of the Interfacial Shear Stress on the Final Mechanical Properties. *Biomolecules* **2020**, *10*, 1549. [CrossRef]
39. Nicolais, L.; Nicodemo, L. Strength of particulate composite. *Polym. Eng. Sci.* **1973**, *13*, 469. [CrossRef]
40. Boeriu, C.G.; Bravo, D.; Gosselink, R.J.A.; van Dam, J.E.G. Characterisation of structure-dependent functional properties of lignin with infrared spectroscopy. *Ind. Crops Prod.* **2004**, *20*, 205–218. [CrossRef]
41. Barczewski, M.; Sałasińska, K.; Szulc, J. Application of sunflower husk, hazelnut shell and walnut shell as waste agricultural fillers for epoxy-based composites: A study into mechanical behavior related to structural and rheological properties. *Polym. Test.* **2019**, *75*, 1–11. [CrossRef]
42. Salasinska, K.; Barczewski, M.; Górny, R.; Kloziński, A. Evaluation of highly filled epoxy composites modified with walnut shell waste filler. *Polym. Bull.* **2018**, *75*, 2511–2528. [CrossRef]
43. Facca, A.G.; Kortschot, M.T.; Yan, N. Predicting the elastic modulus of natural fibre reinforced thermoplastics. *Compos. Part A Appl. Sci. Manuf.* **2006**, *37*, 1660–1671. [CrossRef]
44. Mohanty, A.K.; Misra, M.; Hinrichsen, G. Biofibres, biodegradable polymers and biocomposites: An overview. *Macromol. Mater. Eng.* **2000**, *276–277*, 1–24. [CrossRef]
45. Aliotta, L.; Lazzeri, A. A proposal to modify the Kelly-Tyson equation to calculate the interfacial shear strength (IFSS) of composites with low aspect ratio fibers. *Compos. Sci. Technol.* **2020**, *186*, 107920. [CrossRef]
46. Quiles-Carrillo, L.; Montanes, N.; Sammon, C.; Balart, R.; Torres-Giner, S. Compatibilization of highly sustainable polylactide/almond shell flour composites by reactive extrusion with maleinized linseed oil. *Ind. Crops Prod.* **2018**, *111*, 878–888. [CrossRef]
47. Quiles-Carrillo, L.; Montanes, N.; Garcia-Garcia, D.; Carbonell-Verdu, A.; Balart, R.; Torres-Giner, S. Effect of different compatibilizers on injection-molded green composite pieces based on polylactide filled with almond shell flour. *Compos. Part B Eng.* **2018**, *147*, 76–85. [CrossRef]
48. Álvarez-Chávez, C.R.; Sánchez-Acosta, D.L.; Encinas-Encinas, J.C.; Esquer, J.; Quintana-Owen, P.; Madera-Santana, T.J. Characterization of extruded poly (lactic acid)/pecan nutshell biocomposites. *Int. J. Polym. Sci.* **2017**, *2017*. [CrossRef]
49. Molinari, G.; Gigante, V.; Fiori, S.; Aliotta, L. Dispersion of Micro Fibrillated Cellulose (MFC) in Poly (lactic acid) (PLA) from Lab-Scale to Semi-Industrial Processing Using Biobased Plasticizers as Dispersing Aids. *Chemistry* **2021**, *3*, 896–915. [CrossRef]
50. Taheri, H.; Hietala, M.; Oksman, K. One-step twin-screw extrusion process of cellulose fibers and hydroxyethyl cellulose to produce fibrillated cellulose biocomposite. *Cellulose* **2020**, *27*, 8105–8119. [CrossRef]

Article

Hydrogels Based on Poly([2-(acryloxy)ethyl] Trimethylammonium Chloride) and Nanocellulose Applied to Remove Methyl Orange Dye from Water

Karina Roa [1], Yesid Tapiero [1], Musthafa Ottakam Thotiyl [2] and Julio Sánchez [1,*]

1. Departamento de Ciencias del Ambiente, Facultad de Química y Biología, Universidad de Santiago de Chile (USACH), Santiago 9160000, Chile; karina.roa@usach.cl (K.R.); yesidtm@gmail.com (Y.T.)
2. Department of Chemistry, IISER Pune, Dr. Homi Bhabha Road, Pune 411008, India; musmuhammed@gmail.com
* Correspondence: julio.sanchez@usach.cl

Abstract: Bio-based hydrogels that adsorb contaminant dyes, such as methyl orange (MO), were synthesized and characterized in this study. The synthesis of poly([2-(acryloyloxy)ethyl] trimethylammonium chloride) and poly(ClAETA) hydrogels containing cellulose nanofibrillated (CNF) was carried out by free-radical polymerization based on a factorial experimental design. The hydrogels were characterized by Fourier transformed infrared spectroscopy, scanning electron microscopy, and thermogravimetry. Adsorption studies of MO were performed, varying time, pH, CNF concentration, initial dye concentration and reuse cycles, determining that when the hydrogels were reinforced with CNF, the dye removal values reached approximately 96%, and that the material was stable when the maximum swelling capacity was attained. The maximum amount of MO retained per gram of hydrogel (q = mg MO g^{-1}) was 1379.0 mg g^{-1} for the hydrogel containing 1% (w w^{-1}) CNF. Furthermore, it was found that the absorption capacity of MO dye can be improved when the medium pH tends to be neutral (pH = 7.64). The obtained hydrogels can be applicable for the treatment of water containing anionic dyes.

Keywords: adsorption; fibrillated nanocellulose; hydrogel; methyl orange; polymer

Citation: Roa, K.; Tapiero, Y.; Thotiyl, M.O.; Sánchez, J. Hydrogels Based on Poly([2-(acryloxy)ethyl] Trimethylammonium Chloride) and Nanocellulose Applied to Remove Methyl Orange Dye from Water. *Polymers* **2021**, *13*, 2265. https://doi.org/10.3390/polym13142265

Academic Editors: Andrea Lazzeri, Maria Beatrice Coltelli and Patrizia Cinelli

Received: 2 June 2021
Accepted: 7 July 2021
Published: 10 July 2021

Publisher's Note: MDPI stays neutral with regard to jurisdictional claims in published maps and institutional affiliations.

Copyright: © 2021 by the authors. Licensee MDPI, Basel, Switzerland. This article is an open access article distributed under the terms and conditions of the Creative Commons Attribution (CC BY) license (https://creativecommons.org/licenses/by/4.0/).

1. Introduction

Pollutants released by industrial liquid waste affect the quality of water in water bodies. They can engender serious health effects in plants, animals, and humans. Wastewater pollutants include dyes, surfactants, oils, lubricants, organic solvents, petroleum derivatives, and pharmaceuticals such as antibiotics, anti-allergy, and hormones [1,2]. Artificial dyes, which are largely used in various industries such as textiles, food, cosmetics, leather, paper, and pharmaceuticals, are highly dangerous organic pollutants [3,4]. Dyes are mutagenic agents even at low concentrations and render an undesirable color to water bodies [5]. The presence of dyes in wastewater that drains into water bodies makes it difficult for light to penetrate natural water bodies and negatively impacts photosynthetic activity [6]. A representative artificial dye is methyl orange (MO, dimethylaminoazobenzenesulfonate), which is non-biodegradable in nature; besides, it is a water-soluble carcinogen, azo dye that is widely used in textile industries, printing paper manufacturing, textile laboratories, chemical research, pharmaceuticals and research laboratories [7]. It pollutes water at low concentrations; large volumes of MO are produced as waste.

The MO molecule has a bright orange color when dissolved in water, stable chemical structure due to the presence of azo (–N=N–) and aromatic groups (which are highly toxic, carcinogenic and teratogenic), and is harmful to the environment and organisms since it shows low biodegradability [8]. MO can lead to critical health issues, like cyanosis, vomiting, tachycardia, tissue necrosis, and jaundice, and has been declared carcinogenic and tumorigenic by the International Agency for Research on Cancer (IARC) and the National Institute for Occupational Safety and Health [9]. Additionally, it is an allergenic substance that can cause eczema upon contact with the skin. Its presence in living organisms is considered harmful and can lead to a significant increase in the activity of the azo-nitro-reductase enzymes, producing aromatic amines that may cause intestinal cancer [10,11].

It is difficult to eliminate and control the dye concentration efficiently through traditional treatment methods such as coagulation, sedimentation, chemical oxidation, and biological digestion [12]. In general, conventional water treatment methods generate large volumes of residual sludge, use excessive process times, and consume large amounts of energy [13]. Adsorption technology is considered one of the most effective methods because of its simplicity, low energy consumption, short treatment times, low generation of sludge, high efficiency, flexibility, and insensitivity to toxic substances.

Various materials have been used to adsorb MO, for example, activated carbon from natural sources [14], algae [15], hybrid materials with metal oxides [16], chitosan and hydrogen peroxide–treated anthracite sheets [17], and hypercrosslinked cyclodextrin networks in the form of nanofibrous membranes [18]. For example, Borsagli et al. designed and developed novel three-dimensional porous scaffolds made of N-acyl thiolated chitosan using 11-mercaptoundecanoic acid, with high adsorption capacities for the anionic MO dye in an aqueous medium [19]. Liu et al. prepared hydrogel particles of methacrylateethyltrimethylammonium chloride and acrylamide copolymer, with the ability to eliminate anionic dyes such as amaranth red, orange G, and MO reaching 94% efficiency [20]. Onder et al. prepared copolymer hydrogels of poly([2-(acryloyloxy)ethyl] trimethylammonium chloride-co-1-vinyl-2-pyrrolidinone) (p(AETAC-co-NVP)) which showed the ability to retain the MO and alizarin red S dyes through electrostatic interactions when the test pH values were 7.0 and 5.0, respectively [21], and Dalalibera et al. prepared hydrogels based on polyacrylic acid with the ability to absorb and selectively separate cationic and anionic dyes at a pH of 8.0 to 10.0 [22].

Recently, nanocomposites based on cationic polymers and nanocellulose have been prepared for application in water treatment. These materials have shown remarkable capabilities to remove oxyanions such as chromates and have improved mechanical properties [23], for example, Szekely et al. prepared nanocomposite hydrogels based on cellulose acetate, modified with the addition of small amounts of polymers of intrinsic microporosity and graphene oxide (GO), which demonstrated the ability to absorb neonicotinoid insecticidal pollutants in an aqueous medium [24]. Khan et al. prepared nanocomposite hydrogels with a porous 3D network structure based on cellulose-aluminum oxide nanoparticles-graphene oxide (GO) (Al_2O_3/GO), with application in the removal of fluoride ions from drinking water [25]. Hameed et al. prepared carboxymethyl cellulose/potato starch/amylum starch hydrogels where aluminum sulfate octahydrate was used as a crosslinking agent, which reached high capacity in the retention of heavy metals (cadmium, lead, and iron) from municipal drinking water [26].

In addition, nanocellulose has been used for the synthesis of hydrogels with applications in biomedicine, as in the case of the research work by Chen et al. They prepared fluorescent compound hydrogels of nitrogen-doped carbon points/cellulose nanofibrils (NCD/CNF-gel), where the mechanical properties were highlighted [27].

Several synthetic resins have been studied (such as Amberlite [28]), and the challenge now is to manufacture biomass-derived materials for removal of anionic dyes such as MO. There are not many papers on biomass-derived hydrogels that remove this dye, and that is why in this study hydrogels with ammonium groups reinforced with CNF were developed.

The aim of this study was to synthesize poly([2-(acryloyloxy)ethyl] trimethylammonium chloride) and poly(ClAETA) hydrogels containing fibrillated nanocellulose (CNF). The amount of crosslinkers, CNF, and initiators was optimized by studying the effects on the percentage yield of synthesis, degree of crosslinking, and water adsorption. Developing applicable hydrogels for adsorption and treatment of aqueous MO-containing wastes.

2. Materials and Methods

2.1. Materials

CNF was obtained by oxidation with 2,2,6,6-tetramethylpiperidine-1-oxyl (TEMPO) [29]. The synthesis was performed using a cellulose sample from Norweigan spruce wood, according to the procedure described by Dax et al. [30]. The suspension used in this study had a concentration of 10.79 mg g^{-1}. Through conductometric titration analysis, it was determined that the CNF contained 0.96 mmol COO$^-$ per gram of fiber.

[2-(acryloyloxy)ethyl]trimethylammonium chloride (ClAETA) solution (80 wt% in water; Aldrich, St. Louis, MO, USA), N,N-methylene-bis-acrylamide (MBA) (99%; Aldrich, St. Louis, MO, USA), ammonium persulfate (APS) (98%; Aldrich, St. Louis, MO, USA), NaOH (Merck, St. Louis, MO, USA), HNO$_3$ (70%, Merck, St. Louis, MO, USA), HCl (36 v%; Merck, St. Louis, MO, USA), MO (85 vol% dye content; Sigma-Aldrich, St. Louis, MO, USA), and demineralized diethyl ether (99%, Merck, Milwaukee, WI, USA) were purchased and used in this study.

2.2. Synthesis of Hydrogels

The synthesis of the hydrogels containing CNF was performed via free-radical polymerization. For all experiments, the mass of ClAETA monomer was fixed at 5.0 g in 30 mL of deionized water type I. The synthesis was performed in a polymerization tube (Schlenk) under an inert atmosphere of nitrogen gas, which was immersed in a glycerine bath at 70 °C for 2.5 h. At the end of the reaction, the samples were dried in an oven (BIOBASE, model BOV-T50F, Jinan, Shandong, China) to obtain a constant weight. Subsequently, the samples were lyophilized (LABCONCO FREEZONE, Kansas City, USA).

Figure 1 shows the chemical structures of MO, poly(ClAETA), and the synthesis scheme.

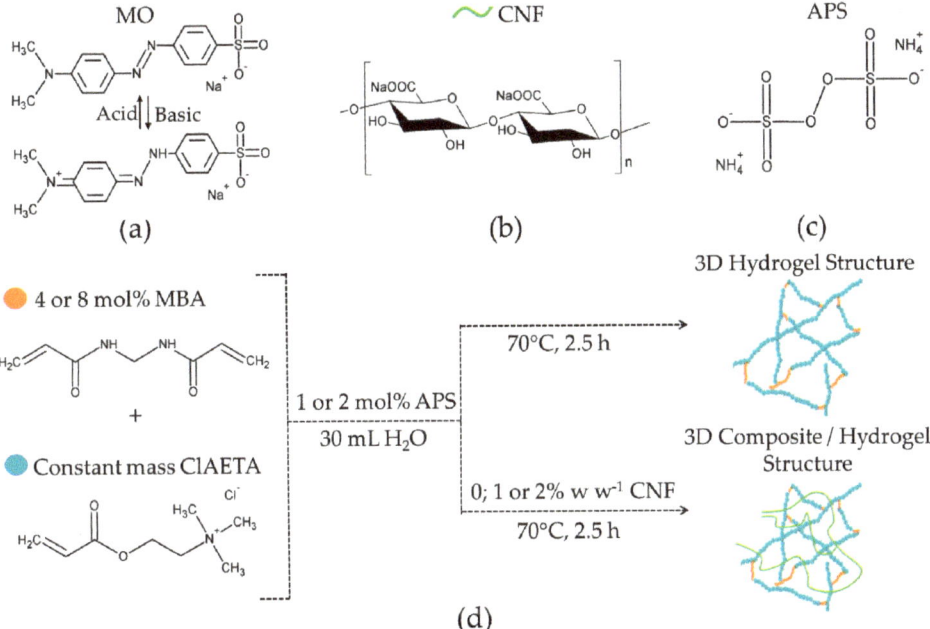

Figure 1. (a) Chemical structure of MO dye (molecular formula: $C_{14}H_{14}N_3O_3SNa$), (b) chemical structure of CNF, (c) chemical structure of poly(ClAETA), and (d) illustrative representation of the synthesis process of the exchange hydrogels obtained from this work.

2.3. Experimental Design

The synthesized materials were evaluated by analysis of variance (ANOVA) using Minitab 19 Statistical Software. The parameters studied in this experimental design were as follows: free-radical initiator (APS) (mol%), cross-linking agent (MBA) (mol%), and amount of CNF (wt%). Each parameter was considered with respect to the amount of ClAETA monomer. For this study, two test levels were used (minimum and maximum), as shown in Table 1. For the preparation, all possible combinations of the three factors were analyzed.

Table 1. Factors to be studied and their respective levels, where −, +, and 0 indicate the minimum, maximum, and control levels respectively.

Levels			−	+	0
	MBA (mol%)	A	4	8	
3 Factors	APS (mol%)	B	1	2	
	CNF (% w w^{-1})	C	1	2	0

The combinations of the factors provide a design matrix of the type $2 \times 2 \times 3 \rightarrow 12$, thus assigning 12 treatments to be evaluated, as shown in Table 2. This method of experimental design is known as "Full Factorial Design (FFD)" [31]. An experiment is defined in which all possible combinations of factor configurations are tested and all possible interactions are determined. Full factorial designs are large compared to screening designs. Generally, a FFD is used when you have a small number of factors and levels, and you search for information on all possible interactions. In a FFD, an experimental run is performed for each combination of factor levels. The sample size is the product of the number of factor levels. FFDs are the most conservative of all design types.

Table 2. Experimental design matrix ($2 \times 2 \times 3 \rightarrow 12$) for the synthesis of ion exchange hydrogels. Treatment "c" indicates that the lowest level of testing has been completed.

Code	Treatment	Factors		
		A MBA (mol%)	B APS (mol%)	C CNF (% w w^{-1})
Hy01	1	4	1	1
Hy02	c	4	1	2
Hy03	0	4	1	0
Hy04	4	4	2	1
Hy05	5	4	2	2
Hy06	6	4	2	0
Hy07	7	8	1	1
Hy08	8	8	1	2
Hy09	9	8	1	0
Hy10	10	8	2	1
Hy11	11	8	2	2
Hy12	12	8	2	0

The factorial experimental design for three factors (A, B, and C) facilitated the investigation of the individual and combined effects of A, B, C, AB, AC, BC, and ABC, based on two levels for each factor. Hence, seven effects were analyzed using ANOVA. In this test, it was assumed that the data followed a trend represented by the F statistic (Ronald

Fisher). To ensure that the uncontrolled factors did not affect the results, the experiments were performed randomly. In addition, we analyzed whether the values found (yield of the reaction, cross-linking degree, and water absorption capacity) generated statistically significant effects, through standardized Pareto charts using the same software.

2.4. Theory Section: Determination of Yield of the Reaction, Cross-Linking Degree, and Water Absorption Capacity

For each of the synthesized materials, the reaction performance, degree of cross-linking, and water absorption were calculated according to the following procedure.

2.4.1. Determination of the Reaction Yield

The yield of the polymerization reaction was determined by weighing the resulting freeze-dried sample (xerogel) and comparing it with the total mass of the reagents using Equation (1).

$$\%Y = \frac{mass_{xerogel}(g)}{mass_{total}(g)} \times 100\% \tag{1}$$

where $mass_{xerogel}$ (g) is the mass of the xerogel in grams, and $mass_{total}$ (g) is the total mass of the reactive compounds.

2.4.2. Determination of Water Absorption Capacity

To determine the maximum water absorption capacity of each of the synthesized hydrogels, 100 mg of xerogel was added to 80 mL of distilled water. The hydrated hydrogel was weighed and deposited in water. The water absorption capacity (%WA) was determined using Equation (2).

$$\%WA = \left(\frac{W_1 - W_0}{W_0}\right) \times 100\% \tag{2}$$

where W_1 and W_0 are the weights in grams of the swollen and initial hydrogels, respectively.

2.4.3. Determination of the Effective Cross-Link Density of a Cross-Linked Structure

To determine the degree of cross-linking, 0.01 g of xerogel was taken in a petri dish, and 1 mL of demineralized ether type I was added. The swollen gels were dried superficially with filter paper and left to stand for 10 min, and the weight of the petri dish with the hydrated polymer material was recorded with XB220 Precisa Analytical Balance. The measurements continued until a constant weight was obtained for each sample. This weight was used to calculate the volume fraction v_{2m} according to Equation (3):

$$w = 1 - \frac{\%WA}{100\%} \tag{3}$$

where w is the weight fraction of polymer in swollen gel [32].

$$C_d = \frac{mass_{xerogel,wet}(g)}{mass_{total,dry}(g)} \tag{4}$$

where C_d is the equilibrium degree of swelling of the polymer in a gel sample that is swollen to equilibrium in water, $mass_{xerogel,wet}$ (g) is the wet mass of the xerogel in grams, and $mass_{total,dry}$ (g) is the total mass of the dry sample. The hydrated material was dried in an oven (BIOBASE model BOV-T50F) at 50 °C for 24 h. The dried mass of the material was measured again.

$$v_{2m} = 1/C_d \tag{5}$$

where v_{2m} is the polymer volume fraction of the cross-linked polymer in the swollen gel.

$$\frac{1}{v_{2m}} - 1 - \frac{\rho_{polymer}}{\rho_{water\ 25\ °C}} * \left(\frac{1}{w} - 1\right) = 0 \tag{6}$$

where $\rho_{polymer}$ is the polymer density in g cm^{-3}, and ρ_{water} is the water density at 25 °C. By using the Excel solver function, it is possible to determine the value of the density of the polymeric part of the hydrogel using Equation (6).

$$\overline{M_C} - \frac{\left(1 - \frac{2}{\phi}\right) \times V_1 \times v_{2m}^{\frac{2}{3}} \times v_{2r}^{\frac{1}{3}}}{\overline{v} \times \left(\ln(1 - v_{2m}) + v_{2m} + \chi \times v_{2m}^2\right)} = 0 \tag{7}$$

where $\overline{M_C}$ is the average molecular weight of the network chains (g mol^{-1}), v_{2r} is the polymer volume fraction in the relaxed state, V_1 (18,07 cm^3 mol^{-1}) 1 is the molar volume of the swelling agent (water, in this study), and $\overline{v}$ is the specific volume of the polymer. In this study, the reference value of cellulose is taken as 0.664 cm^3 g^{-1} [33]. By using the Excel solver function, it is possible to determine the value of $\overline{M_C}$ using Equation (7).

$$\phi = 3 \tag{8}$$

where ϕ is functional at the cross-linking site [34,35].

$$\chi = \frac{1}{2} + \frac{v_{2m}}{3} \tag{9}$$

where χ is the polymer–solvent interaction [35].

$$v_e \left(\frac{mol}{cm^3}\right) = \rho_{polymer}/M_C \tag{10}$$

where v_e is the effective cross-linking density of a cross-linked structure [36].

2.5. Physicochemical Characterization

2.5.1. Fourier Transform Infrared (FTIR) Spectroscopy

Spectrum Two (UATR Two; Perkin Elmer) spectrophotometer with discs of KBr, in the spectral range of 4000–400 cm^{-1}, was used for the FTIR measurements to determine the functional groups of the hydrogels.

2.5.2. Scanning Electron Microscopy (SEM)

The surface morphology of the hydrogels was observed using scanning electron microscopy (SEM; Zeiss EVO MA 10 model, Oberkochen, Germany). The analyzed samples were previously hydrated with water until saturation.

2.5.3. Thermogravimetric Analyses (TGA)

Thermobalance (STARe System Mettler Toledo, Greinfensee, Switzerland) equipment was used for the thermal gravimetric analysis, which was performed under a nitrogen gas atmosphere with a heating rate of 10 °C min^{-1} and a temperature range between 30 °C and 550 °C. A 250 mL min^{-1} flow rate of nitrogen gas was employed with aluminum as a reference material.

2.6. Adsorption Capacity of MO

The hydrogels were analyzed according to their ability to adsorb MO dye. An aqueous solution (water conductivity 3.15 µS/cm) of MO dye (150 mg L^{-1}) was prepared, and 40 mL was used for each test. The hydrogel (50 mg) was placed in contact with 40 mL of aqueous MO dye, and the duration of the experiment was 300 min. The following hydrogels were selected for these tests: Hy01, Hy02, Hy03, Hy04, Hy07, Hy10, and Hy12.

First, the adsorption kinetics were determined using washed and unwashed hydrogel Hy12 after synthesis. Subsequently, the adsorption was studied as a function of time up to 300 min for Hy01 and Hy03; the adsorption capacity of the hydrogel was calculated every 15 min up to 60 min, then every 30 min up to 240 min, and finally at 300 min. Thereafter, the adsorption tests were performed simultaneously as mentioned above with Hy01, Hy02, and Hy03, by varying the amount of CNF in the hydrogel.

The adsorption tests were performed with Hy01 by varying the pH of the colored solution to acidic (pH 3.0) with HCl (0.1 mol L^{-1}) and alkaline (pH 10.0) with NaOH (0.1 mol L^{-1}). The MO dye concentration was measured using a UV-visible spectrometer (model BK-UV1800, BIOBASE) at a wavelength of 466 nm [37].

The amount of MO dye retained by the hydrogel was determined according to Equation (11):

$$q = \frac{(C_i - C_p) \times V_i}{m_x} \quad (11)$$

where q is the adsorption capacity or amount in milligrams of MO retained per gram of sorbent (mg g^{-1}); C_p is the concentration of dye in the supernatant (mg L^{-1}); C_i is the initial concentration of dye in the supernatant (mg L^{-1}); V_i is the volume of the dye solution (mL); and m_x is the mass of the xerogel (g).

The removal percentage %R of MO by the hydrogels was determined using Equation (12):

$$\%R = \frac{C_i - C_p}{C_i} \times 100 \quad (12)$$

where %R is the removal percentage; C_p is the concentration of dye in the supernatant (mg L^{-1}); and C_i is the initial concentration of the dye in the supernatant (mg L^{-1}).

The effect of the initial concentration of MO was studied by varying the concentrations between 50, 100, 200, 350, 500, 750, 1000, 1500, and 2000 mg L^{-1}.

2.7. Adsorption Capacity of MO

To study adsorption and desorption cycles, adsorption was performed under the best conditions (pH, time, and concentration) and desorption was performed using 40 mL of HCl (0.1 mol L^{-1}) as eluent during 30 min of constant agitation at 200 rpm. Each adsorption and desorption was then centrifuged for 10 min at 9000 rpm in a centrifuge, measuring the concentration of the supernatant by a UV-visible spectrometer (model BK-UV1800, BIOBASE, Jinan, Shandong, China).

2.8. Kinetic Model

To understand the dynamics of adsorption as a function of time, kinetic models described in the literature were used [38]. Equations (13) and (14) show the linear expression for pseudo-first order and pseudo-second order respectively.

$$\ln(q_e - q_t) = \ln(q_e) - k_1 t \quad (13)$$

$$\frac{t}{q_t} = \frac{1}{k_2 q^2_e} + \frac{1}{q_e} t \quad (14)$$

where q_e (mg g^{-1}) it is the sorption in equilibrium, q_t (mg g^{-1}) is the sorption at any time t, k_1 and k_2 (min^{-1}) are the sorption constants of the pseudo-first-order and pseudo-second-order model.

3. Results

3.1. Synthesis of Hydrogels Modifying the Amount of Cross-Linker, Initiator, and CNF

Upon freeze-drying, the hydrogels had a granular texture and were slightly light brown. The nanocomposites synthesized with CNF had a slightly translucent appearance and, when freeze-dried, had granular characteristics and were light brown. Table 3 lists the reaction yields, water absorption, and effective cross-link density of the cross-linked structure of the 12 synthesized hydrogels.

Table 3. Characterization results of the synthesized hydrogels: the reaction yields, water absorption, and effective cross-link density of the cross-linked structure of the 12 synthesized hydrogels.

Code	Treatment	%Y	%WA	v_e (mol cm^{-3})
Hy01	1	30.8	3119.0	7.28×10^{-4}
Hy02	c	17.3	1818.0	1.47×10^{-4}
Hy03	0	96.4	9567.7	5.23×10^{-5}
Hy04	4	29.8	4631.0	2.73×10^{-3}
Hy05	5	17.8	918.3	1.84×10^{-3}
Hy06	6	101.4	826.0	4.79×10^{-4}
Hy07	7	28.5	1318.4	5.58×10^{-4}
Hy08	8	18.2	889.5	2.05×10^{-5}
Hy09	9	101.1	2032.0	7.75×10^{-4}
Hy10	10	30.6	5004.0	8.55×10^{-4}
Hy11	11	19.0	882.2	4.06×10^{-4}
Hy12	12	97.8	890.9	4.73×10^{-4}

The responses to the variations in the factors in the experimental design are explained below.

3.1.1. Yield of the Reaction

Table 3 lists the reaction yields, water absorption, and effective cross-link density of the cross-linked structure of the 12 synthesized hydrogels.

The reaction yields of only four synthesized materials exceeded 90%: Hy06 > Hy09 > Hy12 > Hy03. All four hydrogels were prepared without the addition of CNF. Likewise, it can be observed that when the CNF concentration increased in the reactive mixture, the efficiency of the reaction decreased from approximately 30%, when 1% w w^{-1} CNF was added to the mixture, to approximately 17%, when 2% w w^{-1} CNF was added to the mixture. This indicated that CNF had an inhibitory effect on the growth of polymers through free-radical reactions (see Table 3).

3.1.2. Water Absorption Capacity

The water absorption capacity of these hydrogels was high, indicating that they could effectively facilitate the removal of contaminants. From the results obtained for the %WA capacity, the following decreasing order can be observed: Hy03 > Hy10 > Hy04 > Hy01 > Hy09 (see Table 3). The highest water absorption capacity was achieved for the hydrogel that did not contain CNF; however, the remaining hydrogels that attained the highest %WA values contained 1% w w^{-1} CNF in their structures.

3.1.3. Effective Cross-Link Density of a Cross-Linked Structure

The analysis of the effective cross-link density of a cross-linked structure revealed that five hydrogels exceeded the value of 6.0×10^{-4} mol cm^{-3}: Hy04 > Hy05 > Hy10 > Hy09 > Hy01 (see Table 3). Except Hy12 and Hy06 all hydrogels contained CNF. The v_e values were higher when 2% w w^{-1} CNF was added to the mixture.

3.2. Statistical Analysis

The results of the ANOVA on the yield of the reaction response (see Table S1a) indicated that the factors (A, B, and C) individually did not meet the criteria that represented a linear model of the data for the response of interest. Meanwhile, the value of the p statistic (significance value of 0.05) was found to be 0.037, which was less than 0.05. This result confirmed the prior result obtained through the F statistic. Therefore, the performance of the polymerization reaction for hydrogel synthesis was independent of factors A, B, and C. Furthermore, when analyzing factor B (CNF w w^{-1} %), it was observed that the value of the F statistic was greater than that of $F_{critical}$ and the p statistic (0.008 < 0.05), considering the results obtained from the double and triple interactions of the factors. The values of the F statistic were less than the $F_{critical}$ values, and the values of the p statistic were less than 0.05.

The ANOVA results for the response of the water absorption capacity (see Table S1b), including all the factors (individually as well as in their binary and ternary combinations), generated an F statistic value that was less than the $F_{critical}$ statistic value, and the p values were greater than the significance value of 0.05. All parameters significantly influenced the water absorption capacity.

The ANOVA results for the response of the effective cross-link density of a cross-linked structure can be observed in Table S1c, where the F statistic value was greater than the $F_{critical}$ value; similarly, the value of the p-statistic (0.027) was less than the significance value (0.05). Considering the results of the individual factors and their double combinations when comparing the values of the F statistic with their $F_{critical}$ values, F < $F_{critical}$. Furthermore, the values of the p statistic for these factors and their double combinations generated values less than 0.05; therefore, they were significant and influenced the response of the effective cross-link density of a cross-linked structure.

To establish the relationship between the investigated factors and yield of the reaction, water absorption capacity, and the effective cross-link density of the cross-linked structure, the data were graphically displayed using standardized Pareto charts. The changes in the level of a factor that influenced the different system responses were represented as the effects of the factor. The standardized Pareto chart compares the absolute values and the significance of the effects.

Figure 2 shows standardized Pareto charts for the response variables, highlighting that the critical value obtained from the Student's t-test statistic is $t_{0.25,8}$ = 2.776.

From the ANOVA analysis for the three factors, it was found that, for the treatment of the data obtained for the reaction yield, only the CNF factor w w^{-1} % generated a value of the p statistic at 0.008 << 0.05; therefore, this factor significantly influences the performance results. Furthermore, the error had eight degrees of freedom, and we worked with α = 0.05; therefore, the critical value obtained from the Student's t-test was $t_{0.25,8}$ = 2.776. Figure 2a shows that the bar of the CNF factor w w^{-1}% exceeds the critical value of 2.776; therefore, it was concluded that the reaction performance was affected significantly. Other factors did not significantly affect the yield of the reaction. The ANOVA analysis of the results obtained for the effective cross-link density of a cross-linked structure and the water absorption capacity also presented eight degrees of freedom in the error value; therefore, the critical value of the Student's t-test statistic was $t_{0.25,8}$ = 2.776. It can be observed from Figure 2b that none of the factors or the combination of these factors had a significant effect on the water absorption capacity, as none of the bars exceeded the critical value of $t_{0.25,8}$ = 2.776.

The effective cross-link density of the cross-linked structure was not strongly affected by the interactions of the three factors (MBA mol% $\times$ APS mol% $\times$ CNF w w^{-1} %) because the value shown in Figure 2c does not exceed the critical value of $t_{0.25,8}$ = 2.776.

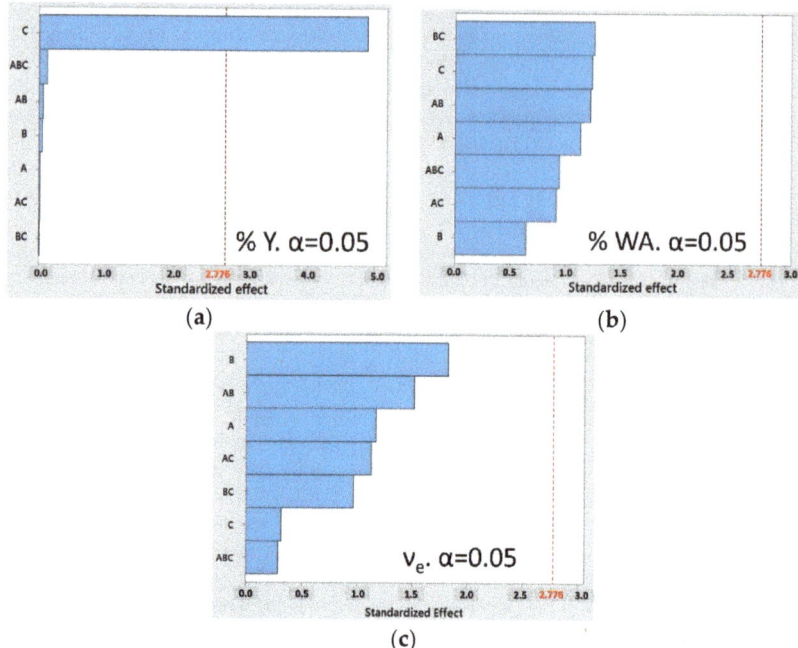

Figure 2. Result of the standardized Pareto analysis for the response variables (parameters A: MBA mol%; B: APS mol%; and C: CNF w w^{-1}%): (**a**) yield of the reaction, (**b**) water absorption capacity, and (**c**) the effective cross-link density of the cross-linked structure.

The analysis of the effect of APS mol% in these tests revealed a random behavior. APS affects the length of the polymer chains, making them longer or shorter depending on the concentration used. Increasing the initiator percentage produces more free radicals that accelerate the chain termination reaction, resulting in shorter chains and thus increasing the possibilities for water entry [39]. The incorporation of CNF helps ensure that the hydration of the polymer does not lead to swelling drastically, since the fibrils become a part of the interpolymeric network when physically mixed and by electrostatic interactions, and continue to absorb significant percentages of water. An increase in hydrogel swelling is observed when the concentration of CNF is increased, which may be explained by the increased number of carboxyl groups in the hydrogels [40].

3.3. Physicochemical Characterization

3.3.1. Fourier Transformed Infrared Spectroscopy

FTIR spectroscopy was performed to identify the functional groups of the polymers and the specific interactions between poly(ClAETA) and CNF, as well as to understand how this reinforcement interferes with the hydrogel base.

Figure 3a shows the characteristic signal of CNF, where the characteristic band of the -OH stretching (3319 cm^{-1}), band for -CH stretching (2900 cm^{-1}), and the band with the signal of the -C=O (1610 cm^{-1}) indicate the occurrence of -COOH groups in CNF, suggesting oxidation in the glucose ring of the hydroxyl group at the C-6 position [41,42]. Figure 3b,c shows the characteristic signals for two synthesized hydrogels, Hy06 and Hy10, corresponding to NH stretching (3357 cm^{-1}) and disappearance of the vinyl group (C=C) (1200 cm^{-1}), respectively. The carbon double bond signal disappeared, and the signals of the functional groups of the 2-(acryloyloxy) ethyl trimethylammonium chloride monomer (ClAETA) remained, corroborating the formation of the polymer [43,44]. The signals indicating the stretching of CH in the polymer (2977 cm^{-1}), the carbonyl bond

(C=O) (1736 cm^{-1} and 1726 cm^{-1}), and the quaternary ammonium group (-N$^+$ (CH$_3$)$_3$) (1500 cm^{-1}) were also observed [44].

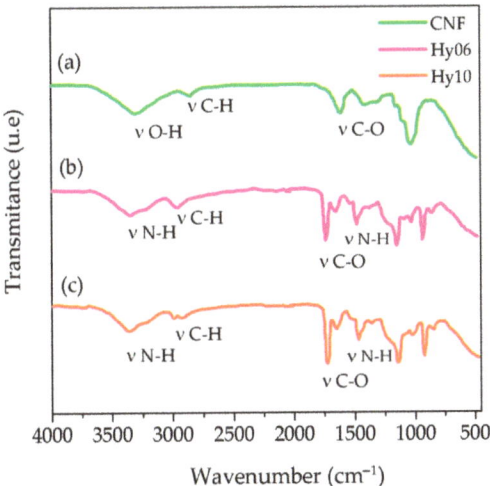

Figure 3. FTIR analysis of (**a**) CNF, (**b**) Hy06 (hydrogel without CNF), and (**c**) Hy10 (hydrogel with CNF).

3.3.2. Morphological Analysis by Scanning Electron Microscopy

The microstructural changes in the hydrogels were analyzed using SEM, and the micrographs are shown in Figure 4.

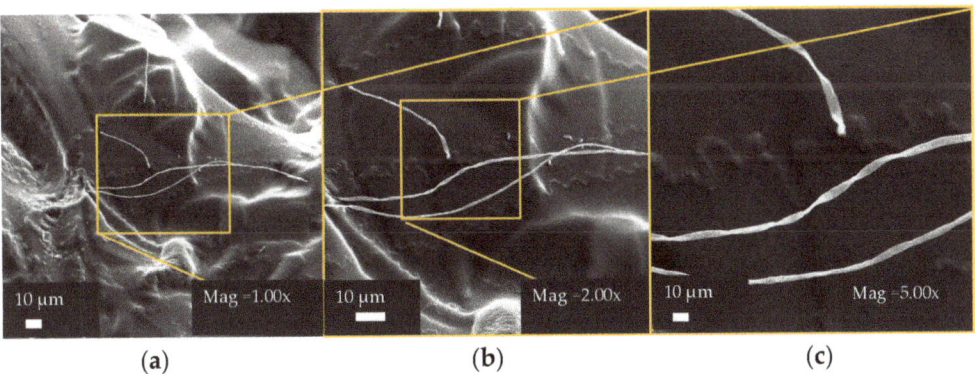

Figure 4. SEM images of Hy05 (2 wt% CNF) samples at (**a**) ×1000, (**b**) ×2000, and (**c**) ×5000 magnification. Scale bar: 10 μm.

Figure 4 shows the structure of Hy05, which contained 2 wt% CNF. Separated fibers were observed on the sample surface; it is known that the surface of poly(ClAETA) (without CNF) has a smooth morphology, as reported in previous studies [43]. CNF was partially carboxylated, where the carboxylate groups interacted electrostatically with the quaternary ammonium groups present in the polymer, according to the SEM images. The CNF fibers were confirmed to be homogeneously distributed in the hydrogel network. Hy05 presented uniform roughness across the surface of the material; in addition, this physical mixture of CNF with the polymer chains helped in the stability and compactness of the structure [45]. In prior research studies, it has been reported that if one of the polymeric components has negatively charged functional groups, the formation of porosities is facilitated because of the repulsion forces between charges [40].

3.3.3. Thermogravimetric Analysis

Thermogravimetric analysis is a technique that allows the evaluation of the thermal stability of synthesized poly(ClAETA) hydrogels and the determination of the effect generated by the addition of CNF in the hydrogels. Figure 5 shows the thermograms of hydrogels Hy03, Hy04, Hy07, and Hy11, which show a typical sigmoidal shape, indicating the weight loss in three stages. The first stage was recorded at temperatures around 100 °C, corresponding to the dehydration of water in the polymer and the elimination of humidity [46]. The second stage occurred at approximately 280–330 °C, corresponding to the thermal decomposition of the groups that protruded from the polymer chain. Similarly, a peak was observed at 390–410 °C, corresponding to exothermic reactions resulting from the decomposition of the ammonium salt [43].

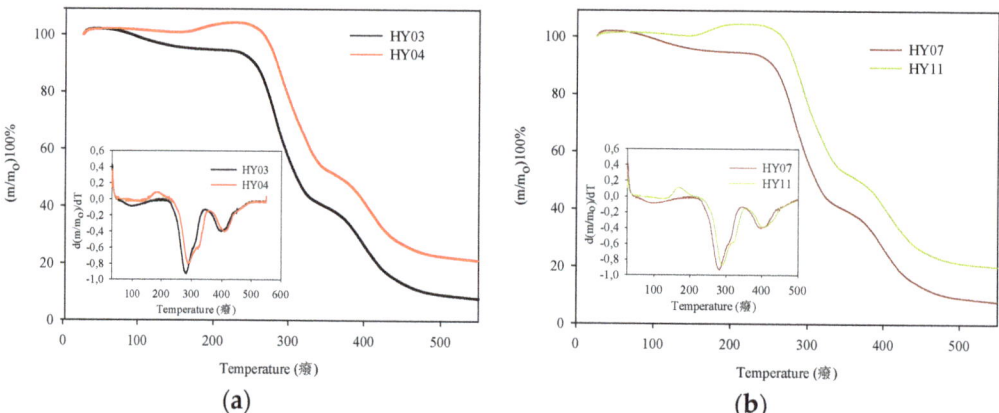

Figure 5. Thermogravimetric analysis TGA of the samples: (**a**) Hy03 and Hy04, (**b**) Hy07 and Hy11.

Upon analysis of Hy03, which did not contain CNF, a decomposition was observed from 248.0 °C to 330.0 °C, resulting in a residual mass percentage of 45.1%; in the second stage, a residual mass percentage of 18.4% was obtained with respect to the initial mass; and finally, at 550 °C, only 8% of the residual mass of the hydrogel remained. In the case of Hy04, a rapid decomposition was observed that generated 58% of the residual mass in the second stage; in the third stage, a residual mass percentage of 26.7% was obtained; and finally, at 550 °C, the resulting value of the residual mass was 21.6%. It can be observed that there was a difference in stability between the hydrogels Hy03 and Hy04 at 18 °C, which could be due to the presence of CNF [47]. In the thermogram of Hy07, the first stage had a temperature range of 238–317.3 °C, which generated a residual mass percentage of 46.7%; in the third stage, 317.3 °C and 426 °C corresponded to a residual mass percentage value of 17.8%; and finally, at 550 °C, only 8% of the residual mass of the hydrogel remained.

Finally, the hydrogel Hy11 presented the second range of decomposition between the temperatures 256 °C and 330.2 °C, and the residual mass was 57.2%; between 330.2 °C and 434.3 °C, the residual mass was 24.8%. For this material, a thermal stability was achieved at approximately 8 °C. Analysis of Hy11 with Hy07 reveals that the main chain decomposition stage had a wider temperature range, which could be caused by the higher concentration of APS and CNF. Finally, at 550 °C, a residual mass percentage of 20.3% was obtained. The higher residual mass may be due to the higher amount of initiator, which can accelerate the cross-linking process during polymerization [48].

3.4. Adsorption Capacity of Methyl Orange Dye by Hydrogels

The functionality of hydrogel as a water dye adsorbent was evaluated. These tests were conducted using MO as the study molecule. For all the tests, the concentration was fixed at 150 mg L^{-1}.

First, we studied the adsorption of dyes when the polymers were washed after synthesis, as well as the effects of this process on the result. For this test, we used Hy12. Figure 6 shows dye retention and capacity of adsorption per gram of resin for the washed (Hy12W) and unwashed (Hy12NW) hydrogel.

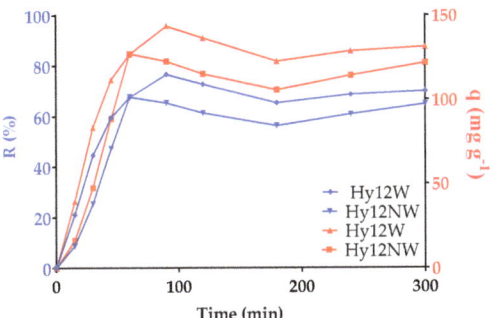

Figure 6. Dye retention and capacity of adsorption per gram of resin by hydrogel, as a function of time for Hy12NW and Hy12W at pH 7.64.

In the case of the Hy12W hydrogel, the minimum concentration of MO that remained in the solution at 90 min was 8.26%, compared to the case of Hy12NW, in which MO concentration was 28.13% at 60 min. Furthermore, as shown in Figure 6, Hy12W attained an optimal MO adsorption capacity of 96% at 90 min after the start of the experiment; however, the MO adsorption capacity was 37.55% at 300 min. In Hy12NW, an optimal MO adsorption capacity of 84.77% was attained at 60 min; at 300 min, the MO adsorption capacity was 81.16%. Thus, we concluded that despite washing, the hydrogels without CNF had a greater instability, which led to the desorption of the previously adsorbed dye.

The removal of MO using two synthesized hydrogels, one with CNF and one without the reinforcement (Hy01 and Hy03, respectively), at a pH of 7.64 was evaluated with the same percentages of cross-linker and initiator. The results for the percentage of adsorption and the retention of the hydrogel are shown in Figure 7.

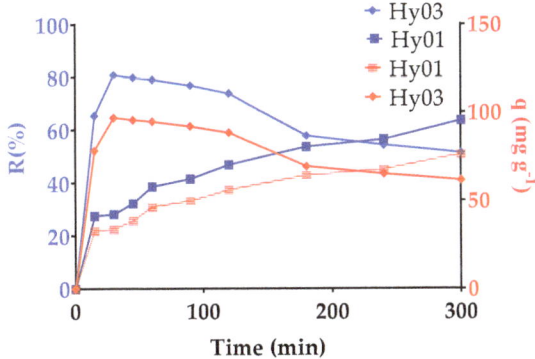

Figure 7. Dye retention and capacity of adsorption per gram of resin by hydrogel as a function of time for Hy01 and Hy03 at pH 7.64.

It was observed that for Hy03, the removal percentage at 30 min after being immersed in the MO solution was 80.92%, and adsorption decreased thereafter. Concurrently, it was observed that the concentration of dissolved dye in the solution increased, which could be due to the low stability of hydrogel. In general, it is observed that hydrogels absorb the contaminant until a plateau of stability or equilibrium is attained in the adsorption curve, which is maintained until the hydrogel reaches its maximum swelling capacity, whereupon the sample sorption starts decreasing [30]. In the case of the hydrogel with CNF, Hy01, a slower response to adsorption is observed, but this response is sustained when nearing 300 min with 76.41 mg of MO per gram of resin.

Comparing the amount of MO retained per gram of resin (q = mg MO g^{-1}) and considering different percentages of CNF (see Figure 8), it can be ascertained that with CNF, the adsorption was slower but sustained with time, except in the case of Hy03 (control sample, without CNF), which retained a greater amount of MO in less time but lost the retained dye and returned it to the solution. In Hy01, containing 1% CNF, and Hy02, containing 2% CNF, we observed 76.41 mg of MO per gram of resin and 72.42 mg of MO per gram of resin after 300 min, respectively, which could be caused by the stability provided by the incorporation of CNF into the matrix.

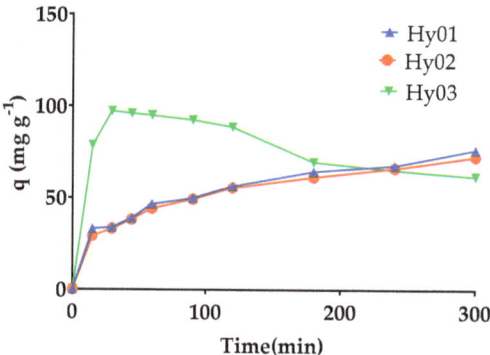

Figure 8. Removal of MO as a function of time when containing different concentrations of CNF at pH 7.64.

3.4.1. Effect of pH on MO Removal

The performance in terms of retention of the MO dye in the acidic and basic environment of Hy01 was analyzed. Figure 9 shows the results of the q response (concentration of MO dye absorbed in the hydrogel structure). There was a higher MO load at pH 7.64, with 76.41 mg of MO for each gram of resin; the MO load at pH 3 and 10 was 59.92 and 65.70 mg for each gram of resin, respectively. MO dye is a weak acid that is widely used as an indicator of pH change. It has a pKa value of 3.47; therefore, at pH 3.0, the sulfonate molecules in the dye are neutralized and the amino groups are protonated, generating a positive charge [28]. The positive charge of the quaternary ammonium group exerts electrostatic repulsion with the positive charge of the hydrogel. In contrast, in a basic medium, the molecules of MO have a completely ionized sulfonate group, and the amino component is uncharged; thus, the molecules in the dyes and the fixed quaternary ammonium groups in poly(ClAETA) can completely interact with each other electrostatically. In addition, CNF has carboxyl groups on its surface and a pKa of approximately 4.6. In an alkaline environment, the carboxylic acid groups in CNFs gradually change to carboxylic anions, leading to the interaction of weakened hydrogen bonds and increased electrostatic repulsion in the hydrogels [40].

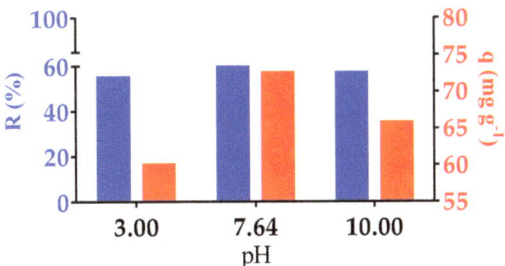

Figure 9. Removal of MO dye at pH 3, 7.64, and 10. *y*-axis: right—MO retention (%R); left—adsorption capacity (q).

3.4.2. Adsorption as a Function of MO Concentration

The initial concentration of the dye affects adsorption because it affects the mass transfer between the aqueous phase and the solid phase. For this reason studies were carried out varying the concentration between 50–2000 mg L^{-1} at pH 7.64 with an initial hydrogel mass of 50 mg (see Figure 10a). The adsorption capacity increased with increasing initial concentration; a similar situation was observed in previous investigations with this dye [49,50]. With respect to the retention percentage, it also increase as the concentration of MO increases. Equilibrium is achieved between 1500 and 2000 mg L^{-1}, obtaining a maximum adsorption capacity of 1379 mg g^{-1} when the initial concentration is 2000 mg L^{-1}. This high dye concentration was also studied by Onder et al., who using their hydrogel of [(2-(acryloyloxy)ethyl]trimethylammonium chloride-*co*-1-vinyl-2-pyrrolidone] hydrogel reached 905.6 mg g^{-1} [21]. Regenerability and reusability of the adsorbent are also very important, as they make the adsorption process economical. Figure 10b shows the adsorption–desorption cycles when hydrochloric acid was used as an eluent [51]. As the number of cycles increases, the adsorption capacity gradually decreases to 5% by the fifth cycle. In general, in our experimental conditions, the reuse is recommended up to the third cycle, since after this adsorption capacity decreases significantly. It is also noted that the hydrogel delivers low concentrations of MO up to the third desorption process.

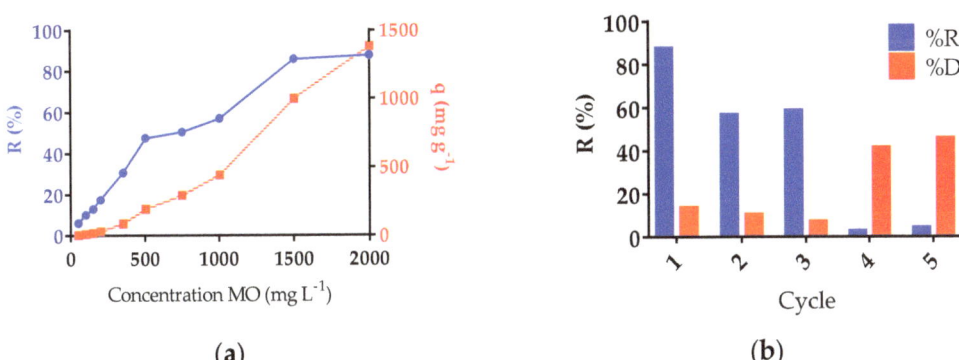

Figure 10. (a) Effect of MO concentration on the adsorption capacity of Hy01 at pH 7.64 with different concentrations of MO (mg L^{-1}). (b) Effect of adsorption–desorption cycles on adsorption capacity (pH = 7.64; initial concentration: 2000 mg/L; and 50 mg adsorbent).

3.4.3. Kinetic Models

A study of adsorption kinetics is desirable because it provides information on the progress of adsorption and whether physical or chemical interactions predominate the process. The pseudo-first and pseudo-second-order kinetic parameter values for MO adsorption are presented in Table 4. The correlation coefficient criterion (highest value of R^2) was used to describe the most suitable kinetic adsorption model [52]. According to the described criteria, MO adsorption for hydrogel Hy01 conforms to the pseudo-second-order kinetic model with an R^2 value greater than 0.9733 (see Figure 11). Similar kinetic results were obtained in previously reported MO adsorption studies [7,19,53].

Table 4. The MO sorption data for the pseudo-first and -second-order kinetic model.

Hydrogel	K_1 (min^{-1})	R^2 Pseudo-First Order	K_2 (g mg^{-1} min^{-1})	R^2 Pseudo-Second Order
Hy01	0.0083	0.9595	0.0004111	0.9733

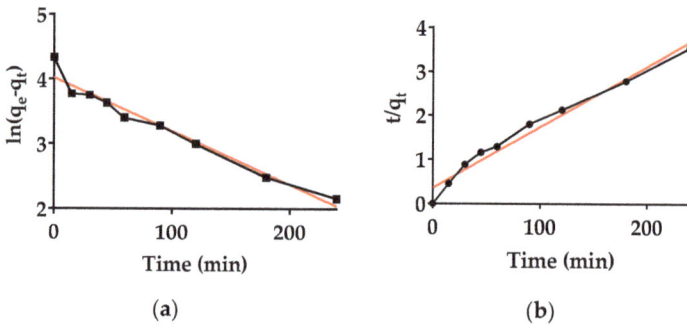

Figure 11. MO sorption kinetics: (**a**) pseudo-first-order and (**b**) pseudo-second-order kinetic model.

From a systematic study of the literature, it is clear that the textile industry is the main industry that generates large volumes of wastewater containing dyes, consuming about 100 L of water to process approximately 1 kg of textile material [54]. These are highly recalcitrant and biocompatible synthetic chemical compounds, considered as potential threats to human and environmental health [4]. About 3500 different types of synthetic dyes are used in the textile industry. The most commonly used dyes are anthraquinone and azo dyes and more than 60% of these dyes are reactive [55]. These chemical species are released through the industrial processes of dyeing and washing, among others [56], resulting in wastewater with high concentrations fluctuating between 350–1000 mg L^{-1} of dyes [4,57,58]. Most synthetic dyes are soluble in water, thanks to the ionizable groups that compose them, such as: -OH, -COOH, and -SO$_3$H in acid dyes, and -NH$_2$, -NHR, and -NR$_2$ in basic dyes. It is estimated between 50% and 70% of the world production of 10,000 synthetic dyes (dyes and distinctive dyes used in the textile industry) corresponds to azo dyes, which represent the class of compounds most used in textile and food processes [10,11]. The pH value of the aqueous medium favors the ionization of the groups, depending on the pKa value of the chemical species in the solution. It is known that the average pH value of wastewater is 8.75 ± 1.29 [4], where the vast majority of dyes are in an ionic state. Regarding the applicability of hydrogels, they are materials that possess a number of functional groups suitable for dyes, and also offer the possibility of reuse/regeneration in sorption–desorption cycles by washing processes with acidic solutions for the anionic hydrogel and basic brine/NaCl for the cationic hydrogel. If it is not possible to regenerate the structures, these materials should be disposed of in the solid waste landfill, following the usual route for hazardous solid waste [59].

Table 5 shows a comparison of adsorption capacities of MO by biopolymer composites, highlighting the possibility of further developing this type of adsorbent materials with natural polymers, such as CNF, which was corroborated to improve the stability at the time of adsorption, giving the possibility of reusing the hydrogel in desorption–adsorption cycles.

Table 5. Comparative table of maximum adsorption results.

Adsorbent	Qmax (mg g^{-1})	Ref.
Ppy@magnetic chitosan	95	[60]
three-dimensional (3D) porous scaffolds made of N-acyl thiolated chitosan using 11-mercaptoundecanoic acid	434.89	[19]
particles of methacrylateethyltrimethylammonium chloride (DMC) and acrylamide (AM) copolymer hydrogel	992.63	[20]
Chitosan/diatomite composite	35	[61]
Banana peel	21	[62]
Chitosan/organic rectorite-Fe$_3$O$_4$	5.6	[53]
Poly([2-(acryloyloxy)ethyl] trimethylammonium chloride), poly(ClAETA), hydrogels containing fibrillated nanocellulose (CNF).	1379.0	This study

Is important to advance in the development of bio-based materials to be tested in real applications in industry. The adsorption is inexpensive, simple, and easy to adapt. In addition, its treatment period is short, causes no pollution to the environment, and has been confirmed as one of the most promising technologies for removing dyes from wastewaters [50].

4. Conclusions

Nanocomposite hydrogels based on ClAETA were successfully synthesized by varying the concentrations of CNF, MBA, and APS. From the ANOVA analysis, it was observed that the concentration of APS significantly affects the performance of the hydrogel synthesis compared to the other factors. It was determined that the combination of the three factors significantly affected the degree of cross-linking because the APS affects the length of the polymeric chains formed, the MBA maintains the solidity and porosity of the hydrogel, and the APS provides stability and rigidity, and an increase in hydrogel swelling is observed when the concentration of CNF is increased, which may be explained by the increased number of carboxyl groups in the hydrogel. In contrast, all individual factors, in double or triple combination with each other, did not significantly affect the water absorption capacity.

In addition, in the microstructural analysis, the texture of the hydrogels was determined, and the CNF fibers were individually identified. The functional groups of the structures of the hydrogels can be determined by FTIR spectroscopic analysis. From TGA it was verified that the hydrogels containing CNF generated greater thermal stability compared to hydrogels with only poly(ClAETA). The surface morphology of the obtained hydrogels was observed by SEM and the incorporated CNF was observed.

In the application of the hydrogels to the absorption of the dye, it was observed that the hydrogels containing only poly(ClAETA) achieved removal values above 80% and then decreased, but these were unstable after reaching the maximum swelling capacity and tended to destabilize. In contrast, hydrogels with CNF, such as Hy01, had lower removal rates than those without CNF but were chemically and mechanically more stable, capturing 1379 mg of MO per gram of resin after 300 min. The reuse/regenerative hydrogel was tested and was found to be satisfactory in up to three cycles. Tests with pH variations indicated that the adsorption of MO was favored under neutral pH. Therefore, it can be concluded that the incorporation of CNF improves the MO adsorption as a function of time.

Supplementary Materials: The following are available online at https://www.mdpi.com/article/10.3390/polym13142265/s1, Table S1: ANOVA results, where the significance is 0.05: (a) yield of the reaction, (b) crosslinking degree, and (c) water absorption capacity. DF (degree of freedom of the data), SS (the sum of the squares of the data), MS (mean sum of the squares of the data), F (F-statistic), value-p (p-value) and $F_{critical}$ (F-statistic critical value).

Author Contributions: K.R., conceptualization, methodology, writing; Y.T., visualization, software, data curation; M.O.T., original draft preparation, visualization, investigation; J.S., writing, reviewing, editing, supervision, project administration. All authors have read and agreed to the published version of the manuscript.

Funding: This research was funded by FONDECYT, grant 1191336.

Institutional Review Board Statement: Not applicable.

Informed Consent Statement: Not applicable.

Data Availability Statement: Exclude this statement.

Acknowledgments: The authors thank project FONDECYT n° 1191336.

Conflicts of Interest: The authors declare no conflict of interest.

References

1. Senthil Kumar, P.; Janet Joshiba, G.; Femina, C.C.; Varshini, P.; Priyadharshini, S.; Arun Karthick, M.S.; Jothirani, R. A critical review on recent developments in the low-cost adsorption of dyes from wastewater. *Desalin. Water Treat.* **2019**, *172*, 395–416. [CrossRef]
2. Wang, J.; Wang, Z.; Vieira, C.L.Z.; Wolfson, J.M.; Pingtian, G.; Huang, S. Review on the treatment of organic pollutants in water by ultrasonic technology. *Ultrason. Sonochem.* **2019**, *55*, 273–278. [CrossRef]
3. Pauletto, P.S.; Gonçalves, J.O.; Pinto, L.A.A.; Dotto, G.L.; Salau, N.P.G. Single and competitive dye adsorption onto chitosan–based hybrid hydrogels using artificial neural network modeling. *J. Colloid Interface Sci.* **2020**, *560*, 722–729. [CrossRef]
4. Kishor, R.; Purchase, D.; Saratale, G.D.; Saratale, R.G.; Ferreira, L.F.R.; Bilal, M.; Chandra, R.; Bharagava, N.R. Ecotoxicological and health concerns of persistent coloring pollutants of textile industry wastewater and treatment approaches for environmental safety. *J. Environ. Chem. Eng.* **2021**, *9*, 105012. [CrossRef]
5. Shakoor, S.; Nasar, A. Removal of methylene blue dye from artificially contaminated water using citrus limetta peel waste as a very low cost adsorbent. *J. Taiwan Inst. Chem. Eng.* **2016**, *66*, 154–163. [CrossRef]
6. Bharathiraja, B.; Aberna, E.S.I.; Iyyappan, J.; Varjani, S. Itaconic acid: An effective sorbent for removal of pollutants from dye industry effluents. *Curr. Opin. Environ. Sci. Health* **2019**, *12*, 6–17. [CrossRef]
7. Sejie, F.P.; Nadiye-tabbiruka, M.S. Removal of Methyl Orange (MO) from Water by adsorption onto Modified Local Clay (Kaolinite). *Phys. Chem.* **2016**, *6*, 39–48.
8. Wu, L.; Liu, X.; Lv, G.; Zhu, R.; Tian, L.; Liu, M.; Li, Y.; Rao, W.; Liu, T.; Liao, L. Study on the adsorption properties of methyl orange by natural one-dimensional nano-mineral materials with different structures. *Sci. Rep.* **2021**, *11*, 10640. [CrossRef] [PubMed]
9. Karthikeyan, P.; Ramkumar, K.; Pandi, K.; Fayyaz, A.; Meenakshi, S.; Park, C.M. Effective removal of Cr (VI) and methyl orange from the aqueous environment using two-dimensional (2D) $Ti_3C_2T_x$ MXene nanosheets. *Ceram. Int.* **2021**, *47*, 3692–3698. [CrossRef]
10. da Silva, R.J.; Mojica-sánchez, L.C.; Gorza, F.D.S.; Pedro, G.C.; Maciel, B.G.; Ratkovski, G.P.; da Rocha, H.D.; do Nascimento, K.T.O.; Medina-llamas, J.C.; Chávez-Guajardo, A.E.; et al. Kinetics and thermodynamic studies of Methyl Orange removal by polyvinylidene fluoride-PEDOT mats. *J. Environ. Sci.* **2021**, *100*, 62–73. [CrossRef]
11. Kgatle, M.; Sikhwivhilu, K.; Ndlovu, G.; Moloto, N. Degradation Kinetics of Methyl Orange Dye in Water Using trimetallic Fe/Cu/Ag nanoparticles. *Catalysts* **2021**, *11*, 428. [CrossRef]
12. Neethu, N.; Choudhury, T. Treatment of Methylene Blue and Methyl Orange Dyes in Wastewater by Grafted Titania Pillared Clay Membranes. *Recent Pat. Nanotechnol.* **2018**, *12*, 200–207. [CrossRef] [PubMed]
13. Katheresan, V.; Kansedo, J.; Lau, S.Y. Efficiency of various recent wastewater dye removal methods: A review. *J. Environ. Chem. Eng.* **2018**, *6*, 4676–4697. [CrossRef]
14. Yönten, V.; Sanyürek, N.K.; Kivanç, M.R. A thermodynamic and kinetic approach to adsorption of methyl orange from aqueous solution using a low cost activated carbon prepared from *Vitis vinifera* L. *Surf. Interfaces* **2020**, *20*, 1–8. [CrossRef]
15. Maruthanayagam, A.; Mani, P.; Kaliappan, K.; Chinnappan, S. In vitro and In silico Studies on the Removal of Methyl Orange from Aqueous Solution Using Oedogonium subplagiostomum AP1. *Water Air Soil Pollut.* **2020**, *231*, 232. [CrossRef]
16. Makeswari, M.; Saraswathi, P. Photo catalytic degradation of methylene blue and methyl orange from aqueous solution using solar light onto chitosan bi-metal oxide composite. *SN Appl. Sci.* **2020**, *2*. [CrossRef]

17. Mohamed, E.A.; Selim, A.Q.; Ahmed, S.A.; Sellaoui, L.; Bonilla-Petriciolet, A.; Erto, A.; Li, Z.; Li, Y.; Seliem, M.K. H_2O_2-activated anthracite impregnated with chitosan as a novel composite for Cr (VI) and methyl orange adsorption in single-compound and binary systems: Modeling and mechanism interpretation. *Chem. Eng. J.* **2020**, *380*, 122445. [CrossRef]
18. Topuz, F.; Holtzl, T.; Szekely, G. Scavenging organic micropollutants from water with nanofibrous hypercrosslinked cyclodextrin membranes derived from green resources. *Chem. Eng. J.* **2021**, *419*, 129443. [CrossRef]
19. Borsagli, F.G.L.M.; Ciminelli, V.S.T.; Ladeira, C.L.; Haas, D.J.; Lage, A.P.; Mansur, H.S. Multi-functional eco-friendly 3D scaffolds based on N-acyl thiolated chitosan for potential adsorption of methyl orange and antibacterial activity against Pseudomonas aeruginosa. *J. Environ. Chem. Eng.* **2019**, *7*, 103286. [CrossRef]
20. Liu, X.; Zhou, Z.; Yin, J.; He, C.; Zhao, W.; Zhao, C. Fast and environmental-friendly approach towards uniform hydrogel particles with ultrahigh and selective removal of anionic dyes. *J. Environ. Chem. Eng.* **2020**, *8*, 104352. [CrossRef]
21. Onder, A.; Ilgin, P.; Ozay, H.; Ozay, O. Removal of dye from aqueous medium with pH-sensitive poly chloride-co-1-vinyl-2-pyrrolidone] cationic hydrogel. *J. Environ. Chem. Eng.* **2020**, *8*, 104436. [CrossRef]
22. Dalalibera, A.; Vilela, B.P.; Vieira, T.; Becegato, A.V.; Paulino, T.A. Removal and selective separation of synthetic dyes from water using a polyacrylic acid-based hydrogel: Characterization, isotherm, kinetic, and thermodynamic data. *J. Environ. Chem. Eng.* **2020**, *8*, 104465. [CrossRef]
23. Dax, D.; Chávez, M.S.; Xu, C.; Willför, S.; Mendonça, R.T.; Sánchez, J. Cationic hemicellulose-based hydrogels for arsenic and chromium removal from aqueous solutions. *Carbohydr. Polym.* **2014**, *111*, 797–805. [CrossRef]
24. Alammar, A.; Park, S.; Ibrahim, I.; Arun, D.; Holtzl, T.; Dumée, L.F.; Ngee, H.; Szekely, G. Architecting neonicotinoid-scavenging nanocomposite hydrogels for environmental remediation. *Appl. Mater. Today* **2020**, *21*, 100878. [CrossRef]
25. Singh, N.; Kumari, S.; Goyal, N.; Khan, S. Al_2O_3/GO cellulose based 3D-hydrogel for efficient fluoride removal from water. *Environ. Nanotechnol. Monit. Manag.* **2021**, *15*, 100444.
26. Hameed, A.; Khurshid, S.; Adnan, A. Synthesis and characterization of carboxymethyl cellulose based hydrogel and its applications on water treatment. *Desalin. Water Treat.* **2020**, *196*, 214–227. [CrossRef]
27. Chen, X.; Song, Z.; Li, S.; Thang, N.T.; Gao, X.; Gong, X.; Guo, M. Facile one-pot synthesis of self-assembled nitrogen-doped carbon dots/cellulose nanofibril hydrogel with enhanced fluorescence and mechanical properties. *Green Chem.* **2020**, *22*, 3296–3308. [CrossRef]
28. Santander, P.; Oyarce, E.; Sánchez, J. New insights in the use of a strong cationic resin in dye adsorption. *Water Sci. Technol.* **2020**, *81*, 773–780. [CrossRef]
29. Liu, J.; Korpinen, R.; Mikkonen, K.S.; Willför, S.; Xu, C. Nanofibrillated cellulose originated from birch sawdust after sequential extractions: A promising polymeric material from waste to films. *Cellulose* **2014**, *21*, 2587–2598. [CrossRef]
30. Dax, D.; Chávez Bastidas, M.S.; Honorato, C.; Liu, J.; Spoljaric, S.; Seppälä, J.; Mendonça, R.T.; Xu, C.; Willför, S.; Sánchez, J. Tailor-made hemicellulose-based hydrogels reinforced with nanofibrillated cellulose. *Nord. Pulp Pap. Res. J.* **2015**, *30*, 373–384. [CrossRef]
31. Das, A.K.; Dewanjee, S. Optimization of Extraction Using Mathematical Models and Computation. In *Computational phytochemistry*; Elsevier: Kolkata, India, 2018; pp. 75–106.
32. Sen, M.; Yakar, A.; Güven, O. Determination of average molecular weight between cross-links ($\overline{M}(c)$) from swelling behaviours of diprotic acid-containing hydrogels. *Polymer (Guildf)* **1999**, *40*, 2969–2974. [CrossRef]
33. Zu, G.; Shen, J.; Zou, L.; Wang, F.; Wang, X.; Zhang, Y.; Yao, X. Nanocellulose-derived highly porous carbon aerogels for supercapacitors. *Carbon N. Y.* **2016**, *99*, 203–211. [CrossRef]
34. Ben Ammar, N.E.; Saied, T.; Barbouche, M.; Hosni, F.; Hamzaoui, A.H.; Şen, M. A comparative study between three different methods of hydrogel network characterization: Effect of composition on the crosslinking properties using sol–gel, rheological and mechanical analyses. *Polym. Bull.* **2018**, *75*, 3825–3841. [CrossRef]
35. Mahmudi, N.; Şen, M.; Rendevski, S.; Güven, O. Radiation synthesis of low swelling acrylamide based hydrogels and determination of average molecular weight between cross-links. *Nucl. Instrum. Methods Phys. Res. Sect. B Beam Interact. Mater. Atoms* **2007**, *265*, 375–378. [CrossRef]
36. Şen, M.; Hayrabolulu, H. Radiation synthesis and characterisation of the network structure of natural/synthetic double-network superabsorbent polymers. *Radiat. Phys. Chem.* **2012**, *81*, 1378–1382. [CrossRef]
37. Wang, B.; Yang, X.; Ma, L.; Zhai, L.; Xuan, J.; Liu, C.; Bai, Z. Ultra-high efficient pH induced selective removal of cationic and anionic dyes from complex coexisted solution by novel amphoteric biocomposite microspheres. *Sep. Purif. Technol.* **2020**, *231*, 115922. [CrossRef]
38. Mouni, L.; Belkhiri, L.; Bollinger, J.; Bouzaza, A.; Assadi, A.; Tirri, A.; Dahmoune, F.; Madani, K.; Remini, H. Removal of Methylene Blue from aqueous solutions by adsorption on Kaolin: Kinetic and equilibrium studies. *Appl. Clay Sci.* **2018**, *153*, 38–45. [CrossRef]
39. Zhou, A.; Chen, W.; Liao, L.; Xie, P.; Zhang, T.C.; Wu, X.; Feng, X. Comparative adsorption of emerging contaminants in water by functional designed magnetic poly(N-isopropylacrylamide)/chitosan hydrogels. *Sci. Total Environ.* **2019**, *671*, 377–387. [CrossRef] [PubMed]
40. Li, J.; Xu, Z.; Wu, W.; Jing, Y.; Dai, H.; Fang, G. Nanocellulose/Poly(2-(dimethylamino)ethyl methacrylate)Interpenetrating polymer network hydrogels for removal of Pb(II) and Cu(II) ions. *Colloids Surf. Physicochem. Eng. Asp.* **2018**, *538*, 474–480. [CrossRef]

41. Li, J.; Zuo, K.; Wu, W.; Xu, Z.; Yi, Y.; Jing, Y.; Dai, H.; Fang, G. Shape memory aerogels from nanocellulose and polyethyleneimine as a novel adsorbent for removal of Cu(II) and Pb(II). *Carbohydr. Polym.* **2018**, *196*, 376–384. [CrossRef]
42. Zhang, F.; Ren, H.; Tong, G.; Deng, Y. Ultra-lightweight poly (sodium acrylate) modified TEMPO-oxidized cellulose nanofibril aerogel spheres and their superabsorbent properties. *Cellulose* **2016**, *23*, 3665–3676. [CrossRef]
43. Sánchez, J.; Mendoza, N.; Rivas, B.L.; Basáez, L.; Santiago-García, J.L. Preparation and characterization of water-soluble polymers and their utilization in chromium sorption. *J. Appl. Polym. Sci.* **2017**, *134*, 45355. [CrossRef]
44. Tapiero, Y.; Sánchez, J.; Rivas, B.L. Ion-selective interpenetrating polymer networks supported inside polypropylene microporous membranes for the removal of chromium ions from aqueous media. *Polym. Bull.* **2016**, *73*, 989–1013. [CrossRef]
45. Wang, W.; Zhang, X.; Teng, A.; Liu, A. Mechanical reinforcement of gelatin hydrogel with nanofiber cellulose as a function of percolation concentration. *Int. J. Biol. Macromol.* **2017**, *103*, 226–233. [CrossRef] [PubMed]
46. Tanan, W.; Panichpakdee, J.; Saengsuwan, S. Novel biodegradable hydrogel based on natural polymers: Synthesis, characterization, swelling/reswelling and biodegradability. *Eur. Polym. J.* **2019**, *112*, 678–687. [CrossRef]
47. Lubis, R.; Wirjosentono, B.; Eddyanto; Septevani, A.A. Preparation, characterization and antimicrobial activity of grafted cellulose fiber from durian rind waste. *Colloids Surf. Physicochem. Eng. Asp.* **2020**, *604*, 125311. [CrossRef]
48. Li, I.; Huang, G.; Tsai, C.; Chen, Y.; Hong, M.; Tsai, T. Ammonium nitrogen adsorption from aqueous solution by poly(sodium acrylate)s: Effect on the amount of crosslinker and initiator. *J. Appl. Polym. Sci.* **2020**, *137*, 49581. [CrossRef]
49. Kang, S.; Qin, L.; Zhao, Y.; Wang, W.; Zhang, T.; Yang, L.; Rao, F.; Song, S. Enhanced removal of methyl orange on exfoliated montmorillonite/chitosan gel in presence of methylene blue. *Chemosphere* **2020**, *238*, 124693. [CrossRef]
50. Zhai, L.; Bai, Z.; Zhu, Y.; Wang, B.; Luo, W. Fabrication of chitosan microspheres for efficient adsorption of methyl orange. *Chin. J. Chem. Eng.* **2018**, *26*, 657–666. [CrossRef]
51. Azam, K.; Raza, R.; Shezad, N.; Shabir, M.; Yang, W.; Ahmad, N.; Shafiq, I.; Akhter, P.; Razzaq, A.; Hussain, M. Development of recoverable magnetic mesoporous carbon adsorbent for removal of methyl blue and methyl orange from wastewater. *J. Environ. Chem. Eng.* **2020**, *8*, 104220. [CrossRef]
52. Oyarce, E.; Pizarro, G.D.C.; Oyarzún, D.P.; Martin-Trasanco, R.; Sánchez, J. Adsorption of methylene blue in aqueous solution using hydrogels based on 2-hydroxyethyl methacrylate copolymerized with itaconic acid or acrylic acid. *Mater. Today Commun.* **2020**, *25*, 101324. [CrossRef]
53. Zeng, L.; Xie, M.; Zhang, Q.; Kang, Y.; Guo, X.; Xiao, H.; Peng, Y.; Luo, J. Chitosan/organic rectorite composite for the magnetic uptake of methylene blue and methyl orange. *Carbohydr. Polym.* **2015**, *123*, 89–98. [CrossRef]
54. Shah, M.P.; Patel, K.A.; Nair, S.S.; Darji, A.M. Microbial decolourization of methyl orange dye by Pseudomonas spp. *OA Biotechnol.* **2013**, *1*, 54–59.
55. Saeed, M.; Usman, M.; ul Haq, A. Catalytic Degradation of Organic Dyes in Aqueous Medium. In *Photochemistry and Photophysics-Fundamentals to Applications*; IntechOpen: London, UK, 2018; Volume 13, p. 197.
56. Sarayu, K.; Sandhya, S. Aerobic Biodegradation Pathway for Remazol Orange by Pseudomonas aeruginosa. *Appl. Biochem. Biotechnol.* **2010**, *160*, 1241–1253. [CrossRef]
57. Aouni, A.; Fersi, C.; Cuartas-uribe, B.; Bes-pía, A.; Alcaina-miranda, M.I.; Dhahbi, M. Reactive dyes rejection and textile effluent treatment study using ultra filtration and nano filtration processes. *Desalination* **2012**, *297*, 87–96. [CrossRef]
58. Bandala, E.R.; Pel, M.A.; Garc, A.J.; Salgado, M.D.J.; Moeller, G. Photocatalytic decolourisation of synthetic and real textile wastewater containing benzidine-based azo dyes. *Chem. Eng. Process. Process Intensif.* **2008**, *47*, 169–176. [CrossRef]
59. Singh, T.; Singhal, R. Reuse of a Waste Adsorbent Poly (AAc/AM/SH)-Cu Superabsorbent Hydrogel, for the Potential Phosphate Ion Removal from Waste Water: Matrix Effects, Adsorption Kinetics, and Thermodynamic Studies. *J. Appl. Polym. Sci.* **2013**, *9*, 3126–3139. [CrossRef]
60. Alsaiari, N.S.; Amari, A.; Katubi, K.; Alzahrani, F.; Rebah, F.B.; Tahoon, M.A. Innovative Magnetite Based Polymeric Nanocomposite for Simultaneous Removal of Methyl Orange and Hexavalent Chromium from Water. *Processes* **2021**, *9*, 576. [CrossRef]
61. Zhao, P.; Zhang, R.; Wang, J. Adsorption of methyl orange from aqueous solution using chitosan/diatomite composite. *Water Sci. Technol.* **2017**, *57*, 1633–1642. [CrossRef] [PubMed]
62. Annadurai, G.; Juang, R.; Lee, D. Use of cellulose-based wastes for adsorption of dyes from aqueous solutions. *J. Hazard. Mater.* **2002**, *B92*, 263–274. [CrossRef]

Article

Phosphorus Release and Adsorption Properties of Polyurethane–Biochar Crosslinked Material as a Filter Additive in Bioretention Systems

Yike Meng [1,*], Yuan Wang [2,*] and Chuanyue Wang [1]

1 College of Civil and Transportation Engineering, Hohai University, Nanjing 210098, China; 19860001@hhu.edu.cn
2 College of Water Conservancy and Hydropower Engineering, Hohai University, Nanjing 210098, China
* Correspondence: mengyike@hhu.edu.cn (Y.M.); wangyuan@hhu.edu.cn (Y.W.)

Citation: Meng, Y.; Wang, Y.; Wang, C. Phosphorus Release and Adsorption Properties of Polyurethane–Biochar Crosslinked Material as a Filter Additive in Bioretention Systems. *Polymers* **2021**, *13*, 283. https://doi.org/10.3390/polym13020283

Received: 11 December 2020
Accepted: 13 January 2021
Published: 17 January 2021

Publisher's Note: MDPI stays neutral with regard to jurisdictional claims in published maps and institutional affiliations.

Copyright: © 2021 by the authors. Licensee MDPI, Basel, Switzerland. This article is an open access article distributed under the terms and conditions of the Creative Commons Attribution (CC BY) license (https://creativecommons.org/licenses/by/4.0/).

Abstract: Bioretention systems are frequently employed in stormwater treatment to reduce phosphorus pollution and prevent eutrophication. To enhance their efficiency, filter additives are required but the currently used traditional materials cannot meet the primary requirements of excellent hydraulic properties as well as outstanding release and adsorption capacities at the same time. In this research, a polyurethane-biochar crosslinked material was produced by mixing the hardwood biochar (HB) with polyurethane to improve the performance of traditional filter additives. Through basic parameter tests, the saturated water content of polyurethane-biochar crosslinked material (PCB) was doubled and the permeability coefficient of PCB increased by two orders of magnitude. Due to the polyurethane, the leaching speed of phosphorus slowed down in the batching experiments and fewer metal cations leached. Moreover, PCB could adsorb 93–206 mg/kg PO_4^{3-} at a typical PO_4^{3-} concentration in stormwater runoff, 1.32–1.58 times more than HB, during isothermal adsorption experiments. In the simulating column experiments, weaker hydropower reduced the PO_4^{3-} leaching quantities of PCB and had a stable removal rate of 93.84% in phosphate treatment. This study demonstrates the potential use of PCB as a filter additive in a bioretention system to achieve hydraulic goals and improve phosphate adsorption capacities.

Keywords: polyurethane-biochar crosslinked material; modified filter additive; phosphorus release and adsorption; bioretention facilities; stormwater treatment

1. Introduction

Phosphorus is the main factor of eutrophication in urban rivers [1] and comes from industry, agriculture and transportation activities [2], being mainly spread by urban stormwater runoff, a kind of non-point pollution of surface water. To manage stormwater runoff, developers typically use bioretention facilities, whose primary goal is to reduce floods by reducing the volume of overland flow during a storm event and reinstating natural stormwater infiltration in the developed area to its pre-developmental capacity [3]. The filtration layer in bioretention facilities takes on the role of purification, which has been proven to be efficient in removing oil [4], heavy metal [5] and pathogenic bacteria indicator species [6] from stormwater runoff. However, the performance of traditional filtration layer is not effective at removing phosphorus [7], mainly due to the leaching of phosphorus from compost (a typical filter additive in the filtration layer) [8].

Attempts have been made to improve phosphorus removal. In recent studies, many natural and artificial materials have been investigated to determine their feasibility as filter additives in the filtration layer of bioretention facilities and they can be generally divided into three types: biological waste materials, mineral materials and biochar. Biological waste materials (e.g., coconut [9], peat [10] and livestock manure [11], etc.) still have high leaching quantities of phosphorus, due to the accumulation of a large number of nitrogen and phosphorus nutrients in the growth process, leading to the limitations of phosphorus removal. Mineral filter additives (e.g., volcanic stone [12], montmorillonite [13] and zeolite [14], etc.) have relatively low removal rates and water retention capacities compared to biological waste materials, in spite of their reduced nutrient-leaching quantities. Biochar, a thermal decomposition product of biomass, is suitable as a filter additive in bioretention facilities due to its cleanness [15] and it can reduce the concentration of both nitrogen and phosphorus in runoff [16]. As a popular soil amendment, biochar can also sequester carbon and retain nutrients [17] and this is significant for additive materials to support the growth of plants in the vegetation layer of bioretention facilities. However, pyrolysis brings brittleness to the pore structure of the biochar, which is destroyed by the hydropower of stormwater during long term operation [18], while the saturated hydraulic conductivity of bioretention facilities is significantly reduced [19]. This does not meet the primary goal of bioretention facilities. If this shortcoming of biochar can be improved, it would be a big step forward for bioretention systems.

Polyurethane materials provide a potentially feasible solution to this problem. On the one hand, the water retention capacity of polyurethane improved under multi-field coupling, due to the broken molecular chain and higher connectivity of the pore structure [20]. When subjected to soil, water and air, the structure of polyurethane showed no significant changes, indicating the good durability of its mechanical properties in the long term [21]. On the other hand, it has been widely recognized that polyurethane foams can be employed as highly efficient adsorbents in removing heavy metals [22], ammonium [23], nitrate [24] and some organic pollutants (e.g., dialkyl phthalates [25], oils and trichloromethane [26], etc.). All these properties meet the requirements of filter additives in bioretention systems: a high hydraulic conductivity to reduce overland stormwater, a high retention volume to minimize peak flow, a good endurance to multi-field coupling effects and a high removal capacity of many contaminants from stormwater. However, due to the limitation of raw material composition, the nutrients needed for vegetation growth cannot be provided by polyurethane alone. Considering the characteristics of biochar, it seems that a combination of polyurethane and biochar may achieve acceptable results.

Additionally, polyurethane, as a coating material, will prolong the nutrient release period of inner fertilizers in agriculture and reduce their leaching quantities [27]. Combined with fertilizers, polyurethane composite material has a high potential to preserve moisture and fertility for the amelioration of desertification [28]. Moreover, this advantage could be applied to biochar in the form of a polyurethane–biochar composite material, helping to release phosphorus more slowly.

Some work has been done regarding polyurethane composites in order to relieve the eutrophication crisis in urban rivers caused by phosphorus—Sasidharan developed silver/silver oxide nanoparticles impregnating polyurethane foam with a 61.24% phosphate and this system was still effective in removing 20.58% of phosphate after 7 cycles of reuse [29]. Nie also conducted a column experiment to purify the septic tank effluent, which was mixed with soil and polyurethane and found that the column had a phosphorus removal rate of 96% [30]. While demonstrating above the positive results in the phosphorus removal of polyurethane composites, these studies are limited and do not estimate the phosphorus leaching quantities of polyurethane composites, nor do they try to apply them to bioretention facilities.

Based on the current research, we assume that, if polyurethane and biochar could be combined together as a composite material with both of their advantages, this composite may have an outstanding hydraulic and environmental performance as a filter additive in bioretention systems. Hence, the present study tried to explore the feasibility of a novel composite material, polyurethane-biochar crosslinked material (PCB), as a filter additive in bioretention systems. We estimated the water retention capacity, phosphorus leaching quantities and adsorption capacity of PCB and aimed to improving the performance of bioretention facilities and avoiding eutrophication. This composite material is a sponge structure in which polyurethane interpenetrates and crosslinks the biochar. Hardwood biochar (HB) was selected as a raw material for the production of PCB because of its low nutrient concentrations [31] and high specific surface area [32] and it was also compared with PCB in this study.

2. Materials and Methods

2.1. Synthesis of Polyurethane–Biochar Crosslinked Material

- Hardwood biochar (HB) production: the raw material for the synthesis of PCB used in this study is commercially common hardwood biochar, which was produced using pine at a 600 °C pyrolysis temperature and was purchased from Jinlian Landscape Engineering Services Co., LTD. (Hangzhou, China).
- Polyurethane-biochar crosslinked material (PCB) preparation: PCB was synthesized with a simple one-shoot method, where the polyol and HB (for modifying polyurethane) were mixed with isocyanate. The polyol source used in this research was glycol and isocyanate was diphenyl-methane-diisocyanate (MDI).
- PCB production: 60 g of glycol, 100 g of deionized water (DW) and 5 g of HB were mixed continuously at 750 rpm and 60 °C for 20 min with a magnetic stirrer (VRera, Nanjing, China). After that, while keeping the same rotating speed and temperature, 250 g of MDI was added dropwise at a constant speed before air bubbles formed. The procedure was continued by pouring the mixture into a $30 \times 30 \times 10$ cm^3 of mold and transferring it to a vacuum oven (Xidebao, Shanghai, China) at 60 °C for 3 h and then curing it for 24 h.
- Cutting: The cured PCB was cut into granules with a particle size of 1–2mm, considering the practical application and consistent research scale of HB.
- The PCB used in this study was produced with the assistance of Jinlian Company. Scanning electron microscopy (SEM) and energy dispersive spectroscopy (EDS) were conducted on a Hitachi SU3500/S4800 High-Resolution Focused Ion Beam and Scanning Electron Microscope (Hitachi, Tokyo, Japan) working at an accelerating voltage of 10 kV, helping to illustrate the microstructure of PCB and HB.

2.2. The Hydraulic Properties and Other Physicochemical Characterizations Tests

The hydraulic property tests include a saturated moisture content test (to evaluate stormwater retention volume) and a permeability coefficient test (to evaluate hydraulic conductivity).

- Saturated moisture content test: The natural bulk densities of the PCB and HB were measured by the cutting ring method (ISO 11272:2017). The samples in the cutting ring were vacuumed by a pump, immersed in deionized water (DW) for 24 h, weighed, dried in an oven at 60 °C for 48 h and weighed again to determine the natural and saturated moisture content (ISO 17892-1:2014).
- Permeability coefficient test: The permeability coefficient of the materials was determined by the constant head method with a Type 70 permeameter (Nanjing Soil Instrument Factory Co., LTD., Nanjing, China) (ISO 17892-11:2019).
- Other physicochemical characterizations influencing the leaching and adsorption capacity of materials were tested:
- The particle size of materials was measured by a sieving method.
- The specific gravity of the materials was measured by the gravity bottle method and the pore ratio of the materials was obtained after conversion with the saturated water content.
- The pH of modifier materials was measured at a material/DW ratio of 1:50 by mass.
- BET surface area was determined by N_2 (77 K) adsorption on an ASAP 2020 Accelerated Surface Area and Porosimetry System (Micromeritics Instrument, Atlanta, GA, USA) after degassing for 12 h via VacPrep^{TM} 061 (Micromeritics Instrument, Atlanta, GA, USA), a Gas Adsorption Sample Preparation Device.
- The cation exchange capacity (CEC) was determined by a hexamminecobalt (III) chloride solution (ISO 23470:2018).

- TP (total phosphorus content) of materials was measured after strong acid digestion and analyzed by ICP-OES (Thermo Fisher Scientific, Waltham, MA, USA).

2.3. Leaching Experiments

The main reason for the unstable efficiency of phosphorus removal in bioretention facilities is the leaching of phosphorus from filter additives, leading to eutrophication in urban rivers. To avoid this pollution and evaluate the phosphorus leaching quantities, polyurethane-biochar crosslinked material (PCB) and hardwood biochar (HB) were continuously rinsed with deionized water (DW) or artificial stormwater (AS) and the release characterizations of phosphorus and metal ions were analyzed. AS was a mixture solution of 120 mg/L $CaCl_2$ and 3 mg/L PO_4-P (Na_2HPO_4) at pH 7.0, referring to the recognized makeup of synthetic urban runoff [4]. In order to reduce the influence of other factors, oils, heavy metals and nitrogen were not added to the AS.

Five grams of material, dried at 60 °C for 48 h, was added to a conical flask containing 100 mL of DW or AS (ISO 21268:2019). At 20 ± 2 °C, the material oscillated at a frequency of 150 rpm for 24 h. After settlement for 30 min, supernatants were aspirated into a centrifuge tube and centrifuged at 5000 rpm (Hitachi CR21 III, Tokyo, Japan) for 20 min. Another 100 mL of DW or AS was added to the conical flask and the leaching-settling-centrifuging steps were repeated another seven times. The supernatants were filtered with 0.45 μm filters and analyzed for phosphate (PO_4-P), total phosphorus (TP-P), metal ions (Na^+, K^+, Mg^{2+}, Ca^{2+}) and water conductivity. The San^{++} Continuous Flow Analyzer (Skalar Dutch) was used for the testing of PO_4-P and TP-P. The samples were mixed with $H_8MoN_2O_4$, $C_8H_4K_2O_{12}Sb_2$ and $C_6H_8O_6$ and a colorimetric analysis was carried out at an 880 nm wavelength for the detection of PO_4-P with a detection limit of 0.001 mg/L (ISO 15681:2018). The method for the detection of TP-P was the same as PO_4-P, only needing a pretreatment (oxidized by $K_2S_2O_8$ solution and digested by UV) before mixing. Metal ions were tested by using a NexION300X inductively coupled plasma mass spectrometer (PerkinElmer, Waltham, MA, USA) with a detection limit of 0.001 mg/L. Conical bottles containing only DW or AS without other materials were used as the control groups. We repeated two sets of tests for each material. After leaching experiments, the DW-rinsed materials were denoted as PCB-DW and HB-DW respectively and then put into a desiccator for later tests.

2.4. Phosphate Adsorption Experiments

As a filter additive in stormwater runoff treatment, the material needs to have a certain phosphate adsorption capacity. In order to evaluate the phosphate adsorption capacity of the materials, adsorption experiments were conducted using PCB-DW and HB-DW with different concentrations of phosphate. Standard solutions of 100 mg/L Na_2HPO_4 were diluted to 0, 0.5, 1, 2, 5, 7 and 10 mg/L with DW and AS, respectively and AS only contained 120 mg/L $CaCl_2$. We placed 0.2 g of PCB-DW and HB-DW into 50 mL conical flasks, added 10 mL of the above solution and oscillated the flasks for 24 h at 150 rpm at 20 ± 2 °C. The extraction method of the supernatants was the same as the leaching test and the concentrations of phosphate in the supernatants were measured. The detection method was the same as above. The experiment was repeated in 2 groups for each material. Additional conical flasks with phosphate solutions but no PCB-DW or HB-DW were used as control groups.

In order to explore the adsorption properties and capacity of phosphate, Langmuir and Freundlich models were used to fit the adsorption equilibrium quantities of phosphate after 24 h. The calculation formula of the equilibrium adsorption quantity q_e (mg/kg) for phosphate at 24 h is [33]:

$$q_e = \frac{(C_0 - C_e)V}{W}, \qquad (1)$$

where, C_0 and C_e are the concentrations (mg/L) of PO_4^{3-} in the solution before and after the adsorption test; V is the solution volume (L); W is the material mass (kg).

The Freundlich model was used to fit the isothermal adsorption results [23]:

$$q_e = K_F C_e^{1/n}, \qquad (2)$$

where, K_F is the volume-affinity parameter (L/mg) of the Freundlich model, which can be regarded as the adsorption capacity at unit a concentration of C_e; n is the Freundlich characteristic constant, the value of $1/n$ is generally between 0 and 1 and its value represents the influence of the concentration on the adsorption capacity. The smaller l/n is, the better the adsorption property is. When $1/n$ is between 0.1 and 0.5, it is easy to absorb; it is difficult to adsorb when $1/n$ is more than 2.

The Langmuir model of single molecular layer physical adsorption was also used to fit the isothermal adsorption results [23]:

$$q_e = q_{\max} \frac{K_L C_e}{1 + K_L C_e} \qquad (3)$$

where q_{max} is the maximum adsorption capacity (mg/kg); K_L is the affinitive parameter of the Langmuir model (L/mg), which is the equilibrium constant of adsorption, also known as the adsorption coefficient. The higher the value of K_L is, the stronger the adsorption capacity is.

The dimensionless coefficient R_L is used to determine whether adsorption easily occurs [34]:

$$R_L = \frac{1}{1 + K_L C_0}, \qquad (4)$$

when $0 < R_L < 1$, adsorption easily occurs; when $R_L > 1$, adsorption does not easily occur; when $R_L = 0$, the adsorption process is reversible. When $R_L = 1$, adsorption is linear.

2.5. Column Experiments

In order to simulate the performance of the filter additives under an actual working situation, column experiments were carried out in the laboratory. The river sand (washed by DW and dried) and filter additives (unwashed) were mixed evenly according to a mass ratio of 10/1. In 3 PVC columns (30 cm in height, 6 cm in diameter and 3 mm holes of sieve tray at the bottom), filter papers were placed at the bottom and evenly mixed geomedia filled the columns, denoted as PCB-Column, HB-Column and Sand-Column. Sand-Column was filled only with river sand as a control group. For each 2 cm of mixed filling, a wooden hammer was employed to drop 10 times from 5 cm above the filling until it was filled to 24 cm. Washed and dried gravels were placed on top of the mixture filling to prevent current scour. After filling, peristaltic pumps (BT01-100) were used at the top of the columns to pump DW for 3 h at a speed of 15 mL/min. The inflow velocity was calculated according to the rainfall intensity formula in China. In this study, rainfall occurred once a year and lasted for 3 h, the catchment ratio was 15 and the infiltration flow per minute was calculated as 15 mL. In total, 50 mL of effluent was collected every 20 min by the effluent tubing at the bottom in order to detect the contents of PO_4-P. The detection methods were the same as above. AS was pumped at the same rate for 3 h after the DW was pumped in for 3 h. AS included 3 mg/L PO_4-P. Effluent was collected and measured as above.

3. Results and Discussion

3.1. Polymerization Process and Microstructure

The Polyurethane-biochar crosslinked material (PCB) was obtained via a one-shoot method. The reaction of PCB and the interaction with the addition of HB are shown in Figure 1. Glycol contained a hydroxyl group that could reacted with an isocyanate group from MDI to obtain a urethane linkage and a monomeric unit was formed by the two constitutional units. Furthermore, the polymer chain was linked through urethane linkages between monomeric units. HB was crosslinked between two monomeric units [22].

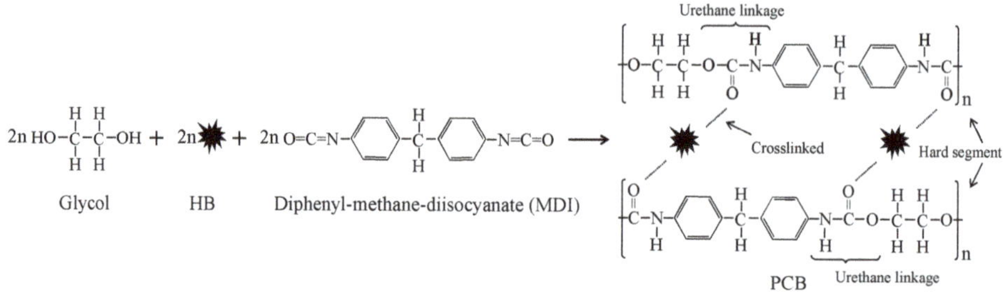

Figure 1. The reaction of polyurethane-biochar crosslinked material (PCB) via the one–shoot method with the addition of HB.

The properties of PCB depend on various factors (chain rigidity, cross-linking degree, intermolecular bonds, etc.) and can be changed in a wide range by the proper selection of raw materials [35]. Considering the application of PCB in bioretention facilities, durability, resilience, porousness and hydrophilia were demanded by the multi-field (water-soil-air) coupling effects. Glycol was chosen as the main polyol source that would reduce the length of carbon chain and improve the hard segment content. The hard segment can improve the initial modulus and tensile strengths of polyurethane materials [36] and crosslinked polymerization would make up for the brittleness of HB to make it resilient against weathering. HB, as an inner material, was blocked in the network of polyurethane by the interaction between carbonyl groups and HB. The blocking way was referred from the FTIR analysis (Figure 2): The interaction between the HB and polyurethane was observed by the shifted wavenumber of carbonyl groups at around 1600 cm^{-1}. As a carbon-rich material, HB possessed abundant hydroxyl groups [37] and was more likely to have this interaction [38].

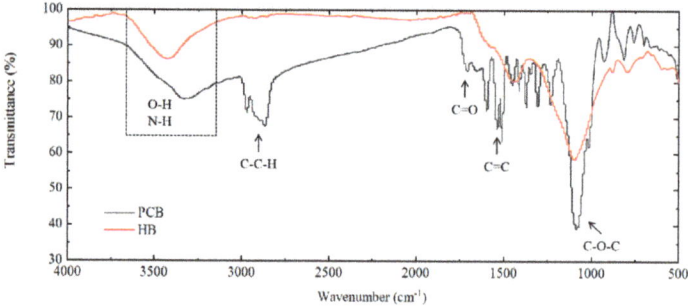

Figure 2. Fourier transform infrared (FTIR) spectra of PCB and hardwood biochar (HB).

The polyol source also influenced the microstructure of polyurethane composites [39]. Small molecular weight polyol has a better smoothness and porousness. As shown in Figure 3, the concave-convex surfaces and throats of PCB can be clearly seen, yet HB has a flat shape and few holes and bumps. This difference in structure may account for the improvements in the hydraulic properties of PCB and HB.

(a) (b)

Figure 3. The scanning electron microscopy (SEM) images of (**a**) PCB and (**b**) HB.

3.2. Hydraulic Properties of Modifiers

Excellent hydraulic performance, including high water retention capacity and permeability, are fundamental criteria for choosing filter additives in bioretention systems, which are also shortcomings of biochar at present [19], when compared to biological waste and mineral materials. Basic experiments were conducted to inspect the improvement of PCB and the results are illustrated in Table 1. After being modified by polyurethane, the material became lighter, with a bulk density of 0.165 g/cm^3, compared to its former bulk density of 0.378 g/cm^3. Such lightweight polyurethane gave PCB a sizable performance boost in water retention capacity [40], whose saturated water content was improved from 195.65% to 383.5%. If added to the filter layer with the same mass ratio (4% of additive materials in traditional bioretention facilities [41]), PCB can reduce 42–63 mm of stormwater within the unit area according to the following water absorption formula:

$$W_{water} = \rho_{filterlayer} \times h \times n \times \omega_{sat}, \qquad (5)$$

where W_{water} is the stormwater retention volume of the filter additive within unit area; $\rho_{filterlayer}$ is the density of the filter layer in bioretention facilities, ranging from 0.8 to 1.2 g/cm^3; h is the filling height, usually calculated as 70 cm; n is the mass ratio of the filter additive; ω_{sat} is the saturated moisture content of the filter additive.

Table 1. Physicochemical properties of PCB and HB.

Media Material	ρ [1] (g/cm^3)	Particle Size (mm)	e [2]	ω_{sat} [3] (%)	K [4] (cm/s)	pH	BET (m^2/g)	CEC (cmol/kg)	TP [5] (g/kg)
PCB	0.165	1–2	3.20	383.50%	8.56×10^{-2}	6.62	83.14	37.5	1.19
HB	0.378	<0.5	3.88	195.65%	6.57×10^{-4}	8.80	118.45	7.4	3.80

[1] ρ is natural bulk density. [2] e is pore ratio. [3] ω_{sat} is saturated moisture content. [4] K is permeability coefficient. [5] TP is total phosphorus content.

This improvement is of great significance and will achieve a high storage volume in order to reduce peak flow and enhance the removal of many contaminants from stormwater [42]. The Permeability of HB was also improved by the polyurethane polymerization, the infiltration coefficient of which changed from 6.57×10^{-4} to 8.56×10^{-2}, with an improvement of more than two orders of magnitude. Apart from the bigger particle size, as an influencing factor on permeability, the internal throats formed in the foaming process also took effect, where stormwater flowed inside the network through the throats, increasing the number of available flowing paths [43].

3.3. Phosphorus Leaching

Before application in bioretention facilities, the quantities of phosphorus that could leach from additive materials should be estimated to prevent potential eutrophication. PO_4-P and TP-P were detected in the successive leaching solution and the results are shown in Table 2 and Figure 4.

Table 2. Phosphorus leaching quantities of PCB and HB in deionized water (DW) or artificial stormwater (AS).

Media Material	PO_4-P			TP-P			
	8 rounds (μmol/g)	1 round (μmol/g)	1 round/ 8-rounds	8 rounds (μmol/g)	8 rounds/ Total	1 round (μmol/g)	1 round/ 8 rounds
PCB-DW	2.68	1.47	54.85%	9.16	23.86%	4.09	44.65%
HB-DW	7.11	0.19	2.67%	8.55	6.98%	0.27	3.16%
PCB-AS	−4.82 [1]	−0.82 [1]	-	0.38	-	1.81	-
HB-AS	−13.67 [1]	−1.79 [1]	-	−11.51 [1]	-	−1.53 [1]	-

[1] The negative value came from the original concentration of AS solution (3 mg/L PO4-P), which was subtracted in a calculation and indicated a decrease in AS.

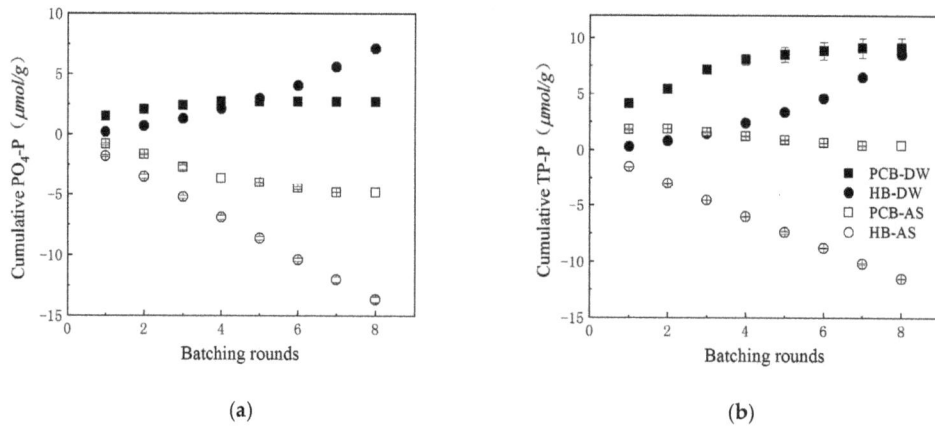

Figure 4. Cumulative phosphorus compounds leached from PCB and HB in DW or AS. (a) PO_4-P; (b) TP-P.

In general, PCB released more than HB in the first DW batching round: 1.47 μmol/g of PO_4-P and 4.09 μmol/g of TP-P for PCB, while 0.19 μmol/g of PO_4-P and 0.27 μmol/g were released for HB. After the first batch, the leaching quantities of phosphorus were in decline in each round and PCB released 2.68 μmol/g of PO_4-P and 9.16 μmol/g of TP-P in total, while these values were 7.11 and 8.55 for HB, respectively. Compared to the compost (around 82 μmol/g of PO_4-P in 6 rounds of leaching) [11] and poultry litter biochar (82.6–146.1 μmol/g of PO_4-P in 10-days leaching) [15] in other studies, the phosphorus leaching quantities of PCB and HB were relatively low.

The inhibition effects of crosslinked polyurethane on the phosphorus leaching of HB can be observed from the different leaching tendencies of the two materials: The leaching quantities of HB increased as the number of rounds increased and it continued to release more and more PO_4-P and TP-P at a nearly constant rate. The leaching tendency of HB was in agreement with former research on biochar leaching properties [44] and it can be predicted that additional phosphorus would be released with further batching. However, after being crosslinked and interpenetrated by the polyurethane, the release of phosphorus from the internal biochar was prevented, as previously reported [27]. PCB's first round of phosphorus leaching accounted for 44.65–54.85% of the total released quantities and the released quantities were only in the range of 2.67–3.16% for HB. The resilient and smooth network of polyurethane could resist scour caused by water and prevent itself from weathering. Polymerization made the HB and polyurethane blend seamlessly, which avoided phosphorus on the surface of HB being washed away by waterpower. The reason for PCB releasing more TP-P than HB could be that the surface of PCB was brushed with organophosphorus flame retardants to meet the storage and transportation conditions [45].

Since the AS contained 3 mg/L of PO_4-P, it was subtracted when calculating the cumulative phosphorus compounds leached from PCB and HB in AS, so the data present negative values. Our assumption from the negative values was that PCB and HB had a certain adsorption capacity to the 3 mg/L of PO_4-P in AS. HB had a better treatment effect on phosphorus, whose adsorption capacity was unimpeded by batching rounds. It is likely that the alkalinity of HB (Table 1) brought this benefit, which could provide an alkaline condition to form hydroxyapatite precipitation with Ca^{2+} and PO_4^{3-} in stormwater runoff [46]. Meanwhile, PCB had a relatively poor treatment performance on phosphorus due to its acidity in the water. It is believed that PCB also had a slight effect on phosphorus adsorption, reducing the leaching quantities of PO_4-P and TP-P in AS. The mechanism of PCB phosphorus adsorption could be ion exchange or physical adsorption but this is inconclusive.

3.4. Leaching of Other Ions

The main commercial processes for removing phosphorus from wastewater effluents are still chemical precipitation with metal ions [46]. The leaching pattern of metal ions could help to elucidate the mechanisms of the removal of phosphorus by PCB and HB. The leaching of low-valence metal ions (Na^+, K^+, Mg^{2+}, Ca^{2+}) as a function of batching rounds is shown in Figure 5 and summarized in Table 3. Generally, PCB leached fewer or nearly equal numbers of metal ions than HB in DW. AS prompted the metal ion-leaching quantities of HB but had no obvious impact on PCB.

Table 3. Cations leaching quantities of PCB and HB in DW or AS.

Media Material	Na^+			K^+			Mg^{2+}			Ca^{2+}		
	8 rounds (μmol/g)	1 round (μmol/g)	1 round/ 8 rounds	8 rounds (μmol/g)	1 round (μmol/g)	1 round/ 8 rounds	8 rounds (μmol/g)	1 round (μmol/g)	1 round/ 8-rounds	8 rounds (μmol/g)	1 round (μmol/g)	1 round/ 8 rounds
PCB-DW	4.28	3.13	73.05%	16.78	8.67	51.69%	25.48	5.92	23.23%	23.27	4.71	20.22%
HB-DW	3.91	1.10	28.04%	90.18	39.16	43.42%	20.55	4.29	20.90%	83.56	12.39	14.83%
PCB-AS	−1.53	−0.72	-	12.07	5.39	44.66%	42.59	4.33	10.17%	−50.65	−12.45	-
HB-AS	28.04	19.87	-	254.03	190.68	75.06%	48.45	18.15	37.46%	83.16	55.63	-

Na^+ was leached at a very low level in DW. From the energy dispersive spectroscopy results (Table 4) of PCB and HB, there was no Na^+ on the analyzed surface of PCB but a small quantity on the surface of HB (0.62%). Hence, the low quantities of Na^+ in DW were reasonable. PCB and HB had similar cumulative release amounts but their release rates and patterns were not consistent: PCB released 3.13 μmol/g Na^+ in the first round, accounting for 73.05% of total release quantities and released less during batching rounds. While HB only released 1.10 μmol/g Na^+ in the first round, after that it leached at an almost constant rate. Because of the valence being the same, the pattern of K^+ leaching was similar to Na^+: the cumulative Na^+/K^+ leaching quantities increased logarithmically, which was consistent with previous research on biochar leaching [47]. AS motivates more metal ions to leach out from HB in the first round but keeps an approximate leaching rate afterwards, as in DW. The leaching quantities of PCB were significantly lower than HB, probably because the K^+ on the surface of HB had been released in the glycol–DW mixing process, most of which was cured in the inner structure of PCB during polymerization process.

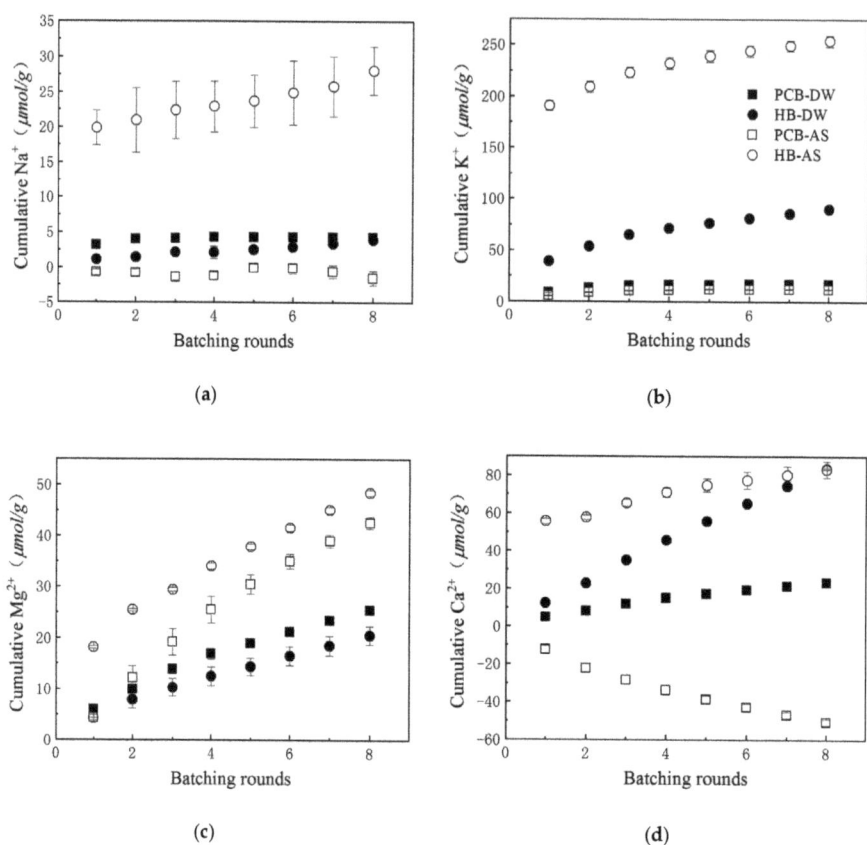

Figure 5. Cumulative low valence metal cations leached from PCB and HB in DW or AS. (**a**) Na$^+$; (**b**) K$^+$; (**c**) Mg^{2+}; (**d**) Ca^{2+}.

Table 4. Energy dispersive spectroscopy (EDS) contents of original and DW-leached materials.

Elements	PCB	PCB-DW	HB	HB-DW
		W_t%		
C	58.74	55.89	38.16	31.93
O	35.18	37.64	32.04	38.17
Na	-	-	0.62	0.27
Mg	0.83	0.73	0.74	-
Al	0.83	0.31	1.56	0.37
Si	1.26	0.24	17.61	25.96
K	0.64	-	3.16	0.86
Ca	2.52	5.19	6.11	2.44
Total		100		

The Mg^{2+} was released from PCB and HB linearly and a nearly equal amount of Mg^{2+} was released in each batching round. EDS results (Table 4) showed that there were 0.74% Mg elements on the surface of HB before leaching which could not be detected after leaching. This indicated that Mg^{2+} adhered to the surface of HB in the form of mineral ions and was washed away. The Mg^{2+} content on the surface of PCB dropped slightly after 8 rounds batching and this suggested that it mostly existed as compounds or was blocked in the polyurethane network. In AS, total Mg^{2+}-releasing quantities and speed increased compared with those of DW: the first-round leaching quantity of PCB in AS was close to that of DW and the release rate of Mg^{2+} in AS increased to 1.22–1.67 times that of in DW in the later leaching process. This increase indicated that the removal of phosphorus was not due to the magnesium–phosphorus precipitation.

The Ca^{2+} leaching rates of PCB and HB seems to be constants whether in DW or AS and are unaffected by the batching rounds. HB leached significantly more Ca^{2+} than PCB, since EDS results showed that HB had a higher Ca^{2+} content than PCB on the surface. Interestingly, HB leached the same Ca^{2+} quantities in AS as in DW, yet with a higher concentration in the first round. The equality of the leaching quantities proved the previous assumption that the reduction of PO_4-P was caused by formation of hydroxyapatite precipitation with Ca^{2+} and PO_4-P in stormwater runoff. On the contrary, the cumulative Ca^{2+}-releasing quantities of PCB were negative and underwent a steady decline, consistent with the tendency of phosphorus leaching but unbalanced in their quantities. This made the mechanism of PCB adsorption to decrease phosphorus concentration uncertain. This could be partly attributed to the calcium–phosphorus precipitation when considering the negative value of Ca^{2+} releasing quantities but other phosphorus-removing approaches coexisted.

Overall, the metal cations of HB, existing as salts on its surface, were easy to washed away, especially for K^+ and Ca^{2+} and this was proved by the SEM and EDS results. After the modification of crosslinked polyurethane, the leaching quantities of metal cations were significantly reduced. Through the correspondence of the leaching quantities between phosphorus and metal ions, it is clear that the mechanism of phosphorus removal by HB is calcium–phosphorus precipitation. PCB has several phosphorus removing approaches, including metal salts precipitation, which requires further research.

3.5. Phosphate Adsorption

Isothermal adsorption experiments can help to estimate the adsorption capacity and properties of additive materials in a bioretention system. The phosphate adsorption results of PCB-DW and HB-DW in DW and AS are shown in Figure 6. PCB-DW had a compelling advantage in phosphate adsorption to contrast to HB-DW: PCB-DW had a stronger equilibrium adsorption capacity, which was 1.32–1.58 times of that of HB-DW. The adsorption rates of PCB-DW were 70–98% under at different concentrations, while HB-DW could adsorb 44–74% of phosphate. PCB-DW and HB-DW adsorbed more in DW than in AS over the tested concentration range but with minimal growth. At a typical phosphate concentration range of 2–5 mg/L in stormwater runoff, the equilibrium adsorption was 93–206 mg/kg for PCB and 60–142 mg/kg for HB.

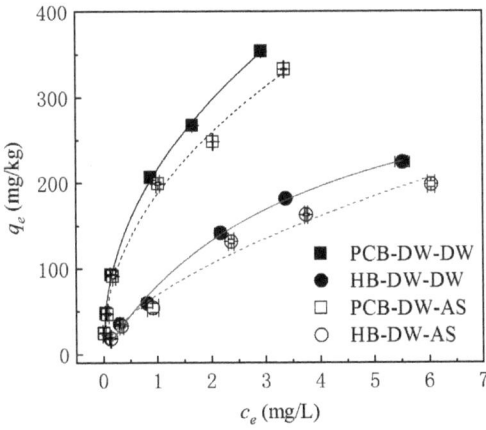

Figure 6. Adsorption isotherms of PCB-DW and HB-DW in DW or AS.

As confirmed by previous research, phosphate was bound to the biochar not only by electrostatic adsorption but also by covalent bonds, forming highly valent cationic-phosphate crystals, including magnesium [48], iron, alum or calcium [46]. There are many factors and complex evolvement courses for PO_4-P adsorption by polyurethane: the main mechanism of phosphate removal is adsorption, which occurs as a result of electrostatic attraction between two oppositely charged ions, where pH plays an important role, preferring to remain around 7 [29]. This explained why PCB (pH = 6.62) showed a poorer adsorption capacity in the leaching experiments but PCB-DW (pH = 6.98) did better in the isothermal adsorption experiments. Adsorption in AS was inferior to that in DW, which indicated that additional Ca^{2+} in AS could not promote the progress of precipitation, instead weakening it and the bivalent and multivalent cations leaching from themselves were adequate for removing phosphorus. In this study, the superiority of PCB-DW was the multiple factors, including the weak acidic conditions with a pH around 7, an abundant supply of bivalent and multivalent cations and a relatively high BET (Table 1). A thorough, quantitative analysis of the factors of phosphate adsorption is still required, however.

The results of the isothermal adsorption of PCB-DW and HB-DW were also fitted to two adsorption models, as shown in Table 5. The Freundlich model fitted the PO_4-P adsorption data of the PCB-DW better, with $R^2 > 0.99$, while the Langmuir model was better for HB-DW. Ahmed also found that the Freundlich model had the best fit for the adsorption of nutrients onto polyurethane materials [23]. With the inverse of the characteristic constants $(1/n) < 1$ in Freundlich models and the Langmuir model coefficient R_L being between 0 and 1, we can draw the conclusion that the adsorption of PO_4-P occurred easily for both PCB-DW and HB-DW. The adsorption of PO_4-P by PCB-DW and HB-DW was nonlinear. With the increase in the PO_4-P concentration in the solution, its adsorption capacity gradually becomes saturated, which was also confirmed by the bending of the fitting curves in Figure 6. K_F (the Freundlich model's volumetric-affinity parameter), to some extent, proved that, compared to HB-DW, PCB-DW had a better adsorption affinity for phosphate. The q_{max} in the Langmuir model reflected the potential maximum adsorption capacity of the materials and PCB-DW had higher q_{max} than HB-DW no matter in DW or AS. Therefore, PCB-DW can be used as an additive with high adsorption performance in stormwater treatment.

Table 5. Parameters for Freundlich and Langmuir isotherms of phosphate adsorption on PCB-DW and HB-DW.

Material	Freundlich			Langmuir			
	K_F (L/mg)	$1/n$	R^2	q_{max} (mg/kg)	K_L (L/mg)	R^2	R_L
PCB-DW-DW	214.978	0.460	0.998	417.833	1.3156	0.959	0.071–0.603
HB-DW-DW	80.500	0.623	0.983	374.176	0.274	0.994	0.267–0.880
PCB-DW-AS	186.782	0.467	0.990	379.160	0.510	0.960	0.164–0.797
HB-DW-AS	69.599	0.606	0.977	319.12	0.276	0.990	0.266–0.879

3.6. Stormwater Infiltration Experiments

In order to explore the actual operation effects after PCB and HB are added into bioretention facilities as additives, column tests were conducted, simulating the rainwater infiltration process and evaluating the suitability and feasibility of the materials. Two simulated rainfall events were carried out, each of which lasted for 3 h. DW was pumped into three columns at a rate of 15 mL/min in the first event. The PO_4-P concentrations of outflows were detected every 20 min and the variations in the outflow concentrations are shown in Figure 7. The outflows from the PCB-Column had higher PO_4-P concentrations compared with the HB-Column. With the increase in DW inflow volume, the effluent concentration decreases linearly, nearly to 0. Traces of PO_4-P were detected in the outflow of the HB-Column, while the content rose slightly during the infiltration process. The tendency of the concentration to grow in the HB-Column outflow was in accordance with the leaching patterns in the batching experiments.

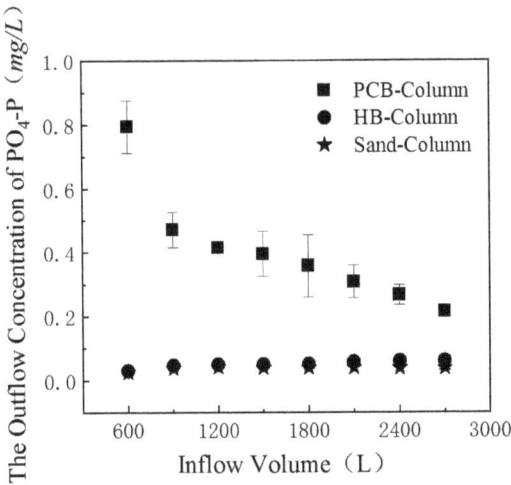

Figure 7. Concentration of PO_4^{3-} in outflows from the three columns in the first simulated rainfall event.

The estimated and detected values of the PO_4-P concentration and total release quantities from PCB-Column and HB-Column are listed in Table 6. The estimate was based on the assumption that the PCB (626.73 g) and HB (834.94 g) in the soil columns would release the same quantities of phosphorus as in the leaching tests. Comparisons in Table 6 illustrated the wide gaps between the estimated and the detected values. These gaps were caused by the differences in the contact ways and the hydrophilicity of materials, which could also be predictable. In column infiltration experiments, the DW inflow had a shorter contact time and a smaller contact surface with the additive PCB and HB and reduced the leaching quantities. Gupta drew a similar conclusion through the observation of heavy metal batch leaching experiments and column experiments [49]. HB was pyrolyzed at 600 °C and oxygenated functional groups on HB's surface made it possess a lower hydrophilicity [50]. The percentage of polyol used as a raw material in polyurethanes is positively associated with hydrophilicity [51] and the molar ratio of urethane linkage:polyol in PCB was 1:1. Combined with the water retention capacity of PCB, we could infer that PCB had a higher hydrophilicity. Additionally, the volume of PCB was double that of HB under same ratio in column experiments and PCB had more contact time and space in the stormwater. Hence, it made sense that there were more PO_4-P leaching quantities in the PCB-Column than in the HB-Column.

Table 6. Comparison of predicted and detected concentrations and total release of PO_4^{3-} in column experiments.

Columns	Concentration of the First 50 mL Effluent (mg/L)		Total Leaching Quantities (mg)	
	Predicted	Detected	Predicted	Detected
PCB-Column	47.60	0.80	52.07	1.22
HB-Column	8.20	0.03	184.03	0.13

DW was replaced by AS in the second simulated rainfall event, while the other conditions remained. Considering the influence of the first rainfall event, the PO_4-P concentration of the outflow was detected after the infiltration for 1 h and the results are shown in Figure 8. The PO$_4$-P concentration of the outflow from the PCB-Column and HB-Column remained stable at lower levels, while it increased sharply from the Sand-Column during the AS infiltration process. MeanPO$_4$-P removal rates for the three columns during the AS flushing are also shown in Figure 8. Compared with the control group (Sand-Column), the experimental groups (PCB-Column and HB-Column) had a significantly higher PO$_4$-P filtration capacity, with removal rates of 93.84% and 90.00%, respectively. This confirmed the feasibility and superiority of PCB as a filter additive in bioretention systems for removing PO$_4$-P in stormwater treatments.

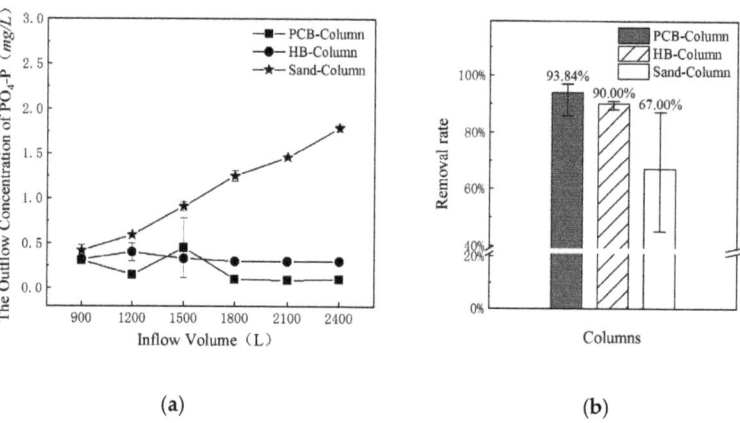

Figure 8. Concentration and mean removal rates of PO$_4^{3-}$ in outflows from the three columns in the second simulated rainfall event. (**a**) The outflow concentrations of PO$_4^{3-}$ from the three columns; (**b**) the mean removal rates of PO$_4$-P from the three columns.

In addition, it should be noted that, in the effluent concentration detection, the concentration of PO$_4$-P was kept at a low and stable state with a downward trend, which confirmed the influence of hydrophilicity on the adsorption effect of filter additives. With the infiltration and scouring of water inflow, oxygenated functional groups carried on the surface of HB were gradually washed away, so its hydrophilicity was improved to some extent. For polyurethane, its surface roughness changed during the infiltration process, which affected its adsorption capacity [52]. After the increase in hydrophilicity and surface roughness, the contact paths and time between stormwater inflow and filter materials increased, leading to the enhancement of adsorption. Although the improvement of hydrophilicity and surface roughness led to the enhancement of the adsorption capacity, the adsorption capacity tended to be saturated as the adsorption process continued. Therefore, the concentration of the outflow did not decrease significantly and it maintained a relatively stable trend in later stages.

4. Conclusions

Polyurethane-biochar crosslinked material (PCB) has been successfully manufactured using polyurethane and hardwood biochar (HB) in order to improve the hydraulic performance of bioretention facilities. This improvement was confirmed through a characteristic analysis of PCB using FTIR spectra, SEM images and hydraulic parameter tests. Biochar was crosslinked through urethane linkages in the polyurethane network. Saturated water content was doubled due to the hydrophilia and porousness of polyurethane. The internal throats, confirmed by the SEM images, increased the permeability coefficient of the filter additive by two orders of magnitude.

From the perspective of phosphorus release and adsorption, PCB is a feasible filter additive in bioretention facilities for stormwater treatment. The network of polyurethane restrained the release of phosphorus from interpenetrated HB with a decreasing cumulative rate of phosphorus leaching and reduced the metal cation leaching quantities compared with HB. The superiority of the adsorption capacity of PCB should be emphasized: for the typical phosphate concentration of stormwater runoff, the equilibrium adsorption quantity of PCB is 93–206 mg/kg for phosphate, which is a result of various factors, including its suitable pH, cation supply and porousness. PCB has a high (93.84%) and stable phosphate removal rate in column experiments, owing to the hydrophilia and porousness of polyurethane.

Overall, the present study offers a feasible filter additive with modified hydraulic properties and environmentally friendly advantages for bioretention facilities to use in stormwater treatment. Changes in the properties via the adjustment of the formula and ratio in PCB preparation lead to a variation in pore size and the functional group should be examined in further research. Meanwhile, the effect of PCB on the removal of other pollutants in stormwater should be investigated in the future.

Author Contributions: Methodology, Y.M.; formal analysis, Y.M.; investigation, Y.M.; resources, Y.W.; data curation, Y.M.; writing—original draft preparation, Y.M.; writing—review and editing, Y.W. and C.W.; visualization, Y.M.; supervision, Y.W. and C.W.; project administration, Y.W. and C.W.; funding acquisition, Y.W. and C.W. All authors have read and agreed to the published version of the manuscript.

Funding: This research was funded by National Natural Science Foundation of China, grant number U1765204 and 41772340.

Institutional Review Board Statement: Not applicable.

Informed Consent Statement: Not applicable.

Data Availability Statement: The data presented in this study are available in this article.

Acknowledgments: The authors would like to acknowledge the instrumentation and technical support of the State Key Laboratory of Hydrology Water Resources and Hydraulic, Hohai University.

Conflicts of Interest: The authors declare no conflict of interest.

References

1. Beretta-Blanco, A.; Carrasco-Letelier, L. Relevant factors in the eutrophication of the Uruguay River and the Río Negro. *Sci. Total Environ.* **2020**, 143299. [CrossRef] [PubMed]
2. Carpenter, S.R.; Caraco, N.F.; Correll, D.L.; Howarth, R.W.; Sharpley, A.N.; Smith, V.H. Nonpoint Pollution of Surface Waters with Phosphorus and Nitrogen. *Ecol. Appl.* **1998**, *8*, 559. [CrossRef]
3. Laurenson, G.; Laurenson, S.; Bolan, N.; Beecham, S.; Clark, I. The Role of Bioretention Systems in the Treatment of Stormwater. *Adv. Agron.* **2013**, *120*, 223–274. [CrossRef]
4. Hsieh, C.-H.; Davis, A.P. Evaluation and Optimization of Bioretention Media for Treatment of Urban Storm Water Runoff. *J. Environ. Eng.* **2005**, *131*, 1521–1531. [CrossRef]
5. Davis, A.P.; Shokouhian, M.; Sharma, H.; Minami, C. Laboratory study of biological retention for urban stormwater management. *Water Environ. Res.* **2001**, *73*, 5–14. [CrossRef]
6. Hunt, W.F.; Smith, J.T.; Jadlocki, S.J.; Hathaway, J.M.; Eubanks, P.R. Pollutant Removal and Peak Flow Mitigation by a Bioretention Cell in Urban Charlotte, N.C. *J. Environ. Eng.* **2008**, *134*, 403–408. [CrossRef]
7. Tirpak, R.A.; Afrooz, A.N.; Winston, R.J.; Valenca, R.; Schiff, K.; Mohanty, S. Conventional and amended bioretention soil media for targeted pollutant treatment: A critical review to guide the state of the practice. *Water Res.* **2021**, *189*, 116648. [CrossRef]
8. Li, L.; Davis, A.P. Urban Stormwater Runoff Nitrogen Composition and Fate in Bioretention Systems. *Environ. Sci. Technol.* **2014**, *48*, 3403–3410. [CrossRef] [PubMed]
9. Hernández-Apaolaza, L.; Guerrero, F. Comparison between pine bark and coconut husk sorption capacity of metals and nitrate when mixed with sewage sludge. *Bioresour. Technol.* **2008**, *99*, 1544–1548. [CrossRef]
10. Gonzales, A.P.S.; Firmino, M.A.; Nomura, C.S.; Rocha, F.R.P.; Oliveira, P.V.; Gaubeur, I. Peat as a natural solid-phase for copper preconcentration and determination in a multicommuted flow system coupled to flame atomic absorption spectrometry. *Anal. Chim. Acta* **2009**, *636*, 198–204. [CrossRef]
11. Li, P.; Lang, M.; Li, C.; Thomas, B.W.; Hao, X. Nutrient Leaching from Soil Amended with Manure and Compost from Cattle Fed Diets Containing Wheat Dried Distillers' Grains with Solubles. *Water, Air, Soil Pollut.* **2016**, *227*, 393. [CrossRef]
12. Jiang, C.; Li, J.; Li, H.; Li, Y. An improved approach to design bioretention system media. *Ecol. Eng.* **2019**, *136*, 125–133. [CrossRef]

13. Cai, M.; Li, F.C.; Chen, S.H.; Gao, L.; Guo, L. The adsorption effect of three minerals on chemical oxygen demand, total nitrogen, total phosphorus and heavy metals in biogas slurry. *IOP Conf. Ser. Earth Environ. Sci.* **2018**, *199*, 042034. [CrossRef]
14. Lin, H.; Ma, R.; Lin, J.; Sun, S.; Liu, X.; Zhang, P. Positive effects of zeolite powder on aerobic granulation: Nitrogen and phosphorus removal and insights into the interaction mechanisms. *Environ. Res.* **2020**, *191*, 110098. [CrossRef] [PubMed]
15. Tian, J.; Miller, V.; Chiu, P.C.; Maresca, J.A.; Guo, M.; Imhoff, P.T. Nutrient release and ammonium sorption by poultry litter and wood biochars in stormwater treatment. *Sci. Total. Environ.* **2016**, *553*, 596–606. [CrossRef] [PubMed]
16. Yao, Y.; Gao, B.; Inyang, M.; Zimmerman, A.R.; Cao, X.; Pullammanappallil, P.; Yang, L. Removal of phosphate from aqueous solution by biochar derived from anaerobically digested sugar beet tailings. *J. Hazard. Mater.* **2011**, *190*, 501–507. [CrossRef] [PubMed]
17. Laird, D.A. The Charcoal Vision: A Win–Win–Win Scenario for Simultaneously Producing Bioenergy, Permanently Sequestering Carbon, while Improving Soil and Water Quality. *Agron. J.* **2008**, *100*, 178–181. [CrossRef]
18. Spokas, K.A.; Novak, J.M.; Masiello, C.A.; Johnson, M.G.; Colosky, E.C.; Ippolito, J.A.; Trigo, C. Physical Disintegration of Biochar: An Overlooked Process. *Environ. Sci. Technol. Lett.* **2014**, *1*, 326–332. [CrossRef]
19. Zhang, J.; Chen, Q.; You, C. Biochar Effect on Water Evaporation and Hydraulic Conductivity in Sandy Soil. *Pedosphere* **2016**, *26*, 265–272. [CrossRef]
20. Ding, Y.; Deng, M.; Dong, J.; Wang, Z.; Huang, S. Relationship between Pore Structure and Thermal Prop-erties of Polyurethane under Multi-Field Coupling. *Eng. Plast. Appl.* **2019**, *47*, 90–94. (In Chinese)
21. Kwiecień, K.; Kwiecień, A.; Stryszewska, T.; Szumera, M.; Dudek, M. Durability of PS-Polyurethane Dedicated for Composite Strengthening Applications in Masonry and Concrete Structures. *Polym.* **2020**, *12*, 2830. [CrossRef] [PubMed]
22. Iqhrammullah, M.; Marlina, M.; Hedwig, R.; Karnadi, I.; Lahna, K.; Olaiya, N.G.; Haafiz, M.K.M.; Hps, A.K.; Abdulmadjid, S.N. Filler-Modified Castor Oil-Based Polyurethane Foam for the Removal of Aqueous Heavy Metals Detected Using Laser-Induced Breakdown Spectroscopy (LIBS) Technique. *Polym.* **2020**, *12*, 903. [CrossRef] [PubMed]
23. Ahmed, Z.; Kim, K.-P.; Shin, J. Kinetic, thermodynamic and equilibrium studies for adsorption of ammonium ion on modified polyurethane. *DESALINATION Water Treat.* **2015**, *57*, 1–9. [CrossRef]
24. Solares, S.B.; Merillas, B.; Cimavilla-Román, P.; Rodriguez-Perez, M.; Pinto, J. Enhanced nitrates-polluted water remediation by polyurethane/sepiolite cellular nanocomposites. *J. Clean. Prod.* **2020**, *254*, 120038. [CrossRef]
25. Okoli, C.P.; Adewuyi, G.O.; Zhang, Q.; Diagboya, P.N.; Qingjun, G. Mechanism of dialkyl phthalates removal from aqueous solution using γ-cyclodextrin and starch based polyurethane polymer adsorbents. *Carbohydr. Polym.* **2014**, *114*, 440–449. [CrossRef]
26. Anju, M.; Renuka, N. Magnetically actuated graphene coated polyurethane foam as potential sorbent for oils and organics. *Arab. J. Chem.* **2020**, *13*, 1752–1762. [CrossRef]
27. Lu, H.; Tian, H.; Zhang, M.; Zhang, M.; Chen, Q.; Guan, R.; Wang, H. Water Polishing improved controlled-release characteristics and fertilizer efficiency of castor oil-based polyurethane coated diammonium phosphate. *Sci. Rep.* **2020**, *10*, 5763. [CrossRef]
28. Yong, Y.; Jie, L. Use of Polyurethane Foams Complex Material to Preserve Moisture and Fertility. *Polym. Technol. Eng.* **2007**, *46*, 943–947. [CrossRef]
29. Sasidharan, A.P.; Meera, V.; Raphael, V.P. Investigations on characteristics of polyurethane foam impregnated with na-nochitosan and nanosilver/silver oxide and its effectiveness in phosphate removal. *Environ. Sci. Pollut. Res.* **2020**. [CrossRef]
30. Nie, J.Y.; Zhu, N.; Lin, K.M.; Song, F.Y. Effect of soil fortified by polyurethane foam on septic tank effluent treatment. *Water Sci. Technol.* **2011**, *63*, 1230–1235. [CrossRef]
31. Mukome, F.N.D.; Zhang, X.; Silva, L.C.R.; Six, J.; Parikh, S.J. Use of Chemical and Physical Characteristics To Investigate Trends in Biochar Feedstocks. *J. Agric. Food Chem.* **2013**, *61*, 2196–2204. [CrossRef] [PubMed]
32. Kizito, S.; Wu, S.; Kirui, W.K.; Lei, M.; Lu, Q.; Bah, H.; Dong, R. Evaluation of slow pyrolyzed wood and rice husks biochar for adsorption of ammonium nitrogen from piggery manure anaerobic digestate slurry. *Sci. Total. Environ.* **2015**, *505*, 102–112. [CrossRef] [PubMed]
33. Hameed, B.; Mahmoud, D.; Ahmad, A. Equilibrium modeling and kinetic studies on the adsorption of basic dye by a low-cost adsorbent: Coconut (Cocos nucifera) bunch waste. *J. Hazard. Mater.* **2008**, *158*, 65–72. [CrossRef] [PubMed]
34. Foo, K.; Hameed, B. Insights into the modeling of adsorption isotherm systems. *Chem. Eng. J.* **2010**, *156*, 2–10. [CrossRef]
35. Teodosiu, C.; Wenkert, R.; Tofan, L.; Paduraru, C. Advances in preconcentration/removal of environmentally relevant heavy metal ions from water and wastewater by sorbents based on polyurethane foam. *Rev. Chem. Eng.* **2014**, *30*, 403–420. [CrossRef]
36. Korley, L.T.J.; Pate, B.D.; Thomas, E.L.; Hammond, P.T. Effect of the degree of soft and hard segment ordering on the morphology and mechanical behavior of semicrystalline segmented polyurethanes. *Polym.* **2006**, *47*, 3073–3082. [CrossRef]
37. Tan, Z.; Yuan, S.; Hong, M.; Zhang, L.; Huang, Q. Mechanism of negative surface charge formation on biochar and its effect on the fixation of soil Cd. *J. Hazard. Mater.* **2020**, *384*, 121370. [CrossRef]
38. Motawie, A.; Madani, M.; Esmail, E.; Dacrorry, A.; Othman, H.; Badr, M.; Abulyazied, D. Electrophysical characteristics of polyurethane/organo-bentonite nanocomposites. *Egypt. J. Pet.* **2014**, *23*, 379–387. [CrossRef]
39. Yan, H.; Zhou, Z.; Peng, C.; Liu, W.; Zhou, H.; Wang, W.; Zhang, Q. Influence of Mass Ratio of Polyols on Properties of Polycaprolactone- Polyethylene Glycol/Methylene Diphenyl Diisocyanate/Diethylene Glycol Hydrogels. *J. Macromol. Sci. Part B* **2017**, *944*, 315–323. [CrossRef]

40. Yang, Y.; Tong, Z.; Geng, Y.; Li, Y.; Zhang, M. Biobased Polymer Composites Derived from Corn Stover and Feather Meals as Double-Coating Materials for Controlled-Release and Water-Retention Urea Fertilizers. *J. Agric. Food Chem.* **2013**, *61*, 8166–8174. [CrossRef]
41. New Jersey Stormwater Best Management Practices Manual. 2007. Available online: https://www.njstormwater.org/bmp_manual2.htm (accessed on 16 January 2021).
42. De Macedo, M.B.; Lago, C.A.F.D.; Mendiondo, E.M. Stormwater volume reduction and water quality improvement by bioretention: Potentials and challenges for water security in a subtropical catchment. *Sci. Total. Environ.* **2019**, *647*, 923–931. [CrossRef] [PubMed]
43. Chunyan, J.; Shunli, H.; Shusheng, G.; Wei, X.; Huaxun, L.; Yuhai, Z. The Characteristics of Lognormal Distribution of Pore and Throat Size of a Low Permeability Core. *Pet. Sci. Technol.* **2013**, *31*, 856–865. [CrossRef]
44. Wang, Y.; Lin, Y.; Chiu, P.C.; Imhoff, P.T.; Guo, M. Phosphorus release behaviors of poultry litter biochar as a soil amendment. *Sci. Total. Environ.* **2015**, *512–513*, 454–463. [CrossRef] [PubMed]
45. Stubbings, W.A.; Harrad, S. Leaching of TCIPP from furniture foam is rapid and substantial. *Chemosphere* **2018**, *193*, 720–725. [CrossRef] [PubMed]
46. De-Bashan, L.E.; Bashan, Y. Recent advances in removing phosphorus from wastewater and its future use as fertilizer (1997–2003). *Water Res.* **2004**, *38*, 4222–4246. [CrossRef] [PubMed]
47. Angst, T.E.; Sohi, S.P. Establishing release dynamics for plant nutrients from biochar. *GCB Bioenergy* **2012**, *5*, 221–226. [CrossRef]
48. Shin, H.; Tiwari, D.; Kim, D.-J. Phosphate adsorption/desorption kinetics and P bioavailability of Mg-biochar from ground coffee waste. *J. Water Process. Eng.* **2020**, *37*, 101484. [CrossRef]
49. Gupta, N.; Gedam, V.V.; Moghe, C.; Labhasetwar, P. Comparative assessment of batch and column leaching studies for heavy metals release from Coal Fly Ash Bricks and Clay Bricks. *Environ. Technol. Innov.* **2019**, *16*, 100461. [CrossRef]
50. Suliman, W.; Harsh, J.B.; Abu-Lail, N.I.; Fortuna, A.-M.; Dallmeyer, I.; Garcia-Perez, M. The role of biochar porosity and surface functionality in augmenting hydrologic properties of a sandy soil. *Sci. Total. Environ.* **2017**, *574*, 139–147. [CrossRef]
51. Brzeska, J.; Tercjak, A.; Sikorska, W.; Kowalczuk, M.; Rutkowska, M. Predicted Studies of Branched and Cross-Linked Polyurethanes Based on Polyhydroxybutyrate with Polycaprolactone Triol in Soft Segments. *Polymers* **2020**, *12*, 1068. [CrossRef]
52. Akkas, T.; Citak, C.; Sirkecioglu, A.; Guner, F.S. Which is more effective for protein adsorption: Surface roughness, surface wettability or swelling? Case study of polyurethane films prepared from castor oil and poly(ethylene glycol). *Polym. Int.* **2013**, *62*, 1202–1209. [CrossRef]

Article

Macroscopic Poly Schiff Base-Coated Bacteria Cellulose with High Adsorption Performance

Lili Ren [1], Zhihui Yang [1,2], Lei Huang [1], Yingjie He [1], Haiying Wang [1,2,*] and Liyuan Zhang [3,4,*]

1. School of Metallurgy and Environment, Central South University, Changsha 410083, China; ren_lili@csu.edu.cn (L.R.); yangzh@csu.edu.cn (Z.Y.); znhuanglei@csu.edu.cn (L.H.); yingjiehe@csu.edu.cn (Y.H.)
2. Chinese National Engineering Research Center for Control and Treatment of Heavy Metal Pollution, Central South University, Changsha 410083, China
3. Department of Colloid Chemistry, Max Planck Institute of Colloids and Interfaces, 14476 Potsdam-Golm, Germany
4. Environmental Engineering Research Centre, Department of Civil Engineering, The University of Hong Kong, Pokfulam, Hong Kong, China
* Correspondence: haiyw25@163.com (H.W.); lyzhang@mpikg.mpg.de or zhang_livyl@csu.edu.cn (L.Z.)

Received: 4 March 2020; Accepted: 17 March 2020; Published: 23 March 2020

Abstract: Here, a nanofiber-exfoliated bacteria cellulose aerogel with improved water affinity and high mass transfer was synthesized. Consequently, poly Schiff base can be uniformly coated within the body of bacteria cellulose aerogel without the traditional dispersion treatment. The composite aerogel has adequate mechanical and thermal stability and high mass transfer efficiency. Such an aerogel can serve as a superior adsorbent for flow through adsorption of pollution. Typically, the adsorption capacity towards Cr(VI), Cu(II), Re(VII), Conga red, and Orange G reaches as high as 321.5, 256.4, 153.8, 333.3, and 370.3 mg g^{-1}, respectively. Moreover, the adsorption by this composite aerogel is very fast, such that, for example, at just 2 s, the adsorption is almost finished with Cr(VI) adsorption. Moreover, the composite aerogel exhibits a good adsorption-desorption capability. This research will hopefully shed light on the preparation of bacteria cellulose-derived macroscopic materials powerful in not only environmental areas, but also other related applications.

Keywords: bacteria cellulose; poly Schiff base; chromium; adsorption

1. Introduction

Hexavalent chromium (Cr(VI)), a typical toxic contaminant, is normally discharged by industrial plants, which has strong teratogenicity and carcinogenicity [1–3]. Various technologies have been developed and, among them, adsorption is a popular choice for Cr(VI) owing to its feasibility of operation and cost effectiveness [4–11]. The adsorbent is very important in the adsorption technique, but the current adsorbent mainly suffers from the trade-off between mass transfer and recoverability [12–15]. Typically, high mass transfer requires a good dispersion of adsorbent, but this would decline the recovery efficiency. While high recoverability needs relatively strong noncovalent interaction between adsorbent building blocks, which in turn brings down mass transfer.

Bacteria cellulose (BC), a naturally nanofiber-arranged hydrogel, is a promising candidate adsorbent, exhibiting high mechanical strength and chemical stability [16–23]. Moreover, the highly porous network is of high advantage in boosting the mass transfer within bacteria cellulose. Most recently, a polymer with rich functional groups was adopted to modify the bacteria cellulose to improve the adsorption performance. For example, Jin, X et al. and Wang, J et al. used polyethyleneimine to graft on the bacteria cellulose skeleton for the adsorption of cations (e.g., Cu(II), Pb(II), Hg(II)) [24,25].

Jahan, K et al. and Yang, Z et al. reported the in-situ fabrication of poly(aromatic amine) on bacteria cellulose for effectively separating Cr(VI) from the aqueous solution via reduction and chelation [26,27]. Unfortunately, bacteria cellulose must be dispersed firstly to allow the ultimate interaction between cellulose and polymers [24,27]. This complicates the procedures and moreover weakens the mechanical strength of the final macroscopic product [28]. Moreover, though the batch adsorption performance of these adsorbents were proven to be fine, the dynamic adsorption using macroscopic adsorbent was satisfied to practically purify continuous-flow industrial wastewater, which was scarcely reported.

Here, we report a facile fabrication of bacteria cellulose uniformly coated with poly Schiff base. The pretreatment by sodium silicate was applied first to modify the bacteria cellulose to increase the mass transfer efficiency without breaking the macroscopic structure. As a result, the polymerization of Schiff base proceeded effectively within the body of bacteria cellulose to achieve a uniform coating of polymer. The macroscopic composite monolith exhibits attracting performance in adsorption of Cr(VI) and other typical pollutants in batch and dynamic adsorption tests.

2. Experimental Section

2.1. Materials

The bacterial strain, Gluconacetobacter hansenii (ATCC 53582), was provided by China Center of Industrial Culture Collection. The m-phenylenediamine was purchased from Aladdin Industrial Co., Ltd. (Hangzhou, China). Glutaraldehyde was obtained from Macklin Biochemical Technology Co., Ltd. (Shanghai, China) and other chemical regents were provided from Sinopharm chemical reagent (Shanghai, China).

2.2. Preparation of Porous Bacteria Cellulose/Poly Schiff Base

- Pretreatment

The bacteria cellulose was treated with the solution of alkaline sodium silicate. Then, HCl was added and sodium silicate transformed to NaCl and SiO_2. After the removal of SiO_2 by NaOH, the pretreated bacteria cellulose was freeze-dried and used for polymer coating.

- Polymer Coating

The HCl-protonated m-phenylenediamine was mixed with glutaraldehyde, which formed the oligomer solution due to protonation. The pretreated bacteria cellulose (pBC) was immersed in precursor mixture to absorb the oligomers into matrix of pBC. The poly Schiff base was deposited on cellulose nanofibers by tuning the pH to base. The detail of procedure was given in Supplementary Sections S1.1 and S1.2.

The relative content of poly Schiff base in pBC was estimated according to the following equation:

$$D\% = \frac{W_{product} - W_{BC}}{W_{product}} \times 100\% \qquad (1)$$

where the D% is the content percentage of poly Schiff base in the composites, $W_{product}$ is the total mass of composites, W_{BC} is the mass of bacteria cellulose or porous bacteria cellulose.

2.3. Characterization

The attenuated total reflection infrared (ATR) spectrum was recorded on a Nicolet IS50 Fourier transform infrared (FTIR) spectrometer (Thermo Fisher scientific instruments, Waltham, MA, USA) at resolution of 4 cm^{-1} in the range of 400~4000 cm^{-1}. A field emission scanning electron microscopy (SEM, JSM-6360LV, JEOL, Tokyo, Japan) and transmission electron microscopy (TEM, JEOL-2011, Tokyo, Japan) was employed to inspect the morphology. The surface area and porosity of samples were calculated by Brunauer-Emmett-Teller (BET) and Barrett-Joyner-Halenda (BJH) method, respectively (analysis gas: nitrogen; analysis time: 670 min; thermal delay: 30 s; outgas time: 12 h; bath temperature: 50 °C and outgas temperature: 150 °C) (Autosorb iQ, Quantachrome, Ashland, Virginia, USA) [29]. The crystallographic analysis of aerogels was performed on X-ray diffractometer (XRD) with Cu-Kα radiation (D/Max-RB diffractometer, Rigaku, Tokyo, Japan) from 5~50° by using the dried film of samples. The chemical composition was characterized on a K-Alpha 1063 X-ray photoelectron spectroscope (XPS) with Al Kα X-ray as the excitation source (Thermo Fisher scientific instruments, Waltham, MA, USA).

2.4. Batch Adsorption

A series of adsorption experiments were conducted in the polyethylene vials at 30 °C under 150 rpm rotary shaking. 15 mg adsorbent was accurately weighed and added into the 50 mL Cr(VI) containing solution. The blank assay was carried out under the same condition without any adsorbent simultaneously. The adsorbents were filtrated from the solution by filter paper (pore size ~0.45 μm) after completion of adsorption. The analysis method of Cr(VI) concentration (and other pollutants) was given in Supplementary Section S1.3. Considering the chemical state of Cr, the concentration of Cr(III) can be calculated as $Cr(III) = Cr_{total} - Cr(VI)$. To survey the effect of pH on adsorption capacity, the initial pH of solution was adjusted from 0~6 by 2 M HCl or 2 M NaOH. The adsorption isotherm experiment was implemented at pH 2 for 6 h with the initial Cr(VI) concentration varied from 100~500 mg L^{-1}. For kinetics experiment, Cr(VI) solution with concentration of 400 mg L^{-1} was experienced 1~240 min adsorption at pH 2. The concentration of Cu(II), Re(VII), Conga red, and Orange G solution varied from 50~500 mg L^{-1} for 6 h adsorption to achieve their adsorption isotherm. The Cu(II), Re(VII) and Orange G was adjusted to pH 2 while the Conga red was adjusted to pH 5.5. The adsorption capacity for batch adsorption experiment was calculated as follows:

$$q_t = \frac{C_0 - C_t}{m} V \quad (2)$$

$$q_e = \frac{C_0 - C_e}{m} V \quad (3)$$

where q_t is the adsorption capacity at time t (mg g^{-1}), C_0 is the initial Cr(VI) concentration in solution (mg L^{-1}), C_t is the Cr(VI) concentration in solution at time t (mg L^{-1}), q_e is the equilibrium adsorption capacity (mg g^{-1}), C_e is the equilibrium concentration (mg L^{-1}), m is the weight of adsorbent (g), and V is the volume of solution (L).

2.5. Dynamic Adsorption

For the dynamic adsorption, 0.3 g adsorbent was compressed and squeezed in a glass column with length of 500 mm and ID of 8 mm. 25 mL Cr(VI) solution with concentration of 5~100 mg L^{-1} was adjusted to pH 2 and fed into the top of column using the peristaltic pump with controlling the flow velocity from 28.42~85.26 μL min^{-1} at ambient temperature. The concentration of Cr(VI) (and other pollutant) in effluent was measured by the same method as batch adsorption experiment. For other pollutant, the influent concentration of Cu(II), Re(VII), Conga red and Orange G were 50, 20, 20, and 50 mg L^{-1}, respectively. The Cu(II), Re(VII), and Orange G was adjusted to pH 2 while the Conga red was adjusted to pH 5.5. The removal rate was calculated as follows:

$$R\% = \frac{C_0 - C_a}{C_0} \times 100\% \quad (4)$$

where $R\%$ is the removal rate of effluent, C_0 is the influent concentration of pollutant (mg L^{-1}), C_a is the concentration of pollutant in effluent (mg L^{-1}).

2.6. Desorption and Reusability

The recycle adsorption experiment was tested with 7 successive cycles of dynamic adsorption of Cr(VI). After each cycle, 100 mL 1 M NaOH served as the desorption solution and flowed through the column with 0.3 g adsorbent at flow velocity of 56.84 µL min^{-1}. After being washed by deionized water to neutrality, the adsorbent was freeze-dried for the next cycle.

3. Result and Discussion

3.1. Pretreatment of Bacteria Cellulose

The overview of pathway to fabricate the porous bacteria cellulose/poly Schiff base was described in Scheme 1. Bacteria cellulose was saturated with the sodium silicate solution (alkaline). Then, HCl acid reacted with sodium silicate to produce SiO_2 aggregation. The formation of SiO_2 could enlarge the inner space of bacteria cellulose and disassemble the large bacteria cellulose fiber bundles into much thinner fibers. The SiO_2 was removed by NaOH solution and the pure modified bacteria cellulose was obtained by water rinse and subsequent freeze-drying. After the pretreatment, the bacteria cellulose was still macro-scale monolith. As shown in Figure 1a,b, the smooth surface of pristine bacteria cellulose consisted of compact fibers. In contrast, the pretreated bacteria cellulose had many macroscopic holes (Figure 1e), which should be generated by the SiO_2 formation. The photograph and high-magnification images of bacteria cellulose/SiO_2 composites can verify this (Figure 1c,d). More interestingly, the original bacteria cellulose fibers with diameter 400 nm was decreased to 30 nm (Figure 1f). The small nanofibers were most possibly free from the aggregated cellulose fibers by alkaline treatment. Moreover, the SiO_2 formation within the gap or pores in the aggregated fiber bundles would also produce small nanofibers.

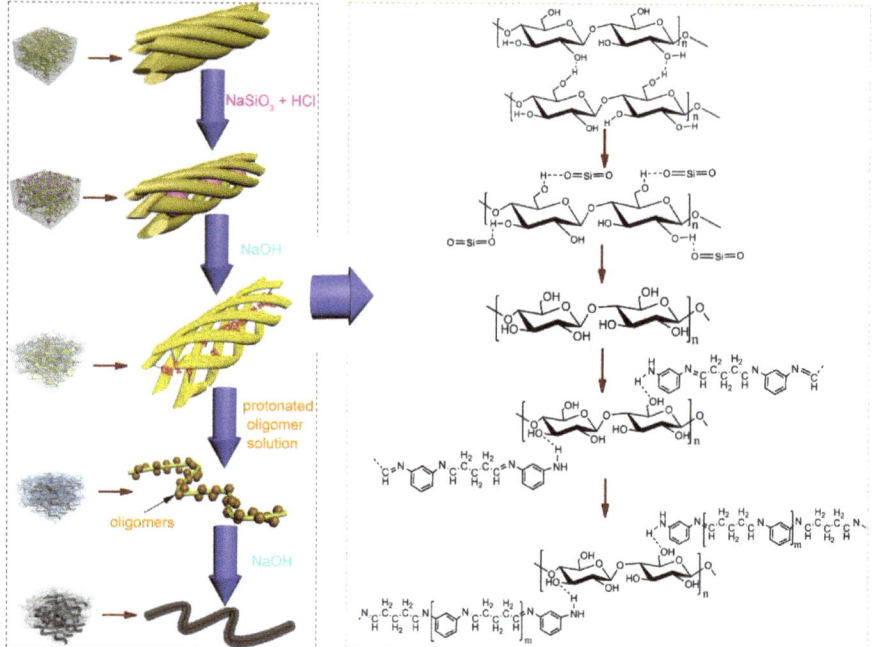

Scheme 1. The pathway of porous bacteria cellulose/poly Schiff base fabrication.

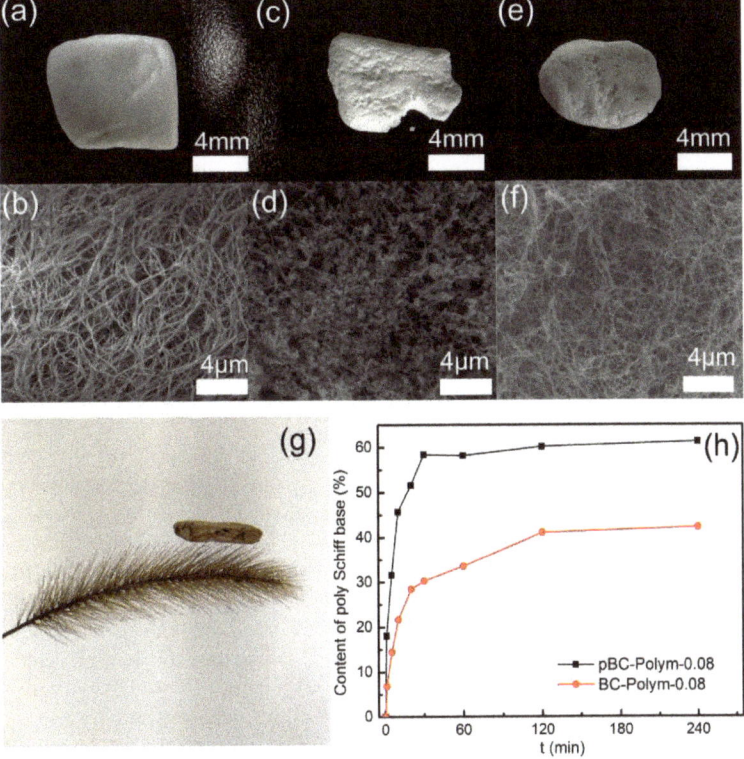

Figure 1. The photograph and of SEM images of bacteria cellulose (**a,b**), bacteria cellulose/SiO$_2$ (**c,d**) and pBC (**e,f**); the photograph of adsorbent with ultra-low density (**g**) and the content of poly Schiff base in pBC-Polym-0.08 and BC-Polym-0.08 with various reaction time (**h**).

XRD was used to validate the variation of bacteria cellulose before and after pretreatment. As seen in Figure 2a, the raw bacteria cellulose exhibited characteristic of typical cellulose I by diffraction peaks at 14.3° and 22.6° [30–32]. Noticeably, after the pretreatment, the peak intensity at 14.3° and 22.6° attenuated obviously, which indicated the decrease of crystallinity. On the other hand, the relative intensity of 16.4° increases. Moreover, a small peak located at 20.0° appeared, indexed as the (110) of cellulose II. It inferred that the structure of semi-crystalline region and amorphous region was altered via pretreatment. This can be explained by the separation of small cellulose nanofibers from original fiber bundles, which inevitably exposed new facet. It must be mentioned that in the raw cellulose the H bonding (by –OH) pushed the formation of cellulose crystal. The decrease of crystallinity of bacteria cellulose demonstrated the breaking of H bonding, which signified that the –OH is free from the crystal. This is conducive to increasing its affinity toward aqueous solution, which can promote the mass transfer within the body of bacteria cellulose.

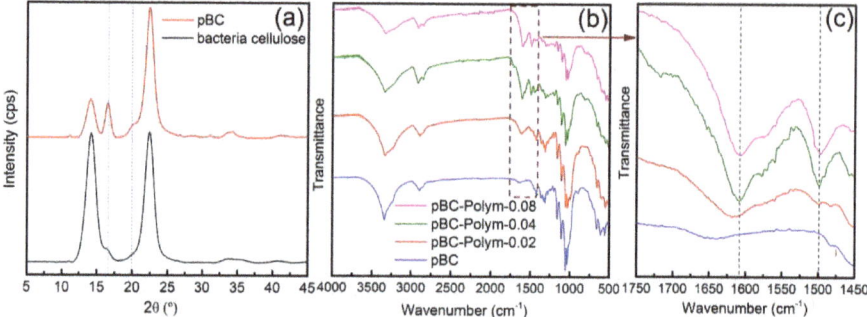

Figure 2. (a) XRD patterns of raw bacteria cellulose and pretreated bacteria cellulose. (b) FTIR spectra of pBC, pBC-Polym-0.02, pBC-Polym-0.04 and pBC-Polym-0.08 and (c) magnified graph in the range of 1750 to 1450 cm^{-1}.

3.2. Schiff Base Loading in Bacteria Cellulose

To allow the precursor coat onto the nanofibers, the bacteria cellulose firstly interacted with acidic aqueous mixture of *m*-phenylenediamine and glutaraldehyde. The pretreated bacteria cellulose immersed into precursor solution as soon as their contact. In contrast, the raw bacteria cellulose took ~30 min to be absolutely wetted by the solution. That should be caused by the pretreatment of bacteria cellulose, which alters its surface properties. After the saturation of precursor, the cellulose was put into the alkaline solution to speed up the polymerization process. Due to the strong interaction between –NH$_2$ and –OH, the poly Schiff base would stably coat onto the nanofibers of bacteria cellulose. The white cellulose became brown after the polymer coating (Figure 1g), which had ultra-low density. The final product was named as pBC-Polym-*x*, where pBC is the pretreated bacteria cellulose and *x* is the concentration of m-phenylenediamine in precursor solution. For comparison, the product without pretreatment was denoted as BC-Polym-*x*.

More importantly, the difference on the wettability of raw and pretreated bacteria cellulose changed the loading mass of poly Schiff base. As shown in Figure 1h, the polymer loading on pretreated bacteria cellulose increased very fast and it took 30 min for saturation. However, for the raw bacteria cellulose, it spent ~120 min for the complete loading. Typically, the content of poly Schiff base on pBC was at least 20% higher than that on raw materials, which can be vividly distinguished by the color of the product (Supplementary Section S2). This should be related to the improvement of the wettability of bacteria cellulose by the pretreatment.

3.3. Characterization

The ATR-FTIR technique was applied to study the structural variation of pBC before and after polymer coating. As seen in Figure 2b and c, functional groups including –OH (3343 cm^{-1}), C–H (2902 cm^{-1}), and –C–O–C– (1109 and 1164 cm^{-1}) can be found in the bare pBC [33–35]. After the polymer coating, two absorption bands at 1609 and 1493 cm^{-1} emerged. These corresponded to the stretching vibration of C=N and N–H shearing vibration in benzenoid amine structure [36,37]. Moreover, these two signals became obvious with the increase of polymer amount coating on the nanofibers.

The morphology of the final product was examined by the SEM technique, taking pBC-Polym-0.04 as an example. The morphology of the surface section was shown in Figure 3a that the gap between the nanofibers was filled with polymer. The result was identical for the cross-section of the product (Figure 3b). That is to say, the poly Schiff base uniformly coated on the nanofibers within the body of bacteria cellulose. The TEM images are given in Figure 3c (agglomeration) and 3d (single fiber). As shown, the incorporated thick polymer evolved to sheath-like structure to encapsulate the cellulose nanofibers, which was consistent with the SEM results. For pBC-Polym-0.02, only nanoparticles can be found on the nanofibers (Supplementary Figure S2). On the other hand, pBC-Polym-0.08 showed a similar morphology but a higher density of coating (Supplementary Figure S3). The N_2 adsorption-desorption isotherm and porosity are shown in Supplementary Section S4, which indicated that the products were mainly composed of mesopores and macropores. In addition, the product exhibited good thermal stability (Supplementary Section S5), and this was beneficial for its practical applications.

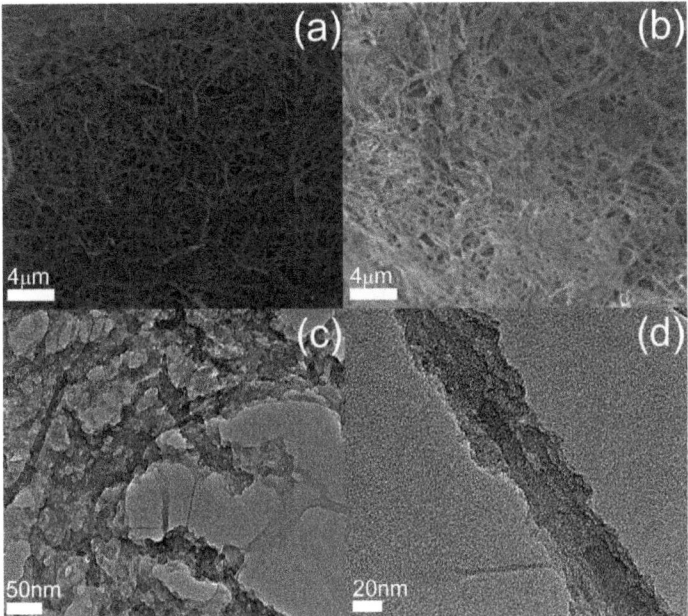

Figure 3. The SEM images of pBC-Polym-0.04: surface (**a**) and cross-section (**b**), and the TEM images of pBC-Polym-0.04: agglomerated fibers (**c**) and single fiber (**d**).

3.4. Adsorption Performance

The effect of pH on the adsorption was investigated and the results were given in Supplementary Section S6. As shown, the optimum pH for adsorption was ~2, which was related to the balance between protonation of the polymer and speciation of Cr ions at different pH. In the following experiments, pH will be used without special caution.

3.4.1. Adsorption Isotherm

Effect of initial concentration of Cr(VI) (100~500 mg L^{-1}) on the adsorption was investigated. As shown in Figure 4a, the adsorption capacity of the series of pBC-Polym-x rapidly increased with the rise of Cr(VI) concentration. Then, the adsorption process tended to be saturated. Similar trend was found for the samples without pretreatment (Figure 4b). However, the adsorption performance is much better for pBC-Polym-x. This can be ascribed to the high polymer loading and the high affinity to aqueous solution of pBC. On the other hand, the saturated capacity of pBC-Polym-0.04 was very close to that of pBC-Polym-0.08. The possible reason was that the thicker polymer coating on the nanofibers (pBC-Polym-0.08) decreased the utilization efficiency of the active materials.

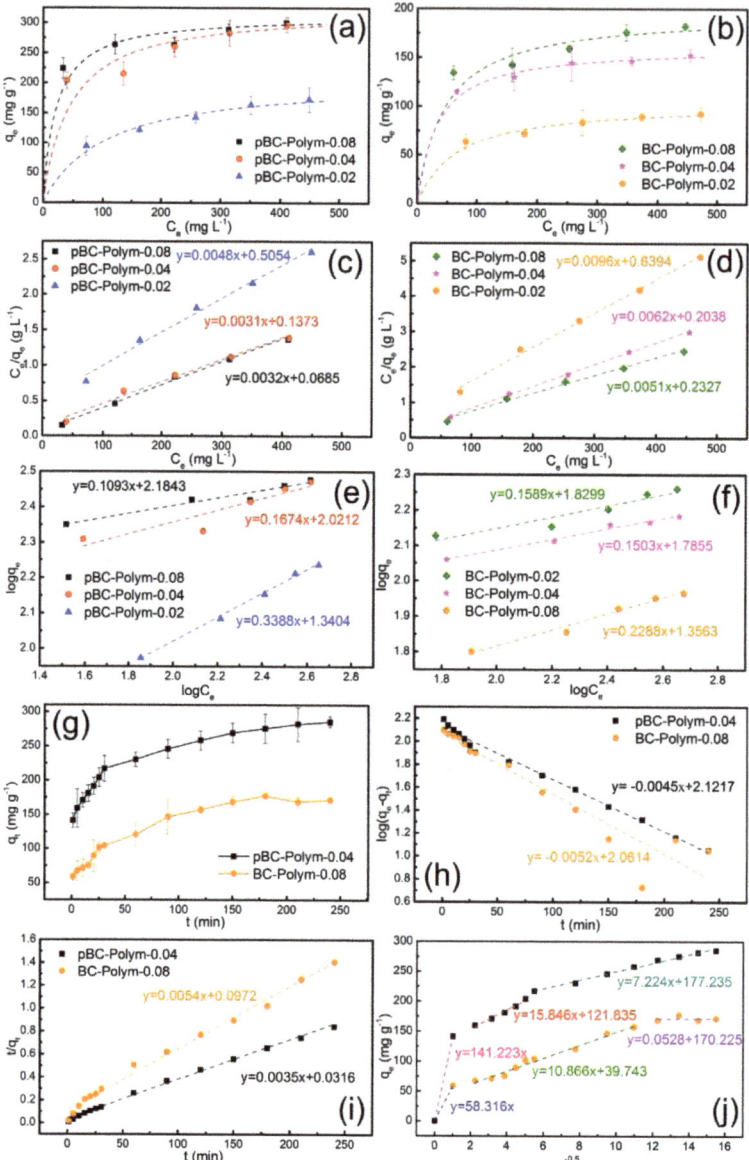

Figure 4. (**a**) Cr(VI) adsorption isotherms of pBC-Polym-0.02, pBC-Polym-0.04 and pBC-Polym-0.08, and (**b**) BC-Polym-0.02, BC-Polym-0.04 and BC-Polym-0.08. Their corresponding fitted Langmuir (**c**,**d**) and Freundlich isotherms (**e**,**f**). Effect of contact time on Cr(VI) adsorption (**g**) and kinetics modeling: (**h**) pseudo-first-order kinetic plots, (**i**) pseudo-second-order kinetic plots and (**j**) plots of intra particle model.

Furthermore, the experimental data (Figure 4c~f) were simulated by Langmuir and Freundlich models. The Langmuir and Freundlich equations are shown below:

$$\frac{C_e}{q_e} = \frac{C_e}{q_m} + \frac{1}{bq_m} \quad (5)$$

$$\log q_e = \log K_f + \frac{1}{n} \log C_e \quad (6)$$

where C_e is the Cr(VI) concentration at equilibrium, q_e is the adsorption capacity at equilibrium, q_m is the maximum adsorption capacity at saturation, b is the isotherm constant for Langmuir model, K_f and n are the constants of isotherm equation. The parameters were listed in Table 1.

Table 1. Isotherms parameter of Cr(VI) adsorption on pBC-Polym-x and BC-Polym-x samples.

Samples	Langmuir Isotherm			Freundlich Isotherm		
	q_{max} (mg g^{-1})	b (L mg^{-1})	R_L^2	K_f (mg g^{-1} (L mg^{-1})$^{1/n}$)	$1/n$	R_f^2
pBC-Polym-0.08	312.5	0.0467	0.995	152.848	0.1093	0.7374
pBC-Polym-0.04	321.5	0.0227	0.986	104.993	0.1674	0.7321
pBC-Polym-0.02	210.1	0.0094	0.991	21.901	0.3388	0.8990
BC-Polym-0.08	197.2	0.0218	0.991	67.590	0.1589	0.9083
BC-Polym-0.04	161.8	0.0303	0.999	61.018	0.1503	0.9830
BC-Polym-0.02	104.6	0.0151	0.995	22.716	0.2288	0.9795

Based on the calculation, adsorption behavior can be better described by the Langmuir model. This demonstrated a homogenous monolayer adsorption on poly Schiff base-coated cellulose nanofibers. The maximum adsorption capacity of pBC-Polym-0.04 was 321.5 mg g^{-1}. Remarkably, the adsorption capacity in this research was higher than most of the macroscopic adsorbents in previous studies (Table 2).

Table 2. Comparison pBC-Polym-0.04 with other macroscopic adsorbents.

Materials	Optimum pH	Kinetics	Adsorption Capacity (mg g^{-1})	Ref
Mesoporous carbon sponge	2~4	Pseudo-second-model	93.9	[38]
Polydopamine and Chitosan cross-linked graphene oxide	3	Pseudo-second-model	312.0	[39]
Magnetic graphene oxide foam	2	Pseudo-second-model	258.6	[40]
3D porous graphene oxide-maize amylopectin composites	5	Pseudo-second-model	13.6	[41]
3D porous cellulose	3	Pseudo second model	220.6	[42]
3D graphene oxide-NiFe LDH composite	3	Pseudo second model	53.6	[43]
Cyclodextrin functionalized 3D-graphene	2	Pseudo second model	107	[44]
Tetraethylenepentamine crosslinked chitosan oligosaccharide hydrogel	3	Pseudo second model	148.1	[45]
Polyaniline-coated bacterial cellulose mat	2	Pseudo second model	128	[26]
pBC-Polym-0.04	2	Pseudo second model	321.5	This study

3.4.2. Adsorption Kinetics

As shown in Figure 4g~i, the adsorption process of pBC-Polym-0.04 was very fast in the initial stage (≤30 min), involving 76% of the total capacity. This was presumably due to the physical adsorption. In this step, the negative charged Cr(VI) was adhered to positive charged adsorbent by electrostatic attraction. Additionally, the fast adsorption process was related to the high affinity to aqueous solution of the pBC-Polym-x, which promised effective mass transfer. The subsequently accumulative adsorption of Cr(VI) ultimately reached the equilibrium within 3 h. The pseudo-first and pseudo-second order models were adopted to analyze the adsorption process. As seen in Table 3, the pseudo-second-order model with the correlation coefficient (R^2) > 0.99 indicated that the Cr(VI) adsorption was mainly controlled by chemical sorption. The adsorption process was deduced as the electrostatic attraction and redox-chelation. The Cr(VI) adsorption was initiated by electrostatic interaction. The Cr(VI) binding on surface of adsorbent was reduced to Cr(III) and chelated by quinoid imine group - the oxidation product of benzenoid amine structure [37,46]. The elaborate verification of adsorption mechanism was stated in Supplementary Section S7.

Table 3. The kinetics parameter of Cr(VI) adsorption on pBC-Polym-0.04 and BC-Polym-0.08.

Samples	q_{exp}	Pseudo-First-Order Model $\log(q_e-q_t)=\log q_e - \frac{k_1}{2.303}t$			Pseudo-Second-Order Model $\frac{t}{q_t}=\frac{1}{k_2 q_e^2}+\frac{1}{q_e}t$			Intra Particles Diffusion $q_t=k_i t^{0.5}+C$					
		q_e	k_1	R^2	q_e	k_2	R^2	k_{id1}	C_1	k_{id2}	C_2	k_{id3}	C_3
pBC-Polym-0.04	285.2	132.3	0.010	0.9903	289.8	0.11	0.9971	141.2	0	15.8	121.8	7.2	177.2
BC-Polym-0.08	170.9	115.1	0.012	0.8993	183.8	0.056	0.9915	58.3	0	10.8	39.7	0.052	170.2

Furthermore, the Weber–Morris intraparticle diffusion model was used to describe the adsorption process (Figure 4j). The whole adsorption can be divided into three regions: (1) Cr(VI) diffusion in aqueous solution onto the exterior surface of adsorbent; (2) Cr(VI) permeation into the inner section of adsorbent and Cr(VI) diffusion into the internal polymers; (3) chemical sorption and reaching equilibrium. Clearly, the surface diffusion of pBC-Polym-0.04 completed within 1 minute. From 0~1 min, Cr(VI) was instantly captured by the poly Schiff base deposited on the surface of aerogel. In the second procedure, faster permeation and mass transfer of pBC-Polym-0.04 was observed. For BC-Polym-0.04, ascribed to the weaker affinity toward solution, the permeation and adsorption of inner polymers lasted from 5~120 min. The k_{id} listed in Table 3 manifested the high adsorption rate of pBC-Polym-0.04.

3.4.3. Dynamic Adsorption

The fast adsorption process is vividly illustrated in Figure 5a. A piece of pBC-Polym-0.04 was directly immersed into the Cr(VI) solution and it was taken out after 2 s. The liquid was squeezed out that the clean water was obtained. Considering this superior property, the pBC-Polym was filled in a column to achieve flow-through adsorption. The practical image is shown in Figure 5b. Here, 25 mL of Cr(VI) solution was pumped to the column by peristaltic pump to flow through the macroscopic adsorbent with a constant velocity.

In the flow-through adsorption, effect of concentration was examined firstly. The flow velocity was set as 28.4 µL min^{-1}. As shown in Figure 6a, the removal rate was persistently remained at above 96% when Cr(VI) concentration < 100 mg L^{-1}. This indicated the excellent performance of pBC-Polym in dynamic process. Figure 6b revealed the relationship between influent velocity and removal efficiency in dynamic adsorption (C_0 = 50 mg L^{-1}). The removal rate hardly decreases with the increased flow rate. This sufficiently demonstrated the high mass transfer efficiency of the ions within the body of adsorbent and strongly verified the high prospect of pBC-Polym in wastewater purification. Moreover, the adsorbent exhibited excellent ability in the treatment of diverse metal ions and organic dyes (Supplementary Section S8).

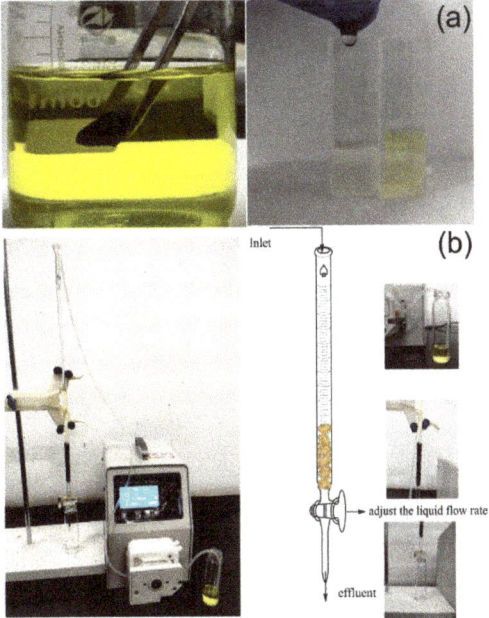

Figure 5. (**a**) Illustration on the fast and convenient adsorption by pBC-Polym-0.04, and (**b**) experimental apparatus for dynamic adsorption.

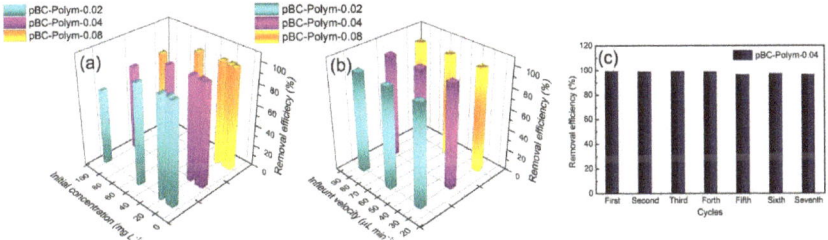

Figure 6. Effect of initial Cr(VI) concentration (**a**) and effect of velocity (**b**) on dynamic adsorption by pBC-Polym-0.02, pBC-Polym-0.04 and pBC-Polym-0.08; recycling behavior of the pBC-Polym-0.04 in Cr(VI) adsorption (**c**).

3.4.4. Recyclability

The repeated adsorption was tested by dynamic adsorption using 25 mL of 20 mg L^{-1} Cr(VI) solution as simulated effluent. 100 mL of 2M NaOH was applied as desorption solution and flow through the adsorbent in the column. The adsorbent was rinsed and freeze-dried after desorption for next cycle. The Cr(VI) removal rate of each cycle was shown in Figure 6c. For seven cycles, no obvious performance attenuation was observed. Typically, the Cr(VI) removal rate of adsorbent remained 96.5%, implying its distinguished reusability.

4. Conclusion

We have successfully fabricated the poly Schiff base-coated bacteria cellulose without the conventional dispersion treatment of the cellulose aerogel. The key is to pretreat the bacteria cellulose with sodium silicate to exfoliate the nanofibers from the big bundles of raw cellulose fibers.

Due to the intact texture of the pretreated bacteria cellulose, the composite aerogel is of good properties (e.g., mechanical and thermal) and high mass transfer efficiency. The adsorption capacity of the aerogel is 321.5 (Cr(VI)), 256.4 (Cu(II)), 153.8 (Re(VII)), 333.3 (Conga red), and 370.3 (Orange G) mg g^{-1}. The adsorption process obeys the pseudo-second order kinetic. The adsorbent can be regenerated easily and used to remove the pollutants without obvious performance attenuation.

Supplementary Materials: The following are available online at http://www.mdpi.com/2073-4360/12/3/714/s1, 1. Detail of Fabrication, 2. Photograph of pBC-Polym-0.04 and BC-Polym-0.04, 3. The morphology of pBC-Polym-0.02 and pBC-Polym-0.08, 4. Specific surface area and porosity, 5. Thermal stability, 7. Adsorption mechanism, 8. Other pollutants

Author Contributions: Conceptualization, L.R. and L.Z.; methodology, L.R.; software, L.R.; validation, Z.Y., and H.W.; formal analysis, L.R; investigation, L.H.; resources, Y.H.; data curation, L.H.; writing—original draft preparation, L.R.; writing—review and editing, L.Z.; visualization, Z.Y.; supervision, H.W.; project administration, Z.Y.; funding acquisition, H.W. All authors have read and agreed to the published version of the manuscript.

Funding: This study was funded by National Key R&D Program of China (2016YFC0403003) and Key R&D Program of Hunan Province (2018SK2026 and 2018SK2043). The authors all declare that they have no conflict of interest.

Acknowledgments: Haiying and Liyuan thank the financial support by Key R&D Program of Hunan Province (2018SK2026 and 2018SK2043) and National Key R&D Program of China (2016YFC0403003).

Conflicts of Interest: The authors all declare that they have no conflict of interest.

References

1. Hausladen, D.M.; Alexander-Ozinskas, A.; McClain, C.; Fendorf, S. Hexavalent Chromium Sources and Distribution in California Groundwater. *Environ. Sci. Technol.* **2018**, *52*, 8242–8251. [CrossRef]
2. Lin, X.; Sun, Z.; Zhao, L.; Ma, J.; Li, X.; He, F.; Hou, H. Toxicity of Exogenous Hexavalent Chromium to Soil-Dwelling Springtail Folsomia Candida in Relation to Soil Properties and Aging Time. *Chemosphere* **2019**, *224*, 734–742. [CrossRef]
3. Yan, X.; Song, M.; Zhou, M.; Ding, C.; Wang, Z.; Wang, Y.; Yang, W.; Yang, Z.; Liao, Q.; Shi, Y. Response of Cupriavidus Basilensis B-8 to Cuo Nanoparticles Enhances Cr(Vi) Reduction. *Sci. Total Environ.* **2019**, *688*, 46–55. [CrossRef] [PubMed]
4. Tran, H.N.; Nguyen, D.T.; Le, G.T.; Tomul, F.; Lima, E.C.; Woo, S.H.; Sarmah, A.K.; Nguyen, H.Q.; Nguyen, P.T.; Nguyen, D.D.; et al. Adsorption Mechanism of Hexavalent Chromium onto Layered Double Hydroxides-Based Adsorbents: A Systematic in-Depth Review. *J. Hazard Mater.* **2019**, *373*, 258–270. [CrossRef] [PubMed]
5. Kalidhasan, S.; Kumar, A.S.K.; Rajesh, V.; Rajesh, N. The Journey Traversed in the Remediation of Hexavalent Chromium and the Road Ahead toward Greener Alternatives—a Perspective. *Coord. Chem. Rev.* **2016**, *317*, 157–166. [CrossRef]
6. Wang, T.; Zhang, L.; Li, C.; Yang, W.; Song, T.; Tang, C.; Meng, Y.; Dai, S.; Wang, H.; Chai, L.; et al. Synthesis of Core-Shell Magnetic Fe3o4@Poly(M-Phenylenediamine) Particles for Chromium Reduction and Adsorption. *Environ. Sci. Technol.* **2015**, *49*, 5654–5662. [CrossRef]
7. Liang, H.; Song, B.; Peng, P.; Jiao, G.; Yan, X.; She, D. Preparation of Three-Dimensional Honeycomb Carbon Materials and Their Adsorption of Cr(Vi). *Chem. Eng. J.* **2019**, *367*, 9–16. [CrossRef]
8. Huang, M.-R.; Lu, H.-J.; Li, X.-G. Synthesis and Strong Heavy-Metal Ion Sorption of Copolymer Microparticles from Phenylenediamine and Its Sulfonate. *J. Mater. Chem.* **2012**, *22*, 34. [CrossRef]
9. Wang, H.; Hou, L.; Shen, Y.; Huang, L.; He, Y.; Yang, W.; Yuan, T.; Jin, L.; Tang, C.J.; Zhang, L. Synthesis of Core-Shell Uio-66-Poly(M-Phenylenediamine) Composites for Removal of Hexavalent Chromium. *Environ Sci. Pollut. Res. Int.* **2020**, *27*, 4115–4126. [CrossRef]
10. Wang, H.; He, Y.; Chai, L.; Lei, H.; Yang, W.; Hou, L.; Yuan, T.; Jin, L.; Tang, C.; Luo, J. Highly-Dispersed Fe2o3@C Electrode Materials for Pb2+ Removal by Capacitive Deionization. *Carbon* **2019**, *153*, 12–20. [CrossRef]
11. Hemavathy, R.V.; Kumar, P.S.; Kanmani, K.; Jahnavi, N. Adsorptive Separation of Cu (Ii) Ions from Aqueous Medium Using Thermally/Chemically Treated Cassia Fistula Based Biochar. *J. Clean. Prod.* **2020**, *249*, 1–12. [CrossRef]

12. Luo, J.; Crittenden, J.C. Nanomaterial Adsorbent Design: From Bench Scale Tests to Engineering Design. *Environ. Sci. Technol.* **2019**, *53*, 10537–10538. [CrossRef] [PubMed]
13. Alfaro, I.; Molina, L.; González, P.; Gaete, J.; Valenzuela, F.; Marco, J.F.; Sáez, C.; Basualto, C. Silica-Coated Magnetite Nanoparticles Functionalized with Betaine and Their Use as an Adsorbent for Mo (Vi) and Re(Vii) Species from Acidic Aqueous Solutions. *J. Ind. Eng. Chem.* **2019**, *78*, 271–283. [CrossRef]
14. Du, Z.; Zheng, T.; Wang, P. Experimental and Modelling Studies on Fixed Bed Adsorption for Cu(Ii) Removal from Aqueous Solution by Carboxyl Modified Jute Fiber. *Powder Technol.* **2018**, *338*, 952–959. [CrossRef]
15. Wang, H.; Li, X.; Chai, L.; Zhang, L. Nano-Functionalized Filamentous Fungus Hyphae with Fast Reversible Macroscopic Assembly & Disassembly Features. *Chem. Commun.* **2015**, *51*, 8524–8527.
16. Wang, D. A Critical Review of Cellulose-Based Nanomaterials for Water Purification in Industrial Processes. *Cellulose* **2018**, *26*, 687–701. [CrossRef]
17. Luo, H.; Xie, J.; Wang, J.; Yao, F.; Yang, Z.; Wan, Y. Step-by-Step Self-Assembly of 2d Few-Layer Reduced Graphene Oxide into 3d Architecture of Bacterial Cellulose for a Robust, Ultralight, and Recyclable All-Carbon Absorbent. *Carbon* **2018**, *139*, 824–832. [CrossRef]
18. Hu, Y.; Liu, F.; Sun, Y.; Xu, X.; Chen, X.; Pan, B.; Sun, D.; Qian, J. Bacterial Cellulose Derived Paper-Like Purifier with Multifunctionality for Water Decontamination. *Chem. Eng. J.* **2019**, *371*, 730–737. [CrossRef]
19. Carpenter, A.W.; de Lannoy, C.F.; Wiesner, M.R. Cellulose Nanomaterials in Water Treatment Technologies. *Environ. Sci. Technol.* **2015**, *49*, 5277–5287. [CrossRef]
20. Cheng, R.; Kang, M.; Zhuang, S.; Shi, L.; Zheng, X.; Wang, J. Adsorption of Sr (Ii) from Water by Mercerized Bacterial Cellulose Membrane Modified with Edta. *J. Hazard Mater.* **2019**, *364*, 645–653. [CrossRef]
21. Hu, Y.; Chen, C.; Yang, L.; Cui, J.; Hao, Q.; Sun, D. Handy Purifier Based on Bacterial Cellulose and Ca-Montmorillonite Composites for Efficient Removal of Dyes and Antibiotics. *Carbohydr. Polym.* **2019**, *222*, 115017. [CrossRef] [PubMed]
22. Zhu, T.; Mao, J.; Cheng, Y.; Liu, H.; Lv, L.; Ge, M.; Li, S.; Huang, J.; Chen, Z.; Li, H.; et al. Recent Progress of Polysaccharide-Based Hydrogel Interfaces for Wound Healing and Tissue Engineering. *Adv. Mater. Interfaces* **2019**, *6*, 17. [CrossRef]
23. Mensah, A.; Lv, P.; Narh, C.; Huang, J.; Wang, D.; Wei, Q. Sequestration of Pb (Ii) Ions from Aqueous Systems with Novel Green Bacterial Cellulose Graphene Oxide Composite. *Materials* **2019**, *12*, 218. [CrossRef] [PubMed]
24. Jin, X.; Xiang, Z.; Liu, Q.; Chen, Y.; Lu, F. Polyethyleneimine-Bacterial Cellulose Bioadsorbent for Effective Removal of Copper and Lead Ions from Aqueous Solution. *Bioresour. Technol.* **2017**, *244*, 844–849. [CrossRef]
25. Wang, J.; Lu, X.; Ng, P.F.; Lee, K.I.; Fei, B.; Xin, J.H.; Wu, J.Y. Polyethyleneimine Coated Bacterial Cellulose Nanofiber Membrane and Application as Adsorbent and Catalyst. *J. Colloid Interface Sci.* **2015**, *440*, 32–38. [CrossRef]
26. Jahan, K.; Kumar, N.; Verma, V. Removal of Hexavalent Chromium from Potable Drinking Using a Polyaniline-Coated Bacterial Cellulose Mat. *Environ. Sci. Water Res. Technol.* **2018**, *4*, 1589–1603. [CrossRef]
27. Yang, Z.; Ren, L.; Jin, L.; Huang, L.; He, Y.; Tang, J.; Yang, W.; Wang, H. In-Situ Functionalization of Poly(M-Phenylenediamine) Nanoparticles on Bacterial Cellulose for Chromium Removal. *Chem. Eng. J.* **2018**, *344*, 441–452. [CrossRef]
28. He, C.; Huang, J.; Li, S.; Meng, K.; Zhang, L.; Chen, Z.; Lai, Y. Mechanically Resistant and Sustainable Cellulose-Based Composite Aerogels with Excellent Flame Retardant, Sound-Absorption, and Superantiwetting Ability for Advanced Engineering Materials. *ACS Sustain. Chem. Eng.* **2017**, *6*, 927–936. [CrossRef]
29. Zhang, H.; Xu, X.; Chen, C.; Chen, X.; Huang, Y.; Sun, D. In Situ Controllable Fabrication of Porous Bacterial Cellulose. *Mater. Lett.* **2019**, *249*, 104–107. [CrossRef]
30. Yan, H.; Chen, X.; Song, H.; Li, J.; Feng, Y.; Shi, Z.; Wang, X.; Lin, Q. Synthesis of Bacterial Cellulose and Bacterial Cellulose Nanocrystals for Their Applications in the Stabilization of Olive Oil Pickering Emulsion. *Food Hydrocoll.* **2017**, *72*, 127–135. [CrossRef]
31. Stoica-Guzun, A.; Stroescu, M.; Jinga, S.I.; Mihalache, N.; Botez, A.; Matei, C.; Berger, D.; Damian, C.M.; Ionita, V. Box-Behnken Experimental Design for Chromium(Vi) Ions Removal by Bacterial Cellulose-Magnetite Composites. *Int. J. Biol. Macromol.* **2016**, *91*, 1062–1072. [CrossRef] [PubMed]

32. Urbina, L.; Guaresti, O.; Requies, J.; Gabilondo, N.; Eceiza, A.; Corcuera, M.A.; Retegi, A. Design of Reusable Novel Membranes Based on Bacterial Cellulose and Chitosan for the Filtration of Copper in Wastewaters. *Carbohydr. Polym.* **2018**, *193*, 362–372. [CrossRef] [PubMed]
33. Oh, S.Y.; Yoo, D.I.; Shin, Y.; Seo, G. Ftir Analysis of Cellulose Treated with Sodium Hydroxide and Carbon Dioxide. *Carbohydr. Res.* **2005**, *340*, 417–428. [CrossRef] [PubMed]
34. Chen, S.; Huang, Y. Bacterial Cellulose Nanofibers Decorated with Phthalocyanine: Preparation, Characterization and Dye Removal Performance. *Mater. Lett.* **2015**, *142*, 235–237. [CrossRef]
35. Jiang, J.; Zhu, J.; Zhang, Q.; Zhan, X.; Chen, F. A Shape Recovery Zwitterionic Bacterial Cellulose Aerogel with Superior Performances for Water Remediation. *Langmuir* **2019**, *35*, 11959–11967. [CrossRef] [PubMed]
36. Zhang, L.; Chai, L.; Liu, J.; Wang, H.; Yu, W.; Sang, P. Ph Manipulation: A Facile Method for Lowering Oxidation State and Keeping Good Yield of Poly (M-Phenylenediamine) and Its Powerful Ag+ Adsorption Ability. *Langmuir* **2011**, *27*, 13729–13738. [CrossRef]
37. Ren, L.; Yang, Z.; Jin, L.; Yang, W.; Shi, Y.; Wang, S.; Yi, H.; Wei, D.; Wang, H.; Zhang, L. Hydrothermal Synthesis of Chemically Stable Cross-Linked Poly-Schiff Base for Efficient Cr (Vi) Removal. *J. Mater. Sci.* **2019**, *55*, 3259–3278. [CrossRef]
38. Liu, Y.; Liu, S.; Li, Z.; Ma, M.; Wang, B. A Microwave Synthesized Mesoporous Carbon Sponge as an Efficient Adsorbent for Cr (Vi) Removal. *RSC Adv.* **2018**, *8*, 7892–7898. [CrossRef]
39. Li, L.; Wei, Z.; Liu, X.; Yang, Y.; Deng, C.; Yu, Z.; Guo, Z.; Shi, J.; Zhu, C.; Guo, W.; et al. Biomaterials Cross-Linked Graphene Oxide Composite Aerogel with a Macro–Nanoporous Network Structure for Efficient Cr (Vi) Removal. *Int. J. Biol. Macromol.* **2019**, (in press). [CrossRef]
40. Lei, Y.; Chen, F.; Luo, Y.; Zhang, L. Three-Dimensional Magnetic Graphene Oxide Foam/Fe3o4 Nanocomposite as an Efficient Absorbent for Cr (Vi) Removal. *J. Mater. Sci.* **2014**, *49*, 4236–4245. [CrossRef]
41. Zhao, X.; Xu, X.; Teng, J.; Zhou, N.; Zhou, Z.; Jiang, X.; Jiao, F.; Yu, J. Three-Dimensional Porous Graphene Oxide-Maize Amylopectin Composites with Controllable Pore-Sizes and Good Adsorption-Desorption Properties: Facile Fabrication and Reutilization, and the Adsorption Mechanism. *Ecotoxicol. Environ. Saf.* **2019**, *176*, 11–19. [CrossRef] [PubMed]
42. Liu, L.; Liao, Q.; Xie, J.; Qian, Z.; Zhu, W.; Chen, X.; Su, X.; Meng, R.; Yao, J. Synthetic Control of Three-Dimensional Porous Cellulose-Based Bioadsorbents: Correlation between Structural Feature and Metal Ion Removal Capability. *Cellulose* **2016**, *23*, 3819–3835. [CrossRef]
43. Zheng, Y.; Cheng, B.; You, W.; Yu, J.; Ho, W. 3D Hierarchical Graphene Oxide-Nife Ldh Composite with Enhanced Adsorption Affinity to Congo Red, Methyl Orange and Cr (Vi) Ions. *J. Hazard Mater.* **2019**, *369*, 214–225. [CrossRef] [PubMed]
44. Wang, Z.; Lin, F.; Huang, L.; Chang, Z.; Yang, B.; Liu, S.; Zheng, M.; Lu, Y.; Chen, J. Cyclodextrin Functionalized 3d-Graphene for the Removal of Cr (Vi) with the Easy and Rapid Separation Strategy. *Environ. Pollut.* **2019**, *254*, 112854. [CrossRef]
45. Kim, M.K.; Sundaram, K.S.; Iyengar, G.A.; Lee, K.-P. A Novel Chitosan Functional Gel Included with Multiwall Carbon Nanotube and Substituted Polyaniline as Adsorbent for Efficient Removal of Chromium Ion. *Chem. Eng. J.* **2015**, *267*, 51–64. [CrossRef]
46. Cheng, Z.; Dai, Z.; Li, J.; Wang, H.; Huang, M.-R.; Li, X.-G.; Liao, Y. Template-Free Synthesis of Tunable Hollow Microspheres of Aniline and Aminocarbazole Copolymers Emitting Colorful Fluorescence for Ultrasensitive Sensors. *Chem. Eng. J.* **2019**, *357*, 776–786. [CrossRef]

© 2020 by the authors. Licensee MDPI, Basel, Switzerland. This article is an open access article distributed under the terms and conditions of the Creative Commons Attribution (CC BY) license (http://creativecommons.org/licenses/by/4.0/).

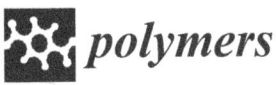

Article

Analysis of Mechanical Properties and Mechanism of Natural Rubber Waterstop after Aging in Low-Temperature Environment

Lin Yu [1,*], Shiman Liu [2], Weiwei Yang [1] and Mengying Liu [1]

[1] College of Mechanics and Materials, Hohai University, Nanjing 210098, China; ntyangweiwei@163.com (W.Y.); liumengying71@163.com (M.L.)
[2] Jianke Engineering Consulting Co., Ltd., Shanghai 200032, China; liushiman0@163.com
* Correspondence: yulin@hhu.edu.cn

Abstract: In order to elucidate the aging performance and aging mechanism of a rubber waterstop in low-temperature environments, the rubber waterstops were placed in the freezing test chamber to accelerate aging, and then we tested its tensile strength, elongation, tear strength, compression permanent deformation and hardness at different times. Additionally, the damaged specimens were tested by scanning electron microscope, Fourier transform infrared spectroscopy and energy dispersive spectrometry. The results showed that with the growth of aging time, the mechanical properties of the rubber waterstop are reduced. At the same time, many protrusions appeared on the surface of the rubber waterstop, the C element gradually decreased, and the O element gradually increased. During the period of 72–90 days, the content of the C element in the low-temperature air environment significantly decreased compared with that in low-temperature water, while the content of O element increased significantly.

Keywords: rubber waterstop; low temperature; aging performance; microscopic examination; mechanism analysis

1. Introduction

In water conservancy and hydro-power projects, considering that concrete cannot be continuously poured, construction joints, settlement joints and deformation joints are needed in order to adapt to the deformation of foundation and the deformation caused by the change in temperature. The rubber waterstop, as the most commonly used waterstop material, can effectively prevent the leakage and seepage of building joints, and play the role of shock absorption and buffer, so it is widely used in projects [1–4]. In applications, rubber waterstop is often exposed to various environments, such as oxygen, ozone, light and temperature, so it often changes in composition and structure [5,6].

At the same time, natural rubber has a huge advantage over fossil-based polymers in terms of environmental friendliness. First of all, natural rubber is derived from rubber trees, while fossil-based polymers are derived from petroleum by-products, which means that natural rubber is inexhaustible and fossil-based polymers are limited. Meanwhile, fossil-based polymers easily burn and release gases that are harmful to the environment, such as toluene. Secondly, the performance of fossil-based polymers in degradability and recycling is worse than that of natural rubber. As fossil-based polymers are difficult to degrade [7], many fossil-based polymers products are discarded and flow into the sea [8]. This causes serious pollution to the marine environment and kills a large number of marine organisms, but natural rubber products can be recycled, and many countries have established the corresponding regulations [9].

Deng Jun et al. [10] studied the effects of different aging temperatures and aging times on the properties of different types of rubber. The results showed that with the increase in

aging temperature and aging time, the rubber hardness increased and its tensile strength decreased. Li Bo et al. [11] accelerated the aging of rubber in a hot oxygen environment and investigated the changes in the mechanical properties of rubber before and after aging and the degree of aging with aging temperature and time. The results showed that the elongation and fracture stress of rubber gradually decreased with increasing aging time and aging temperature, and a small increase in temperature would lead to a significant decrease in mechanical properties. J.R. Beatty et al. [12] studied the influence of time and pressure on rubber in a low-temperature environment. The results showed that the hardness of rubber material at low-temperature increased with the gradual increase in time and pressure and proposed that one of the reasons for the increase in rubber hardness was crystallization.

In the above studies, the environment of rubber aging was mostly set to high temperature and high oxygen or lack of microscopic testing and mechanism analysis. However, the temperature in Xigaze is low all year round, with an average temperature of 6.5 °C [13]. The average temperature in the coldest month (January) is −3.2 °C while the average temperature in the hottest month (July) is 14.6 °C [14]. Additionally, the rubber waterstop used in water conservancy projects is generally poured inside the concrete, which greatly reduces the contact with oxygen and ozone, and also shelters from light [15]. Therefore, we put the rubber waterstop in the freezing test chamber for accelerated aging, and tested its tensile strength, elongation, tear strength, compression set and hardness under different aging times, and carried out microscopic detection on the damaged specimen, trying to find the aging mechanism of rubber in the freezing environment.

2. Experiment Flow Chart

Figure 1 shows the flow chart of our experiments.

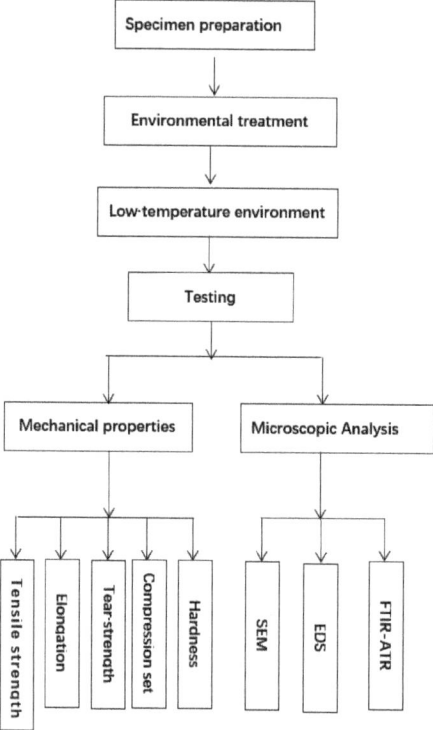

Figure 1. Experiment flow chart.

3. Experimental Program

Materials

The BW5 rubber waterstop with the specification of 300 mm × 8 mm used in the construction joint of the real project was employed in this study. The vulcanized rubber used in this study mainly consisted of polyisoprene (92–95% of cis1,4-polyisoprene) which was produced by Hebei Jingjia Rubber products Co., Ltd. (China). The main components of the rubber waterstop are natural rubber, insoluble sulfur(S), accelerator, zinc oxide (ZnO), stearic acid (SA) and carbon black.

4. Sample Preparation and Aging Environment

Five groups of specimens were prepared in two environments, and each group consisted of three tensile specimens, five tear specimens, three permanent compression deformation specimens and one hardness specimen. Tensile specimens were I type dumbbell shaped with a thickness of 2.0 mm ± 0.2 mm as shown in Figure 2. The tear specimen was of right-angle type with a thickness of 2.0 mm ± 0.2 mm as shown in Figure 3. The compression permanent deformation specimen was a B-type cylinder with a diameter of 13.0 mm ± 0.5 mm.

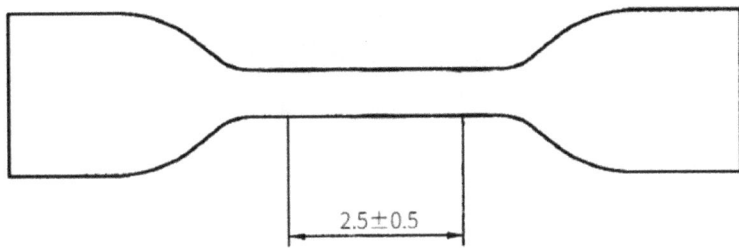

Figure 2. Dumbbell type specimen.

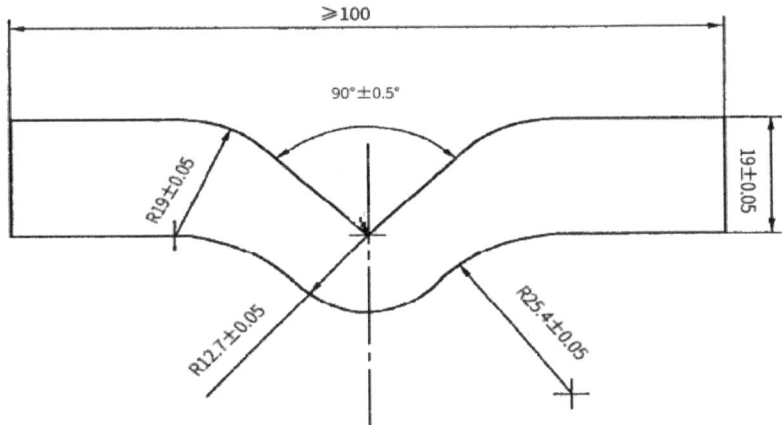

Figure 3. Right-angle specimen.

Considering the constant change of the dam water level in this project [14], the rubber waterstop buried in the dam may be above or below the water surface in different seasons, so the low-temperature environment is divided into low-temperature air environment and low-temperature water environment. Since the lowest annual temperature in Xigaze is −25 °C [16], the low-temperature environment in this paper was set at −25 °C.

5. Equipment

The DW40 refrigerating test chamber was produced by Nanjing Sibenke Experiment Co (China). The KSL-10KN electronic universal testing machine was produced by YangZhou KaiDe Experiment Co (China). The YF-8426 compression permanent deformation machine was produced by YangZhou YuanFeng Experiment Co (China). The Quanta 250F SEM was produced by FEI (Hillsboro, OR, USA). The Nexus 470 Fourier infrared spectrometer was produced by Nicolet (Waltham, MA, USA).

6. Testing Methods

6.1. Mechanical Properties Test

The temperature of the freezer test chamber was set to $-25\,^\circ C$, and then the 5 groups of specimens were aged in the freezing test chamber for 18 days, 36 days, 54 days, 72 days and 90 days, respectively. After reaching the aging time, the specimens were taken out and placed indoors for 24 h. The tensile strength, elongation and tear strength were tested by an electronic universal tensile testing machine. The loading speed of the testing machine was set to 500 mm/min [17,18]. The shore hardness tester was used to test the hardness of the specimen [19]. The compression specimen was put into the compression permanent deformer to make its compression rate reach 23–28%. After 168 h of compression, the specimen was immediately released and allowed to recover at room temperature for 27–33 min, and then measured the height of the specimen [20]. The compression permanent deformation was calculated according to Equation (1):

$$C = \frac{h_0 - h_1}{h_0 - h_s} \times 100\% \qquad (1)$$

where C—compression set (%); h_0—initial height (mm); h_1—height after recovery (mm); and h_s—limiter height (mm).

6.2. Scanning Electron Microscopy with X-Ray Microanalysis (SEM-EDS) Test

The microstructure and elemental analysis of the rubber waterstop after accelerated aging treatment were investigated using SEM (FEI quanta 250F) equipped with an EDS at an accelerating voltage of 30 kV. The test pieces were then observed at 500, 1000 and 6000 magnifications, and the C and O element contents were analyzed by EDS at a magnification of 1000.

6.3. Fourier Infrared Spectroscopy-Attenuated Total Reflection Total Reflection (FTIR-ATR) Test

The FTIR spectra were recorded using a Nicolet Nexus 470 spectrometer equipped with an ATR attachment. The FTIR was operated within the scan angle range of 5°–90° and a scan speed of 5°/min.

7. Results and Discussion

Mechanical Properties

After aging for different times in the low-temperature air environment and low-temperature water environment, the mechanical properties of the rubber waterstop are shown in Tables 1 and 2.

It can be seen from Tables 1 and 2, Figures 4 and 5 that the tensile strength, elongation, tear strength and compression permanent deformation of the rubber waterstop decreased in the low-temperature environment. Tensile strength and tear strength decreased by 34.9% and 24.7% in the low-temperature air environment. Tensile strength and tear strength decreased by 19.1% and 23.3% in the low-temperature water environment. This shows that the brittleness of rubber increases with the aging time, and the adverse effect is more obvious in the low-temperature air environment. The hardness increases slowly, the material becomes hard and gradually loses its elasticity. The decrease in compression permanent deformation indicates that the deformation capacity of vulcanized rubber gradually decreases with the increase in aging time at low temperature, that is, the plasticity weakens. The reason for this is that the rubber main chain and side chain of the molecular chain and crosslinking chain fracture happens, simultaneously creating new crosslinking, whereas in the low-temperature environment, the rubber molecular chain is in a new crosslinking reaction, thus showing surface hardening after aging and brittle crack, namely due to its decreasing tensile strength, elongation, tear strength and compression permanent deformation.

Table 1. Mechanical properties of aging rubber waterstop in low-temperature air environment.

Time(d)	Tensile Strength (MPa)	Elongation (%)	Tear Strength (kN/m)	Hardness (HA)	Compression Set (%)
0	15.2	494.9	55.9	56	20
18	12.8	482.1	52.2	60	19
36	11.8	391.8	51.0	60	17
54	11.3	342.0	48.6	60	15
72	10.6	339.4	46.9	61	13
90	9.9	335.2	42.1	61	13

Table 2. Mechanical properties of aging rubber waterstop in low-temperature water environment.

Time(d)	Tensile Strength (MPa)	Elongation (%)	Tear Strength (kN/m)	Hardness (HA)	Compression Set (%)
0	15.2	494.9	55.9	56	20
18	12.9	466.7	47.8	59	17
36	12.5	414.7	46.9	59	15
54	12.1	398.0	45.4	59	12
72	11.7	360.1	44.6	59	14
90	12.3	359.4	42.9	60	13

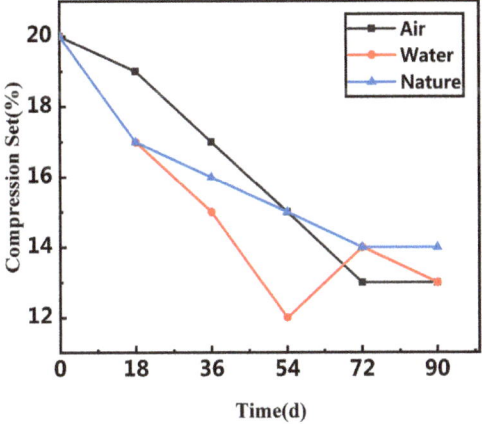

Figure 4. Compression set of waterstop in three different environments.

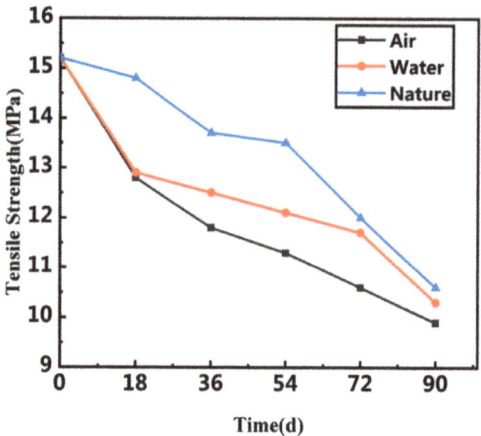

Figure 5. Tensile strength of waterstop in three different environments.

8. Failure Modes

Surface Damage

After aging at low temperature, the surface of the specimen significantly changed. The surface of the specimens without aging was relatively flat, and the micro-pits visible to the naked eye appeared on the surface of the specimens after 36 days of low-temperature treatment, and the number increased with time. At 90 days, pits and tiny holes appeared on the surface of the specimen in the water. At low temperature, the surface of the specimen became rough while the specimen became hard. The apparent morphology of the rubber is shown in Figure 6.

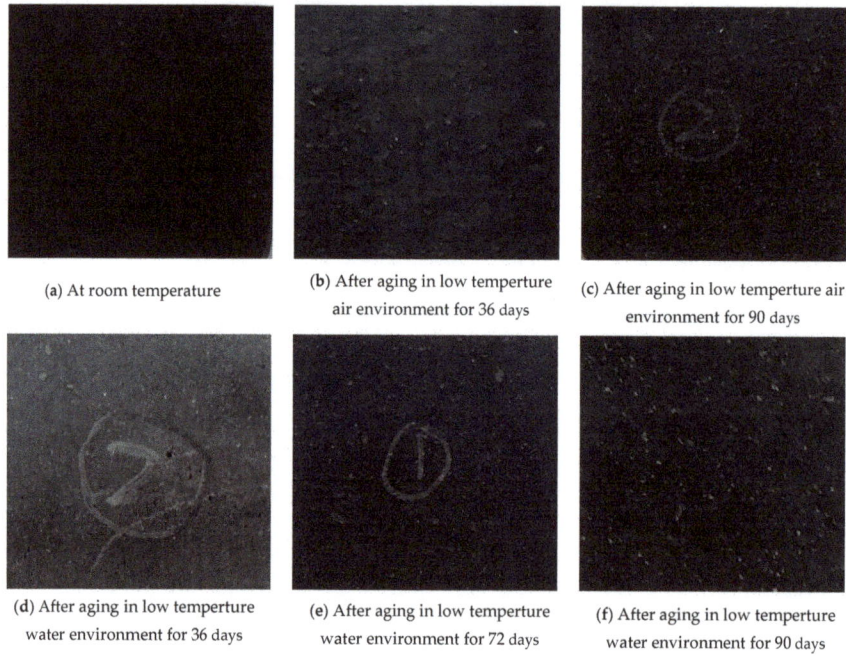

(a) At room temperature

(b) After aging in low temperture air environment for 36 days

(c) After aging in low temperture air environment for 90 days

(d) After aging in low temperture water environment for 36 days

(e) After aging in low temperture water environment for 72 days

(f) After aging in low temperture water environment for 90 days

Figure 6. Apparent morphology of the specimen after aging at low temperature.

The surface morphology of the rubber waterstop before and after aging is further examined by SEM and is shown in Figure 7.

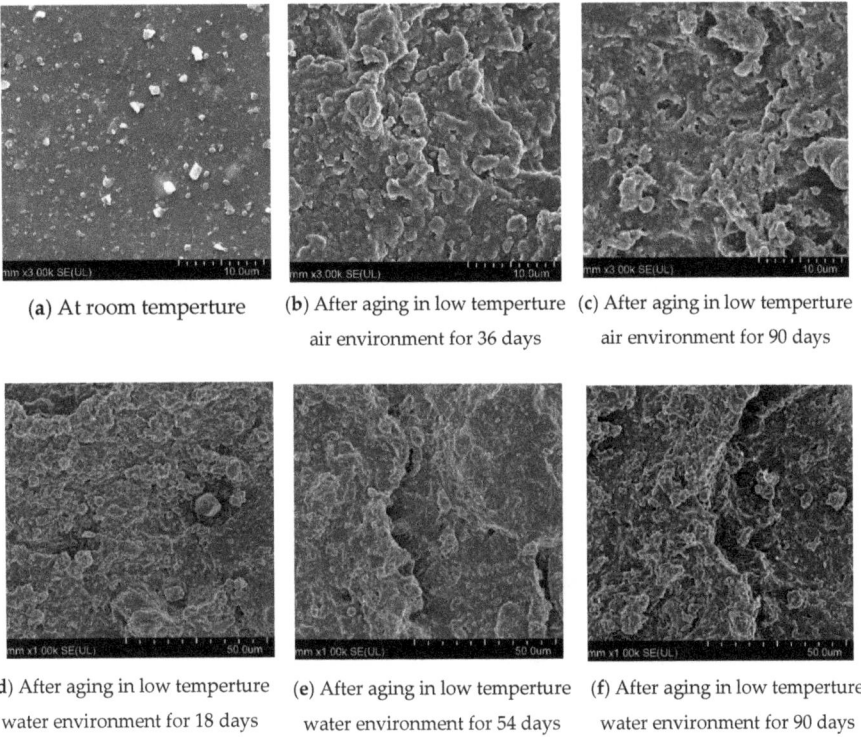

(a) At room temperture (b) After aging in low temperture air environment for 36 days (c) After aging in low temperture air environment for 90 days

(d) After aging in low temperture water environment for 18 days (e) After aging in low temperture water environment for 54 days (f) After aging in low temperture water environment for 90 days

Figure 7. Surface micromorphology.

As shown in Figure 7a, the virgin surface of the rubber waterstop was flat, with white and black additive particles uniformly distributed. As shown in Figure 7b, after 36 days aging under low-temperature air environment, the additive particles on the surface completely disappeared, but many protrusions appeared, and the surface of the specimen became abnormally uneven. With the increase in aging time, the protrusions increased. After 90 days, tiny holes appeared on the surface of the specimen, as shown in Figure 7c. As shown in Figure 7d–f, in the low-temperature water environment, the surface morphology of the specimens protruded, and after 54 days, the rubber surface appeared stepped stratification. After 90 days, the surface became rougher as the protrusion on the surface increased, and the stepped stratification phenomenon became more obvious.

9. Cross-Sectional Damage of Specimens after Tensile Test

Figure 8 shows the cross-sectional morphologies of the tensile test pieces in the low-temperature air environment and low-temperature water environment. The cross-sectional morphologies after 18 days and 90 days in low-temperature air environment are shown in Figure 8b,c, as there were more step-by-step undulations on the surface, and no additive particles. The cross-sectional morphologies changed little in the aging cycle, indicating that the environment has little effect on the adhesion of the rubber matrix. The cross-sectional morphologies after 18 days, 54 days and 90 days in the low-temperature water environment are shown in Figure 8d–f, where a small number of additive particles appeared after 90 days, indicating that this environment can reduce the adhesion of the rubber matrix.

Thus, this environment has an effect on the strength of the rubber waterstop, but the effect is far less than that in the low-temperature air environment.

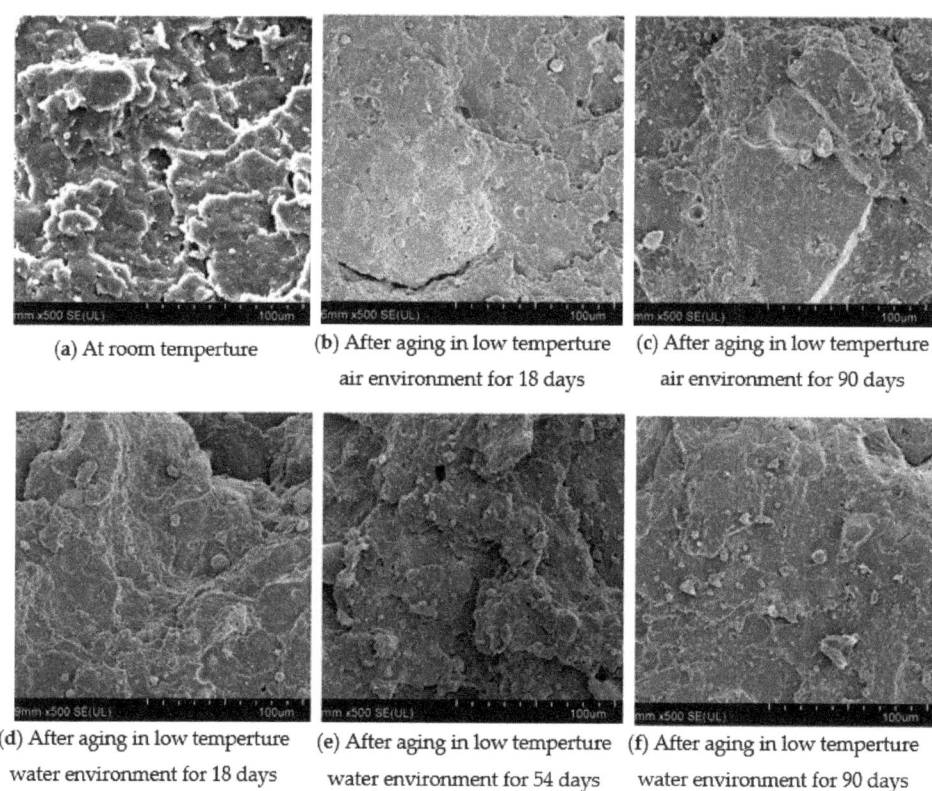

Figure 8. Micromorphology of section.

10. Chemical Analysis

10.1. Energy-Dispersive Spectrometer Analysis

The changes in the content of C and O elements (wt %) in the rubber waterstop samples are shown in Table 3.

Table 3. Changes in C and O element content of different samples.

Element	Time (d) 0	18	36	54	72	90
$C_{(Air)}$	88.0	87.3	86.0	85.6	83.8	75.7
$O_{(Air)}$	8.0	8.5	9.8	9.8	11.6	20.0
$C_{(Water)}$	88.9	88.2	85.3	85.9	85.7	85.2
$O_{(Water)}$	6.7	7.1	10.9	14.6	10.2	10.8

It can be seen that increasing the aging time led to a decrease in the C element and a concurrent increase in the O element in the rubber waterstop. In the low-temperature air environment, the content of the C element decreased by 4.77% while the content of O element increased by 45.0% after 72 days; the oxidation reaction suddenly became faster from 72 days to 90 days, the content of C element decreased by 13.98% and the content of the

O element increased by 105.0% after 90 days. In the low-temperature water environment, the content of C element decreased by 4.16% while the content of the O element increased by 61.19% after 90 days. Intramolecular reactions can occur in vulcanized rubber polymers due to the polyunsaturated property and the short distance between the double bonds. In the aging process of these polymers, the C–C double bond breaks and forms hyperoxide-hydroperoxides, which leads to the decrease in the C element and the increase in the O element on the rubber surface.

Figure 9 shows the change curve of the C and O element content with aging time under three different environments. The x axis represents the aging time, and the y axis represents the content of the elements. Compared with the natural environment, the oxidation reaction of the rubber waterstop in the low-temperature environment was slower. This indicates that low temperature is helpful to alleviate the aging of rubber and the weakening of the adhesion of the rubber matrix.

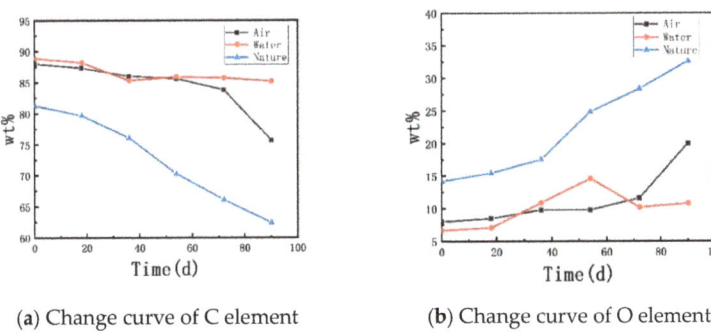

(a) Change curve of C element   (b) Change curve of O element

Figure 9. Change curve of element content.

10.2. Fourier Infrared Spectroscopy-Attenuated Total Refection Analysis

The infrared spectra of the rubber waterstop in low-temperature air environment and low-temperature water environment after 18 days, 36 days, 54 days, 72 days and 90 days are shown in Figures 10 and 11.

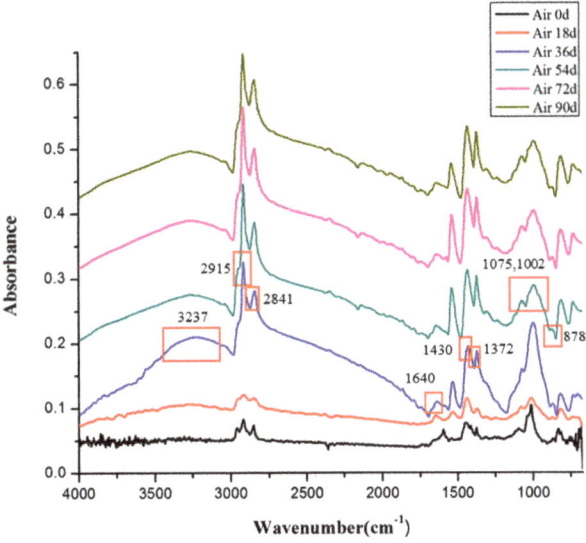

Figure 10. Infrared spectra of low-temperature air environment.

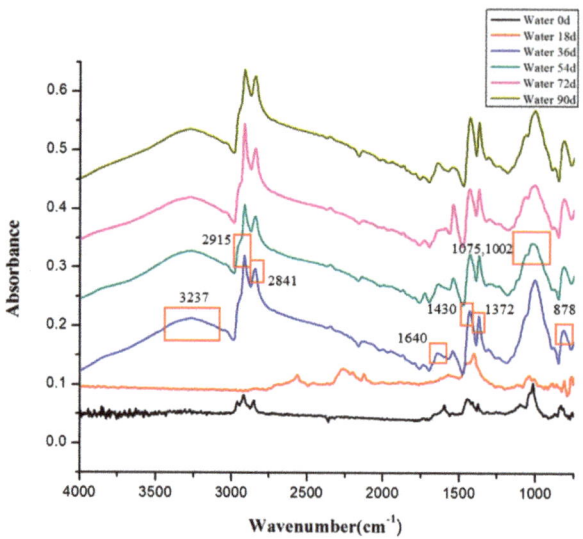

Figure 11. Infrared spectra of low-temperature water environment.

Figure 10 presents the infrared spectra of the rubber waterstop under a low-temperature air environment. In the ATR-FTIR spectra, the peak at 2900–2800 cm^{-1}, 1430 cm^{-1} and 1372 cm^{-1} was assigned to C–H stretching vibrations of –CH$_2$–, –CH$_3$–. The peaks at 1640 cm^{-1} and 878 cm^{-1} were defined as =C–H stretching vibration of the cis-1,4 structure, while the peak at 1443 cm^{-1} was assigned to deformation vibration absorption of –C–O–. It shows that the C–H bond and =CH slowly weaken, and –CO– appears on the molecular chain. The most obvious change was the stretching vibration peak of hydroxyl (O–H) at 3500–3200 cm^{-1}. After 54 days, the peak intensity sharply increased, and then slowly increased. According to the changes of the above spectral peaks, it can be concluded that the double bond of the rubber waterstop slowly decreases and the product of the vicinal diol obviously increases after aging in the low-temperature air environment, which is consistent with the aging law in the natural environment.

In the low-temperature water environment, the infrared spectrum of rubber after aging is similar to that in the low-temperature air environment. With the increase in aging time, the peak intensities of –CH$_3$– at 2915 cm^{-1} and 2841 cm^{-1} and C=C at 1640 cm^{-1} and 878 cm^{-1} decreased slightly, while the peak intensities of –CH$_2$– at 1430 cm^{-1} and CH$_3$ at 1372 cm^{-1} increased slightly. The peak value of –C–O– group at 1002 cm^{-1} obviously decreased, and the change in peak value at 3237 cm^{-1} was the same as that in the dry environment.

From the analysis of Figures 10 and 11, it can be found that during the low-temperature aging test, the basic structure of the rubber and the position of the characteristic peak do not significantly change before and after aging, which indicates that during the low-temperature treatment process of 0–90 days, the basic chemical structure of the rubber waterstop was not obviously damaged or the damage is relatively small.

11. Conclusions

Based on the findings from this study, the following conclusions can be drawn:

1. Increase in low-temperature aging time leads to a decrease in the tensile strength, elongation, tear strength and compression permanent deformation and an increase in the hardness of the rubber waterstop.
2. SEM observation reveals that with the increase in aging time, the surface additive particles disappear, many protrusions appear and gradually increase in the low-

temperature air environment. In the low-temperature water environment, the surface protrusions become rough, and there are only a small number of additive particles in the cross-section at 90 days, which reduces the adhesion of the rubber matrix and affects the strength of the rubber waterstop.
3. FTIR-ATR analysis shows that the infrared spectra of the low-temperature air environment and low-temperature water environment are almost the same, the decrease in functional groups used to characterize the basic structure of polyisoprene is not obvious during the aging process, and the peak amplitude produced by the hydroxyl carboxyl group and other functional groups during the oxidation process is small, and the oxidation is slow.
4. EDS results show that with the increase in aging time, there is an increase in O element and a decrease in C element in the low-temperature air environment. In the low-temperature water environment, the rubber matrix still maintains good adhesion.

Author Contributions: Conceptualization, L.Y. and S.L.; methodology, S.L.; software, S.L.; validation, S.L., W.Y. and M.L.; formal analysis, S.L.; investigation, W.Y.; resources, L.Y.; data curation, S.L.; writing—original draft preparation, S.L.; writing—review and editing, L.Y.; visualization, W.Y.; project administration, L.Y.; funding acquisition, Project of Jiangsu Provincial Construction Engineering Administration. All authors have read and agreed to the published version of the manuscript.

Funding: This research received no external funding.

Institutional Review Board Statement: Not applicable.

Informed Consent Statement: Not applicable.

Data Availability Statement: The data presented in this study are available on request from the corresponding author.

Conflicts of Interest: The authors declare no conflict of interest.

References

1. Huang, L.Q. Application of waterstop in water conservancy engineering. *Hebei Water Resour.* **2018**, *5*, 41.
2. Zhang, L.B. Talking about the application of rubber waterstop in hydraulic engineering. *Technol. Enterp.* **2012**, *9*, 241.
3. Shen, H.Q. Application of rubber waterstop in water conservancy project construction. *Technol. Enterp. Henan Water Resour. South North Water Diversion* **2018**, *8*, 91–92.
4. Tan, L.Y. Talking about the application of rubber waterstop in water conservancy projects. *ChengShi Jianshe LiLun Yan Jiu* **2011**, *24*, 1–2.
5. Thorsten, P.; Ines, J.; Werner, M. Hydrolytic degradation and functional stability of a segmented shape memory poly (ester urethane). *Polym. Degrad. Stab.* **2008**, *1*, 61–73.
6. Liu, S.M.; Yu, L.; Gao, J.X.; Zhang, X.D. Durability of rubber waterstop in extreme environment: Effect and mechanisms of ultraviolet aging. *Polym. Bull.* **2020**, *1*, 14. [CrossRef]
7. Xue, Z.H.; Liu, P. Research on the recycling and reuse of waste plastics. *Plast. Sci. Technol.* **2021**, *49*, 107–110.
8. Liu, C.W. *How to Reduce Marine Plastic Waste and Reduce Economic and Social Costs*; China Finance Press: Beijing, China, 2021.
9. Zeng, X.; Huang, H.S. Development status and Prospect of natural rubber technology in China. *Trop. Agric. China* **2021**, *1*, 25–30.
10. Deng, J.; Zheng, Y.; Pan, Z.C. Study on thermal oxygen aging properties of different rubber materials. *Spec. Rubber Prod.* **2019**, *40*, 17–20.
11. Li, B.; Li, S.X.; Zhang, Z.N. Effect of thermal oxygen aging on mechanical properties and tribological behavior of nitrile butadiene rubber. *J. Mater. Eng.* **2021**, *49*, 114–121.
12. Beatty, J.R.; Davies, J.M. Time and Stress Effects in the Behavior of Rubber at Low Temperature. *J. Rubber Chem. Technol.* **1950**, *23*, 54–66. [CrossRef]
13. GeSang, Z.M.; NiMa, L.Z. Analysis on climate characteristics of Shigatse in 2014. *Tibet Sci. Technol.* **2017**, *12*, 66–70.
14. YangJin, Z.M.; JiaYong, C.C. Study on cloud and precipitation probability in flood season in Xigaze, Xizang. *J. Beijing Agric.* **2015**, *15*, 153–156.
15. Li, S.J.; Liu, L.H.; Gao, Y.F. *Architecture Material*; Chongqing University Press: Chongqing, China, 2014.
16. Yan, Y.Y. Reflection on the present situation and development of highland barley production in Xigaze area. *Tibet J. Agric. Sci.* **2011**, *33*, 10–13.
17. *GB/T 528–2009 Petroleum and Chemical Industry Association of China, Rubber, Vulcanizedor Thermoplastic-Determination of Tensile Stress-Strain Properties*; Chinese Code Press: Beijing, China, 2009.

18. GB/T 529–2008 Petroleum and Chemical Industry Association of China, Rubber, Vulcanizedor Thermoplastic-Determination of Tear Strength (Trouser, Angle and Crescent Test Pieces); ChineseCode Press: Beijing, China, 2008.
19. GB/T 531.1-2008 Petroleum and Chemical Industry Association of China, Rubber, Vulcanizedor Thermoplastic-Determination of Indentation Hardness-Part 1: Duromerer Method (Shore Hardness); Chinese Code Press: Beijing, China, 2008.
20. GB/T 7759.1-2015 Petroleum and Chemical Industry Association of China, Rubber, Vulcanized or Thermoplastic-Determination of Compression Set-Part 1: At Ambient or Elevated Temperatures; Chinese Code Press: Beijing, China, 2015.

MDPI
St. Alban-Anlage 66
4052 Basel
Switzerland
Tel. +41 61 683 77 34
Fax +41 61 302 89 18
www.mdpi.com

Polymers Editorial Office
E-mail: polymers@mdpi.com
www.mdpi.com/journal/polymers

www.ingramcontent.com/pod-product-compliance
Lightning Source LLC
LaVergne TN
LVHW070716100526
838202LV00013B/1104